Constrained Principal Component Analysis and Related Techniques

MONOGRAPHS ON STATISTICS AND APPLIED PROBABILITY

General Editors

F. Bunea, V. Isham, N. Keiding, T. Louis, R. L. Smith, and H. Tong

1. Stochastic Population Models in Ecology and Epidemiology *M.S. Barlett* (1960)
2. Queues *D.R. Cox and W.L. Smith* (1961)
3. Monte Carlo Methods *J.M. Hammersley and D.C. Handscomb* (1964)
4. The Statistical Analysis of Series of Events *D.R. Cox and P.A.W. Lewis* (1966)
5. Population Genetics *W.J. Ewens* (1969)
6. Probability, Statistics and Time *M.S. Barlett* (1975)
7. Statistical Inference *S.D. Silvey* (1975)
8. The Analysis of Contingency Tables *B.S. Everitt* (1977)
9. Multivariate Analysis in Behavioural Research *A.E. Maxwell* (1977)
10. Stochastic Abundance Models *S. Engen* (1978)
11. Some Basic Theory for Statistical Inference *E.J.G. Pitman* (1979)
12. Point Processes *D.R. Cox and V. Isham* (1980)
13. Identification of Outliers *D.M. Hawkins* (1980)
14. Optimal Design *S.D. Silvey* (1980)
15. Finite Mixture Distributions *B.S. Everitt and D.J. Hand* (1981)
16. Classification *A.D. Gordon* (1981)
17. Distribution-Free Statistical Methods, 2nd edition *J.S. Maritz* (1995)
18. Residuals and Influence in Regression *R.D. Cook and S. Weisberg* (1982)
19. Applications of Queueing Theory, 2nd edition *G.F. Newell* (1982)
20. Risk Theory, 3rd edition *R.E. Beard, T. Pentikäinen and E. Pesonen* (1984)
21. Analysis of Survival Data *D.R. Cox and D. Oakes* (1984)
22. An Introduction to Latent Variable Models *B.S. Everitt* (1984)
23. Bandit Problems *D.A. Berry and B. Fristedt* (1985)
24. Stochastic Modelling and Control *M.H.A. Davis and R. Vinter* (1985)
25. The Statistical Analysis of Composition Data *J. Aitchison* (1986)
26. Density Estimation for Statistics and Data Analysis *B.W. Silverman* (1986)
27. Regression Analysis with Applications *G.B. Wetherill* (1986)
28. Sequential Methods in Statistics, 3rd edition *G.B. Wetherill and K.D. Glazebrook* (1986)
29. Tensor Methods in Statistics *P. McCullagh* (1987)
30. Transformation and Weighting in Regression *R.J. Carroll and D. Ruppert* (1988)
31. Asymptotic Techniques for Use in Statistics *O.E. Bandorff-Nielsen and D.R. Cox* (1989)
32. Analysis of Binary Data, 2nd edition *D.R. Cox and E.J. Snell* (1989)
33. Analysis of Infectious Disease Data *N.G. Becker* (1989)
34. Design and Analysis of Cross-Over Trials *B. Jones and M.G. Kenward* (1989)
35. Empirical Bayes Methods, 2nd edition *J.S. Maritz and T. Lwin* (1989)
36. Symmetric Multivariate and Related Distributions *K.T. Fang, S. Kotz and K.W. Ng* (1990)
37. Generalized Linear Models, 2nd edition *P. McCullagh and J.A. Nelder* (1989)
38. Cyclic and Computer Generated Designs, 2nd edition *J.A. John and E.R. Williams* (1995)
39. Analog Estimation Methods in Econometrics *C.F. Manski* (1988)
40. Subset Selection in Regression *A.J. Miller* (1990)
41. Analysis of Repeated Measures *M.J. Crowder and D.J. Hand* (1990)
42. Statistical Reasoning with Imprecise Probabilities *P. Walley* (1991)
43. Generalized Additive Models *T.J. Hastie and R.J. Tibshirani* (1990)
44. Inspection Errors for Attributes in Quality Control *N.L. Johnson, S. Kotz and X. Wu* (1991)
45. The Analysis of Contingency Tables, 2nd edition *B.S. Everitt* (1992)

46. The Analysis of Quantal Response Data *B.J.T. Morgan* (1992)
47. Longitudinal Data with Serial Correlation—A State-Space Approach *R.H. Jones* (1993)
48. Differential Geometry and Statistics *M.K. Murray and J.W. Rice* (1993)
49. Markov Models and Optimization *M.H.A. Davis* (1993)
50. Networks and Chaos—Statistical and Probabilistic Aspects *O.E. Barndorff-Nielsen, J.L. Jensen and W.S. Kendall* (1993)
51. Number-Theoretic Methods in Statistics *K.-T. Fang and Y. Wang* (1994)
52. Inference and Asymptotics *O.E. Barndorff-Nielsen and D.R. Cox* (1994)
53. Practical Risk Theory for Actuaries *C.D. Daykin, T. Pentikäinen and M. Pesonen* (1994)
54. Biplots *J.C. Gower and D.J. Hand* (1996)
55. Predictive Inference—An Introduction *S. Geisser* (1993)
56. Model-Free Curve Estimation *M.E. Tarter and M.D. Lock* (1993)
57. An Introduction to the Bootstrap *B. Efron and R.J. Tibshirani* (1993)
58. Nonparametric Regression and Generalized Linear Models *P.J. Green and B.W. Silverman* (1994)
59. Multidimensional Scaling *T.F. Cox and M.A.A. Cox* (1994)
60. Kernel Smoothing *M.P. Wand and M.C. Jones* (1995)
61. Statistics for Long Memory Processes *J. Beran* (1995)
62. Nonlinear Models for Repeated Measurement Data *M. Davidian and D.M. Giltinan* (1995)
63. Measurement Error in Nonlinear Models *R.J. Carroll, D. Rupert and L.A. Stefanski* (1995)
64. Analyzing and Modeling Rank Data *J.J. Marden* (1995)
65. Time Series Models—In Econometrics, Finance and Other Fields *D.R. Cox, D.V. Hinkley and O.E. Barndorff-Nielsen* (1996)
66. Local Polynomial Modeling and its Applications *J. Fan and I. Gijbels* (1996)
67. Multivariate Dependencies—Models, Analysis and Interpretation *D.R. Cox and N. Wermuth* (1996)
68. Statistical Inference—Based on the Likelihood *A. Azzalini* (1996)
69. Bayes and Empirical Bayes Methods for Data Analysis *B.P. Carlin and T.A Louis* (1996)
70. Hidden Markov and Other Models for Discrete-Valued Time Series *I.L. MacDonald and W. Zucchini* (1997)
71. Statistical Evidence—A Likelihood Paradigm *R. Royall* (1997)
72. Analysis of Incomplete Multivariate Data *J.L. Schafer* (1997)
73. Multivariate Models and Dependence Concepts *H. Joe* (1997)
74. Theory of Sample Surveys *M.E. Thompson* (1997)
75. Retrial Queues *G. Falin and J.G.C. Templeton* (1997)
76. Theory of Dispersion Models *B. Jørgensen* (1997)
77. Mixed Poisson Processes *J. Grandell* (1997)
78. Variance Components Estimation—Mixed Models, Methodologies and Applications *P.S.R.S. Rao* (1997)
79. Bayesian Methods for Finite Population Sampling *G. Meeden and M. Ghosh* (1997)
80. Stochastic Geometry—Likelihood and computation *O.E. Barndorff-Nielsen, W.S. Kendall and M.N.M. van Lieshout* (1998)
81. Computer-Assisted Analysis of Mixtures and Applications—Meta-Analysis, Disease Mapping and Others *D. Böhning* (1999)
82. Classification, 2nd edition *A.D. Gordon* (1999)
83. Semimartingales and their Statistical Inference *B.L.S. Prakasa Rao* (1999)
84. Statistical Aspects of BSE and vCJD—Models for Epidemics *C.A. Donnelly and N.M. Ferguson* (1999)
85. Set-Indexed Martingales *G. Ivanoff and E. Merzbach* (2000)
86. The Theory of the Design of Experiments *D.R. Cox and N. Reid* (2000)
87. Complex Stochastic Systems *O.E. Barndorff-Nielsen, D.R. Cox and C. Klüppelberg* (2001)
88. Multidimensional Scaling, 2nd edition *T.F. Cox and M.A.A. Cox* (2001)
89. Algebraic Statistics—Computational Commutative Algebra in Statistics *G. Pistone, E. Riccomagno and H.P. Wynn* (2001)
90. Analysis of Time Series Structure—SSA and Related Techniques *N. Golyandina, V. Nekrutkin and A.A. Zhigljavsky* (2001)
91. Subjective Probability Models for Lifetimes *Fabio Spizzichino* (2001)
92. Empirical Likelihood *Art B. Owen* (2001)

93. Statistics in the 21st Century *Adrian E. Raftery, Martin A. Tanner, and Martin T. Wells* (2001)
94. Accelerated Life Models: Modeling and Statistical Analysis
Vilijandas Bagdonavicius and Mikhail Nikulin (2001)
95. Subset Selection in Regression, Second Edition *Alan Miller* (2002)
96. Topics in Modelling of Clustered Data *Marc Aerts, Helena Geys, Geert Molenberghs, and Louise M. Ryan* (2002)
97. Components of Variance *D.R. Cox and P.J. Solomon* (2002)
98. Design and Analysis of Cross-Over Trials, 2nd Edition *Byron Jones and Michael G. Kenward* (2003)
99. Extreme Values in Finance, Telecommunications, and the Environment
Bärbel Finkenstädt and Holger Rootzén (2003)
100. Statistical Inference and Simulation for Spatial Point Processes
Jesper Møller and Rasmus Plenge Waagepetersen (2004)
101. Hierarchical Modeling and Analysis for Spatial Data
Sudipto Banerjee, Bradley P. Carlin, and Alan E. Gelfand (2004)
102. Diagnostic Checks in Time Series *Wai Keung Li* (2004)
103. Stereology for Statisticians *Adrian Baddeley and Eva B. Vedel Jensen* (2004)
104. Gaussian Markov Random Fields: Theory and Applications *Håvard Rue and Leonhard Held* (2005)
105. Measurement Error in Nonlinear Models: A Modern Perspective, Second Edition
Raymond J. Carroll, David Ruppert, Leonard A. Stefanski, and Ciprian M. Crainiceanu (2006)
106. Generalized Linear Models with Random Effects: Unified Analysis via H-likelihood
Youngjo Lee, John A. Nelder, and Yudi Pawitan (2006)
107. Statistical Methods for Spatio-Temporal Systems
Bärbel Finkenstädt, Leonhard Held, and Valerie Isham (2007)
108. Nonlinear Time Series: Semiparametric and Nonparametric Methods *Jiti Gao* (2007)
109. Missing Data in Longitudinal Studies: Strategies for Bayesian Modeling and Sensitivity Analysis
Michael J. Daniels and Joseph W. Hogan (2008)
110. Hidden Markov Models for Time Series: An Introduction Using R
Walter Zucchini and Iain L. MacDonald (2009)
111. ROC Curves for Continuous Data *Wojtek J. Krzanowski and David J. Hand* (2009)
112. Antedependence Models for Longitudinal Data *Dale L. Zimmerman and Vicente A. Núñez-Antón* (2009)
113. Mixed Effects Models for Complex Data *Lang Wu* (2010)
114. Intoduction to Time Series Modeling *Genshiro Kitagawa* (2010)
115. Expansions and Asymptotics for Statistics *Christopher G. Small* (2010)
116. Statistical Inference: An Integrated Bayesian/Likelihood Approach *Murray Aitkin* (2010)
117. Circular and Linear Regression: Fitting Circles and Lines by Least Squares *Nikolai Chernov* (2010)
118. Simultaneous Inference in Regression *Wei Liu* (2010)
119. Robust Nonparametric Statistical Methods, Second Edition
Thomas P. Hettmansperger and Joseph W. McKean (2011)
120. Statistical Inference: The Minimum Distance Approach
Ayanendranath Basu, Hiroyuki Shioya, and Chanseok Park (2011)
121. Smoothing Splines: Methods and Applications *Yuedong Wang* (2011)
122. Extreme Value Methods with Applications to Finance *Serguei Y. Novak* (2012)
123. Dynamic Prediction in Clinical Survival Analysis *Hans C. van Houwelingen and Hein Putter* (2012)
124. Statistical Methods for Stochastic Differential Equations
Mathieu Kessler, Alexander Lindner, and Michael Sørensen (2012)
125. Maximum Likelihood Estimation for Sample Surveys
R. L. Chambers, D. G. Steel, Suojin Wang, and A. H. Welsh (2012)
126. Mean Field Simulation for Monte Carlo Integration *Pierre Del Moral* (2013)
127. Analysis of Variance for Functional Data *Jin-Ting Zhang* (2013)
128. Statistical Analysis of Spatial and Spatio-Temporal Point Patterns, Third Edition *Peter J. Diggle* (2013)
129. Constrained Principal Component Analysis and Related Techniques *Yoshio Takane* (2014)

Monographs on Statistics and Applied Probability 129

Constrained Principal Component Analysis and Related Techniques

Yoshio Takane

Professor Emeritus, McGill University
Montreal, Quebec, Canada
and
Adjunct Professor at University of Victoria
British Columbia, Canada

CRC Press
Taylor & Francis Group
Boca Raton London New York

CRC Press is an imprint of the
Taylor & Francis Group, an **informa** business
A CHAPMAN & HALL BOOK

CRC Press
Taylor & Francis Group
6000 Broken Sound Parkway NW, Suite 300
Boca Raton, FL 33487-2742

First issued in paperback 2020

Version Date: 20130923

ISBN 13: 978-0-367-57628-8 (pbk)
ISBN 13: 978-1-4665-5666-9 (hbk)

Visit the Taylor & Francis Web site at
http://www.taylorandfrancis.com

and the CRC Press Web site at
http://www.crcpress.com

Contents

List of Figures

List of Tables

Preface

Regression analysis and principal component analysis (PCA) are the two most frequently used multivariate data analysis techniques. Constrained PCA (CPCA) combines them in a unified framework. Regression analysis predicts one set of variables from another. PCA, on the other hand, finds a subspace of minimal dimensionality that captures the largest variability in the data. How can they be combined in a beneficial way? Why and when is it a good idea to combine them? What kind of benefits are we getting? This book attempts to answer these questions.

PCA has been primarily used in situations where there is little prior knowledge about the data. While no one would deny the role of PCA as an exploratory technique, I also believe that much more can be done with it. In many practical data analysis situations, information regarding rows and columns of a data matrix is available. For example, the data matrix represents preference judgments on a set of stimuli (corresponding to the columns) by a group of subjects (corresponding to the rows). The stimuli may be created by manipulating a set of prescribed attributes, and similarly the subjects may be characterized by their demographic information, such as their age, gender, and level of education. In such cases, we are naturally interested in how this additional information is related to the preference judgments. We may apply regression analysis to decompose the data matrix into what can and cannot be predicted by the additional information. We may further apply PCA to the decomposed parts to explore additional structures within the parts. Since the data matrix is already structured by the external information, the results of PCA tend to be more easily interpretable. This is essentially the idea behind CPCA.

A main attraction of PCA is its computational simplicity. We tried our best to maintain this attractive feature in CPCA. In this book, the use of techniques requiring complicated iterative methods is kept to a minimum. Also, to avoid narrowing down the range of applicability, no distributional assumptions are made. Although this may reduce the inferential capability of CPCA, we feel that the wider applicability of the technique more than offsets such inconvenience. We use some nonparametric procedures such as the bootstrap method to assess the reliability of solutions.

We begin with four concrete examples of CPCA (Chapter 1), designed to provide the reader with a basic understanding of the technique and its applications. Chapter 2 gives a detailed account of two key mathematical ingredients in CPCA, projection and singular value decomposition (SVD). The former underlies regression analysis, which is essentially a projection of the criterion space onto the predictor space. The latter serves as a strong theoretical and computational workforce for data reduction in PCA. Chapter 3 presents the main subject matter of this book. This chapter lays out basic data requirements, models, and analytical tools for CPCA, and their immediate extensions. Chapter 4 introduces techniques that are either special cases of, or closely related to, CPCA. This chapter is of interest particularly for methodologically oriented readers, who have experience in developing statistical techniques of their own. Chapter 5 discusses several topics relevant to practical uses of CPCA, including permutation tests for dimensionality selection, the bootstrap for reliability assessment, and cross validation methods for fine-tuning the values of regularization parameters. Chapter 6 introduces a technique which imposes different constraints on different dimensions (DCDD), and its analytical extensions. The appendix provides a brief overview of computer software for CPCA.

We gladly acknowledge the use of MATLAB® for many of the computational results and figures presented in this book. This software provides an environment extremely conducive to developments of new statistical techniques. For product information please contact:

The Math Works, Inc
3 Apple Hill Drive
Natick, MA 01760-2098 USA
Tel: 508-647-7000
Fax: 508-647-7001
E-mail: info@mathworks.com
Web: www.mathworks.com

I am indebted to many people in writing this book. First and foremost, I would like to thank Professor Michael Hunter, my colleague at University of Victoria, who wrote four key papers on CPCA with me. His expertise in applications of statistical techniques to answer empirical questions has been a great asset throughout the development of CPCA. I also thank Professor Heungsun Hwang, a former colleague at McGill, for many hours of stimulating discussions on developments of various psychometric methods, and Professor Todd Woodward of the University of British Columbia for demonstrating the usefulness of CPCA in brain imaging studies, and developing user friendly programs for CPCA for fMRI data. I am also grateful to Professors Kwanghee Jung and Hye Won Suk, former graduate students at McGill, who read all or parts of the drafts of this book and suggested numerous improvements, to Marina Takane, my daughter, for her editorial assistance, and to Yuriko Oshima-Takane, my wife, for her encouragement throughout this project.

Last but not least, I am deeply indebted to Professor Haruo Yanai of St. Luke College of Nursing for his guidance and mentorship throughout my career. Professor Yanai was in fact instrumental in introducing me to research in statistics and psychometrics when I was an undergraduate student at the University of Tokyo. As has been noted above, CPCA consists of two major ingredients, projection and SVD. These are the main subject matters of a recent book by Yanai et al. (2011), of which this book may be considered a more application-oriented version. I would like to dedicate this book to him to express my sincere appreciation for his long-standing mentorship.

Yoshio Takane

About the Author

Yoshio Takane earned his DL (Doctor of Letters) from the University of Tokyo in 1976, and PhD in quantitative psychology from the University of North Carolina at Chapel Hill in 1977. He was a professor at McGill University, Montreal, until he retired in 2011. He is now an emeritus professor at McGill and an adjunct professor at the University of Victoria, British Columbia, Canada. He is a former president of the Psychometric Society (1986–87), and a recipient of a Career Award (1986) from the Behaviormetric Society of Japan and a Special Award (2013) from the Japanese Psychological Association. He has a long-standing interest in developing statistical techniques for analysis of psychological data, and has published numerous articles on nonmetric multivariate analysis methods, maximum likelihood (ML) multidimensional scaling, choice models incorporating systematic individual differences, constrained principal component analysis, analysis of knowledge representations in neural network models, etc. His most recent interests include regularization techniques for various multivariate data analysis methods, acceleration methods for iterative model fitting, development of structural equation models (SEM) for analysis of brain connectivity, various kinds of singular value decompositions (e.g., sparse SVD, robust SVD) among others.

Chapter 1

Introduction

Multivariate data matrices analyzed by principal component analysis (PCA) are often accompanied by auxiliary information about the rows and columns of the data matrices. For example, the rows of a data matrix may represent subjects for whom some demographic information (e.g., gender, age, level of education, etc.) is available. The columns may represent stimuli constructed by manipulating several attributes with known values. Constrained principal component analysis (CPCA) incorporates such information in PCA of the main data matrix.

CPCA first decomposes the main data matrix into several components according to the external information (External Analysis). Columnwise and/or rowwise regression analyses are used for this purpose with the external information as predictor variables. CPCA then applies PCA to the decomposed components to investigate structures within the components (Internal Analysis). PCA of the portions of the data matrix that can be explained by the external information can be used to extract the most important variation in known structures, while PCA of the portions that cannot be explained by the external information can be used to extract the variation that may be most easily structured among unknown structures. In this introductory chapter, we briefly discuss four examples of analysis by CPCA. The reader is encouraged to pay special attention to what kinds of data are required, what kinds of analysis are possible, what kinds of potential benefits are obtained by applying CPCA, etc. while going through these examples. See Takane and Shibayama (1991), Takane and Hunter (2001, 2011), Hunter and Takane (1998, 2002), Woodward et al. (2006, 2013), and Metzak et al. (2011) for other applications of CPCA.

1.1 Analysis of Mezzich's Data

Table 1.1 presents the data collected from 11 psychiatrists rating four archetypal patients representing four diagnostic categories, 1) manic-depressive depressed (MDD), 2) manic-depressive manic (MDM), 3) simple schizophrenia (SSP), and 4) paranoid schizophrenia (PSP), using 17 Brief Psychiatric Rating Scales (BPRS) by Overall and Gorham (1962). Each of the 17 scales have seven ordered response categories (0: does not apply at all, to 6: applies very well), and pertain to the following aspects of behavior and conditions:

A. Somatic concern

Table 1.1 *Mezzich's (1978) data: Ratings of four psychiatric diseases by 11 psychiatrists using 17 rating scales**.

Cat.	Psy.	A	B	C	D	E	F	G	H	I	J	K	L	M	N	O	P	Q
		Rating Scales																
MDD	1	4	3	3	0	4	3	0	0	6	3	2	0	5	2	2	2	1
	2	5	5	6	2	6	1	0	0	6	1	0	1	6	4	1	4	0
	3	6	5	6	5	6	3	2	0	6	0	5	3	6	5	5	0	0
	4	5	5	1	0	6	1	0	0	6	0	1	2	6	0	3	0	2
	5	6	6	5	0	6	0	0	0	6	0	4	3	5	3	2	0	0
	6	3	3	5	1	4	2	1	0	6	2	1	1	5	2	2	1	1
	7	5	5	5	2	5	4	1	1	6	2	3	0	6	3	5	2	3
	8	4	5	5	1	6	1	1	0	6	1	1	0	5	2	1	1	0
	9	5	3	5	1	6	3	1	0	6	2	1	1	6	2	5	5	0
	10	3	5	5	3	2	4	2	0	6	3	2	0	6	1	4	5	1
	11	5	6	6	4	6	3	1	0	6	2	0	0	6	4	4	6	0
MDM	1	2	2	1	2	0	3	1	6	2	3	3	2	1	4	4	0	6
	2	0	0	0	4	1	5	0	6	0	5	4	4	0	5	5	0	6
	3	0	3	0	5	0	6	0	6	0	3	2	0	0	3	4	0	6
	4	0	0	0	3	0	6	0	6	1	3	1	1	0	2	3	0	6
	5	3	4	0	0	0	5	0	6	0	6	0	0	0	5	0	0	6
	6	2	4	0	3	1	5	1	6	2	5	3	0	0	5	3	0	6
	7	1	2	0	2	1	4	1	5	1	5	1	1	0	4	1	0	6
	8	0	2	0	2	1	5	1	5	0	2	1	1	0	3	1	0	6
	9	0	0	0	6	0	5	1	6	0	5	5	4	0	5	6	0	6
	10	5	5	1	4	0	5	5	6	0	4	4	3	0	5	5	0	6
	11	1	3	0	4	1	4	2	6	3	3	2	0	0	4	3	0	6
SSP	1	3	2	5	2	0	2	2	1	2	1	2	0	1	2	2	4	0
	2	4	4	5	4	3	3	1	0	4	2	3	0	3	2	4	5	0
	3	2	0	6	3	0	0	5	0	0	3	3	2	3	5	3	6	0
	4	1	1	6	2	0	0	1	0	0	3	0	1	0	1	1	6	0
	5	3	3	5	6	3	2	5	0	3	0	2	5	3	3	5	6	2
	6	3	0	5	4	0	0	3	0	2	1	1	1	2	3	3	6	0
	7	3	3	5	4	2	4	2	1	3	1	1	1	4	2	2	5	2
	8	3	2	5	2	2	2	2	1	2	2	3	1	2	2	3	5	0
	9	3	3	6	6	1	3	5	1	3	2	2	5	3	3	6	6	1
	10	1	1	5	3	1	1	3	0	1	1	1	0	5	1	2	6	0
	11	2	3	5	4	2	3	0	0	3	2	2	0	0	2	4	5	0
PSP	1	2	4	3	5	0	3	1	4	2	5	6	5	0	5	6	3	3
	2	2	4	1	1	0	3	1	6	0	6	6	4	0	6	5	0	4
	3	5	5	5	6	0	5	5	6	2	5	6	6	0	5	6	0	2
	4	1	4	2	1	1	1	0	5	1	5	6	5	0	6	6	0	1
	5	4	5	6	3	1	6	3	5	2	6	6	4	0	5	6	0	5
	6	4	5	4	6	2	4	2	4	1	5	6	5	1	5	6	2	4
	7	3	4	3	4	1	5	2	5	2	5	5	3	1	5	5	1	5
	8	2	5	4	3	1	4	3	4	2	5	5	4	0	5	4	1	4
	9	3	3	4	4	1	5	5	5	0	5	6	5	1	5	5	3	4
	10	4	4	2	6	1	4	1	5	3	5	6	5	1	5	6	2	4
	11	3	5	5	5	2	5	4	5	2	4	6	5	0	5	6	5	5

*Reproduced from Table 4.1 of Mezzich and Solomon (1980) with permission from Elsevier.

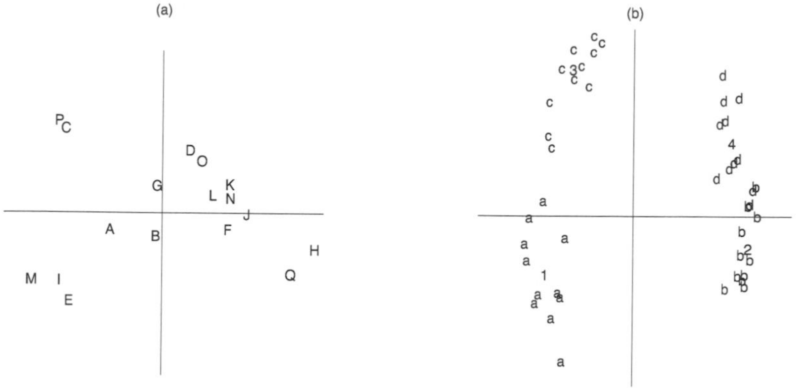

Figure 1.1 *Plots of component loadings (a) and component scores (b) obtained by unconstrained PCA of Mezzich's data. (In (b), symbols "a," "b," "c," and "d" indicate MDD, MDM, SSP, and PSP, respectively.)*

B. Anxiety
C. Emotional withdrawal
D. Conceptual disorganization
E. Guilt feeling
F. Tension
G. Mannerism and posturing
H. Grandiosity
I. Depressive mood
J. Hostility
K. Suspiciousness
L. Hallucinatory behavior
M. Motor retardation
N. Uncooperativeness
O. Unusual thought content
P. Blunted affect
Q. Excitement

The data given in Table 1.1 were first rowwise standardized, and then subjected to unconstrained PCA. While the rowwise standardization of the data is somewhat unconventional in PCA, it has the effect of pronouncing the contrast among the columns of the table (i.e., variables in the data), and is most suitable in the present context. A two-component solution was extracted, and component loadings indicating correlations between the components and observed variables were plotted in Figure 1.1a. The most dominant component is indicated by the horizontal axis, and the second most dominant component by the vertical axis. This figure shows how the 17 BPRS scales are related to each other in reference to these components. Scales pointing in similar directions from the origin measure similar contents. For example, C and P are located in the second quadrant, indicating attributes characteristic of simple

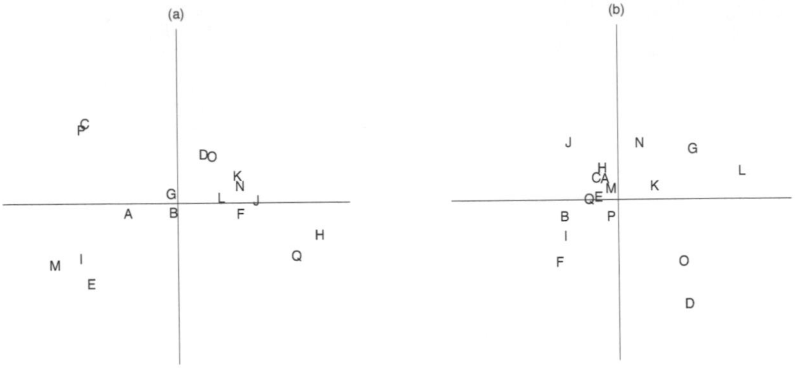

Figure 1.2 *Plots of component loadings by CPCA of Mezzich's data for between-group differences (a) and within-group differences (b).*

schizophrenia (SSP), and E, I, and M are in the third quadrant, indicative of the depressed state of manic-depressive (MDD). Similarly, H and Q are in the fourth quadrant, measuring typical symptoms in the manic state of manic-depressive (MDM), and K, L, and N are in the first quadrant, measuring attributes typical in paranoid schizophrenia (PSP). The distance from the origin indicates how well the particular scale can be represented by the particular direction in the space. For example, C and P are located far away from the origin, which implies that these two scales are relatively well represented in the space defined by the two most dominant components, whereas K, L, and N are located fairly close to the origin, indicating that more components are necessary to capture the contents of these scales.

Component scores, indicating where the 44 sets (11 psychiatrists by 4 diagnostic categories) of ratings stand in terms of the components extracted, were plotted in Figure 1.1b. In the figure, four integers (1 through 4) indicate the centroids of the four diagnostic categories. The 11 sets of ratings by individual psychiatrists are indicated by a unique symbol for each category (a for MDD, b for MDM, c for SSP, and d for PSP). Although there are variations among the psychiatrists within diagnostic categories, there are also fairly good agreements in their ratings, as indicated by the proximity of the component scores for the 11 sets of ratings within the categories. Ratings on SSP are all located in the second quadrant, consistent with our earlier observation that P and C are characteristic of SSP. Likewise, component scores corresponding to MDD are mostly located in the third quadrant along with M, I, and E, those corresponding to MDM are mostly located in the fourth quadrant along with H and Q, and those corresponding to PSP are located in the first quadrant along with K, L, and N. It may be noted that MDM and PSP are much less distinct than MDD and SSP. (There is even some overlap between MDM and PSP.) It seems that the horizontal axis (the first principal component) contrasts between diagnostic groups with and without wild fancies, whereas the vertical axis (the second principal component) contrasts between manic-depressive and schizophrenia.

In the analysis above, external information (which rows represent which diag-

Table 1.2 *A breakdown of the total sum of squares (SS) for Mezzich's data.*

Source	External Analysis	Internal Analysis Comp. 1	Comp. 2	Comp. 3
T (total)	748.0 (100.0%)	364.0 (48.7%)	133.5 (17.4%)	
B (between)	558.5 (74.8%)	356.1 (47.6%)	111.5 (14.9%)	50.7 (8.0%)
W (within)	189.5 (25.3%)	37.2 (5.0%)	26.0 (3.5%)	

nostic groups) was used in an informal way for interpretation. What happens if such information is incorporated in the analysis itself? Let

$$\mathbf{G} = \begin{bmatrix} \mathbf{1}_{11} & \mathbf{0} & \mathbf{0} & \mathbf{0} \\ \mathbf{0} & \mathbf{1}_{11} & \mathbf{0} & \mathbf{0} \\ \mathbf{0} & \mathbf{0} & \mathbf{1}_{11} & \mathbf{0} \\ \mathbf{0} & \mathbf{0} & \mathbf{0} & \mathbf{1}_{11} \end{bmatrix} \tag{1.1}$$

be the matrix of dummy variables, indicating the rows representing the ratings of the four diagnostic categories. Here, $\mathbf{1}_{11}$ denotes an 11-component vector of ones, and $\mathbf{0}$ a vector of zeroes of the same size. The four columns in **G** represent the four diagnostic categories, while the rows represent the 44 sets of ratings. There is only one element equal to unity in each row of **G**, and the column containing the unity indicates the diagnostic group. For example, if the first row of the data matrix pertains to MDD, **G** has a unity in the first column. CPCA was applied to the data with **G** given above as the row constraint matrix. It only analyzes the portions of the data that reflect the differences among the four diagnostic categories, while ignoring those among the psychiatrists. Component loadings obtained in this analysis are plotted in Figure 1.2a. It can be observed that this figure is strikingly similar to Figure 1.1a, indicating that the unconstrained solution in Figure 1.1a mostly reflects the differences among the diagnostic categories (the between-group differences). Figure 1.2b, on the other hand, depicts the within-group component loadings extracted from the portions of the data matrix unrelated to the differences among the diagnostic groups. Figure 1.2b looks completely different from either Figure 1.1a or 1.2a, which is another indication that the variations in the original data are dominated by the between-group variations.

Table 1.2 shows a breakdown of the sum of squares (SS). In CPCA, the decomposition of the original data matrix into between-group and within-group data matrices by external information is called External Analysis, and the subsequent PCA of decomposed matrices Internal Analysis. The two-dimensional solution derived from the original data accounts for approximately 2/3 of the total SS. This means that a majority of SS in the original data can be explained by the two most dominant components. When the original data are decomposed into the between-group data (the part that can be explained by **G**) and the within-group data (the part that cannot be explained by **G**), the former can account for approximately 3/4 of the total SS. The two-dimensional solution extracted from the between-group data can account for 62.5% of the total SS, which is not much of a reduction from 66.1% that can be accounted for by the two-dimensional solution derived from the original data. The within-group data, on the other hand, can account for only 1/4 of the total SS in the original data, and the two-dimensional solution derived from them can account for only 8.5% of the total SS. This is not much larger than the SS that can be accounted for by the third component (not depicted here) derived from the between-group data matrix. It may be safely concluded that the component loadings in Figure 1.2b are perhaps not worth interpreting. (This conclusion, of course, holds only for this particular example, and does not deny the usefulness of residual analyses in general. See Section 6.5.) It may be noted in passing that in this example, the within-group data can further be decomposed into between-psychiatrist data and interactions between psychiatrists and diagnostic groups. However, considering the already rather small effects of the within-group SS, a further decomposition is deemed worthless.

1.2 Analysis of Food and Cancer Data

Here is another example. The data displayed in Table 1.3 were compiled by Segi (1979) based on food data collected by FAO (Food and Agriculture Organization) and health data collected by WHO (World Health Organization). Four variables related to foods are 1) average daily intake of calories (Calorie), and supplies of 2) meat (Meat), 3) milk products (Milk), and 4) alcohol (Alcohol). Two health variables are mortality rates by 1) lower intestine cancer (L-Intes), and 2) rectum cancer (Rectum). Records are available on these six variables in 47 countries in the world. We are interested in investigating how the cancer mortality rates are related to the food variables.

Table 1.4 shows correlations among the six variables in the data set. All four food variables are fairly highly correlated, so it may be supposed that there is something in common underlying these variables. Let's call this common variable the western style diet prevalent in Western Europe and North America, and characterized by high calorie and fatty foods. We may apply a PCA to the four food variables to extract the most dominant component, and use this component as an indication of western style diet. It turns out that the first principal component provides a fairly good summary of the four food variables because it explains over 70% of the total SS. Also, component loadings, indicating correlations between the component and the observed variables, are all high (.943, .898, .819, and .662). Similarly, the two cancer variables are highly correlated. So again, it may be supposed that there must be a common cause for the two cancer variables, which may tentatively be called the "proneness to cancer in lower digestive organs." As before, we may apply a PCA and extract the component most representative of the cancer variables to indicate this common variable. It turns out that the extracted principal component explains over 87% of the total SS. Component loadings are both high (they are both .936; when a PCA is applied to two positively correlated variables, the most dominant component is proportional to the average of the two, and consequently, the two loadings are identical). The correla-

Table 1.3 *Food and cancer data (Segi 1979).*

No.	Calorie	Meat	Milk	Alcohol	L-Intes.	Rectum
1	2336	40	89	38	2.98	2.53
2	3190	575	388	109	14.60	5.91
3	2578	168	127	128	3.92	1.90
4	2200	120	142	23	3.99	0.71
5	2431	205	117	42	6.75	2.84
6	1940	79	101	37	1.28	0.59
7	1895	98	111	41	3.51	0.88
8	1819	55	94	26	0.89	0.57
9	2352	178	100	192	13.76	2.66
10	2570	144	93	68	1.94	0.66
11	2342	140	92	44	4.12	1.61
12	2436	151	133	12	8.53	5.72
13	3349	672	413	121	14.95	4.38
14	2938	720	364	91	12.13	6.07
15	2225	175	167	61	3.27	1.39
16	2468	291	47	25	8.58	4.02
17	2869	187	218	56	9.69	3.89
18	2551	55	47	75	5.01	6.00
19	1915	99	20	19	2.74	1.11
20	2416	146	112	19	9.05	6.18
21	2135	88	17	11	1.84	0.13
22	3366	487	337	218	13.51	9.99
23	3364	486	272	194	13.53	7.88
24	3242	229	136	100	4.07	5.15
25	3398	443	254	192	11.02	11.92
26	3409	474	402	131	11.59	10.99
27	3192	291	590	70	5.48	5.32
28	3351	450	254	317	13.32	7.45
29	3264	485	293	211	13.35	9.74
30	2862	166	181	77	5.69	0.66
31	3246	311	190	142	11.35	9.12
32	3093	459	730	60	9.96	2.73
33	3457	490	438	95	14.67	7.42
34	3079	208	186	262	10.58	6.38
35	3364	486	272	194	13.12	8.72
36	3239	406	386	72	11.79	6.90
37	3107	285	483	49	7.93	6.34
38	3260	327	385	131	5.46	5.49
39	2861	179	91	228	7.86	6.46
40	2903	189	199	74	3.33	3.45
41	2848	192	159	202	6.86	4.27
42	3177	416	444	118	12.17	6.87
43	3521	447	436	227	12.81	7.03
44	3396	529	371	144	13.04	8.56
45	3201	161	182	104	4.11	4.91
46	3256	683	345	146	15.32	6.34
47	3517	753	422	128	16.08	9.59

Table 1.4 *The matrix of correlations among the six variables in Table 1.3.*

	Calorie	Meat	Milk	Alcohol	L-Intes.	Rectum
Calorie	1	.773	.717	.634	.7405	.810
Meat	.773	1	.723	.452	.853	.672
Milk	.717	.723	1	.232	.579	.479
Alcohol	.634	.452	.232	1	.588	.610
L-Intes.	.740	.853	.579	.588	1	.752
Rectum	.810	.672	.479	.610	.752	1

tion between the western style diet and the proneness to cancer in lower digestive organs is found to be .855, which indicates that a high proportion of the variability ($.855^2 \simeq 70.1\%$) in the latter can be predicted by the former. Caution must be exercised in interpreting this number because some portions of the high predictability can also be due to other variables not included in the current data set. For example, accessibility to health care systems, overall wealth and other living conditions, etc. in these countries may also be responsible for cancer mortality rates, and these variables may to some extent explain away some of the high correlations among the food variables and the cancer variables.

The use of principal components in the above analysis may be criticized on the ground that they were extracted completely separately from respective subsets of observed variables without taking into account other subsets of variables. The principal component which captures the most dominant variations in each set may not be a good predictor of other data sets. This situation is analogous to principal component regression (PCR), in which predictor variables in regression analysis are replaced by their principal components to reduce the number of predictor variables or to avoid multicollinearity (high correlations) among the predictor variables. PCR has been under attack for similar reasons as above; the dimension reduction, while preserving the greatest variability in predictor variables, may nonetheless result in important loss in predictive power of the response variables. Note, however, that this criticism does not apply to the present case. The two components extracted are highly correlated, and the component extracted from the food variables is indeed highly predictive of the component extracted from the cancer variables.

The result seen above can be more directly verified by applying a technique similar to the one used in the previous section, in which we use the food variables as the explanatory variables $\mathbf{G}$. This technique is formally called redundancy analysis (RA), which is a special case of CPCA. (Note that in the previous section, $\mathbf{G}$ was a set of dummy variables, whereas it now comprises continuous variables. Note also that in the previous section, the focus of analysis lay in how much of the SS in the main data matrix $\mathbf{Z}$ can be explained by $\mathbf{G}$, whereas in the present case, our interest lies in how well the variability in $\mathbf{Z}$ can be predicted by $\mathbf{G}$. These differences, however, do not affect the method of analysis.) RA seeks to find the subspace in the space of predictor variables most predictive of criterion variables. We applied

RA to the food and cancer data and extracted one (redundancy) component. This component is highly correlated with the two criterion variables (.862 and .811; these correlations are called criterion loadings), and also with the four predictor variables (.924, .914, .633, and .715; these correlations are called predictor loadings). It also correlates highly with the two components extracted in the previous analysis (.957 with the food component, and .894 with the cancer component), confirming our remark above that the previous analysis involving PCA is not unreasonable. One may rightfully wonder, though, why we should even bother with the previous analysis if RA is sufficient. RA is limited to two sets of variables, however. It has been extended to more than two sets of variables (Takane and Hwang 2005), but a special kind of iterative method must be used in the extended RA, whereas the group PCA method used in the previous analysis is easy to apply, and most often provides good approximations to the extended RA.

1.3 Analysis of Greenacre's Data

The two example data sets analyzed so far both consisted of continuous variables. The next two examples involve discrete variables in the form of cross classification tables or contingency tables.

Table 1.5 displays a cross-tabulation of cars purchased in the United States in the fourth quarter of the year 1988 in terms of size classes of the cars, and age and income of purchasers (Greenacre 1993). As given in the panel below, cars are classified into 14 size classes, and the age and income variables were classified into 7 and 9 groups, respectively.

Variables and categories used to create Table 1.5

Size classes		Age groups		Income groups	
A	Full-size, Standard	a1	18 to 24 yrs.	i1	$\geq$ $75,000
B	Full-size, Luxurious	a2	25 to 34 yrs.	i2	$ 50,000 to $ 74,999
C	Personal, Luxurious	a3	35 to 44 yrs.	i3	$ 35,000 to $ 49,999
D	Intermediate, Regular	a4	45 to 54 yrs.	i4	$ 25,000 to $ 34,999
E	Intermediate, Specialty	a5	55 to 64 yrs.	i5	$ 20,000 to $ 24,999
F	Compact, Regular	a6	65 to 74 yrs.	i6	$ 15,000 to $ 19,999
G	Compact, Specialty	a7	$\geq$ 75 yrs	i7	$ 10,000 to $ 14,999
H	Subcompact, Regular			i8	$ 8,000 to $ 9,999
I	Subcompact, Specialty			i9	< $8,000
J	Passenger, Utility				
K	Imports, Economy				
L	Imports, Standard				
M	Imports, Sport				
N	Imports, Luxurious				

In the analysis of a cross classification table, we are typically interested in investigating how the rows and columns are related. In the present case, we are more

Table 1.5: *Greenacre's data: Three-way cross-tabulation of size classes of cars by age groups by income groups**.

		Size class													
		Full-size		Per.	Interm.		Compact		Subc.		Pas.	Imports			
		Std.	Lux.	Lux.	Reg.	Spe.	Reg.	Spe.	Reg.	Spe.	Uti.	Eco.	Std.	Spo.	Lux.
Ag.	Ic.	A	B	C	D	E	F	G	H	I	J	K	L	M	N
a1	i1	36	61	26	47	62	26	156	49	32	2	279	111	128	130
	i2	88	70	24	89	97	47	251	93	65	1	396	122	114	80
	i3	216	149	30	229	230	130	558	256	150	5	716	260	135	104
	i4	55	42	11	75	97	61	305	145	109	0	453	112	78	36
	i5	49	19	4	62	60	44	178	93	77	0	338	69	48	22
	i6	14	7	4	31	50	38	127	70	47	2	253	50	32	17
	i7	32	15	4	47	40	40	179	127	64	2	387	74	39	17
	i8	5	4	1	9	10	6	33	26	18	0	98	19	7	4
	i9	9	5	3	20	17	21	50	50	20	0	147	32	11	6
a2	i1	468	532	208	605	557	175	816	290	171	15	1675	1089	601	992
	i2	659	521	155	941	867	364	1311	564	306	18	2535	1479	677	946
	i3	1140	611	155	1849	1424	744	2553	1078	735	32	4131	2127	821	784
	i4	477	256	68	880	764	482	1607	786	506	18	2647	1129	554	338
	i5	374	170	39	678	563	413	1134	691	399	11	2091	707	329	197
	i6	156	69	22	293	293	250	655	430	256	6	1356	463	185	145
	i7	290	144	28	433	377	319	876	531	328	2	1718	544	238	135
	i8	55	21	5	62	70	57	154	119	50	0	391	107	46	31
	i9	23	22	5	78	65	65	213	147	62	2	547	126	59	34
a3	i1	1091	1236	482	1387	1193	365	1718	633	362	37	2832	1842	1039	2823
	i2	1479	1195	527	1858	1754	680	2703	1168	779	46	4172	2217	1065	1990
	i3	2417	1365	429	3521	3122	1491	4845	2499	1501	57	7032	3322	1226	1675
	i4	953	564	157	1412	1293	829	2127	1223	747	36	3420	1345	522	535
	i5	745	345	96	1129	928	686	1665	1006	558	17	2759	1037	356	294
	6	366	198	38	551	433	357	800	536	287	14	1570	558	243	208
	i7	380	174	43	611	514	392	964	715	386	20	1939	596	242	226
	i8	62	34	9	121	100	82	184	157	75	0	481	129	55	49
	i9	91	43	10	158	137	115	272	199	85	2	619	188	71	67
	i1	1576	2017	682	1507	1699	606	3144	1306	820	16	4770	2505	1888	2809
	i2	2003	2022	614	2026	2137	966	3924	1723	1197	21	5516	2342	1473	1698

a4	i3	3260	2254	635	3488	3242	1690	5774	2988	1988	36	7724	3079	1570	1454
	i4	1231	758	203	1432	1287	795	2189	1248	812	15	3339	1077	526	449
	i5	780	457	82	937	852	616	1601	1023	636	12	2460	741	304	262
	i6	300	207	48	411	379	288	682	448	240	6	1256	365	173	136
	i7	497	323	67	637	470	400	993	663	429	14	1699	446	245	189
	i8	74	63	12	122	100	81	165	134	65	0	346	87	29	31
	i9	101	66	16	135	112	97	208	145	88	4	446	111	43	44
a5	i1	1771	2379	531	1481	1251	551	1787	805	477	18	2975	1702	1033	1722
	i2	2549	2421	449	1931	1431	763	2062	995	623	11	3098	1486	751	1014
	i3	4611	3100	451	3778	2333	1426	3163	1973	1140	19	4749	2108	841	867
	i4	1968	1173	166	1629	1009	778	1413	914	494	7	2021	786	270	278
	i5	1575	804	112	1591	848	734	1250	799	503	8	1813	678	244	199
	i6	637	386	49	610	301	326	525	358	229	4	866	268	99	95
	i7	870	495	57	901	479	497	761	584	389	6	1277	413	166	146
	i8	105	75	14	129	75	88	101	87	56	0	196	54	26	19
	i9	147	102	12	172	100	100	160	140	81	2	305	78	28	26
a6	i1	1462	1963	360	1168	693	247	634	322	158	3	986	596	294	692
	i2	2390	2548	358	1581	782	396	743	405	264	3	1099	644	261	550
	i3	4854	3441	412	3506	1477	960	1392	944	486	11	2136	1085	312	641
	i4	3390	2036	224	2736	1018	871	976	720	347	5	1410	673	156	303
	i5	3141	1457	167	2647	996	954	1041	770	426	3	1422	569	172	220
	i6	1446	661	100	1257	427	483	479	435	229	4	752	295	77	96
	i7	2114	947	111	2171	690	903	853	779	425	2	1278	407	117	123
	i8	236	141	8	250	90	133	127	104	59	1	192	58	8	17
	i9	181	113	10	244	71	138	109	127	62	0	190	55	11	21
a7	i1	600	933	276	567	336	138	374	173	61	3	985	453	321	595
	i2	1099	1246	218	806	429	230	384	245	96	1	894	370	212	357
	i3	1951	1697	249	1818	756	522	634	491	227	3	1332	585	256	352
	i4	1656	1144	129	1705	499	568	515	418	191	1	900	358	111	155
	i5	1504	818	100	1686	497	553	430	471	194	4	854	326	87	118
	i6	782	388	42	899	259	373	219	281	132	0	432	160	39	54
	i7	1134	545	61	1544	345	702	435	502	226	4	775	227	61	69
	i8	139	81	9	218	43	104	57	81	30	0	121	23	8	13
	i9	113	60	9	186	45	115	47	92	35	0	111	29	8	10
Ag.	Ic.	A	B	C	D	E	F	G	H	I	J	K	L	M	N
		Std.	Lux.	Lux.	Reg.	Spe.	Reg.	Spe.	Reg.	Spe.	Uti.	Eco.	Std.	Spo.	Lux.

*Reproduced from Exhibit 15.4 of Greenacre (1993) with permission from Elsevier.

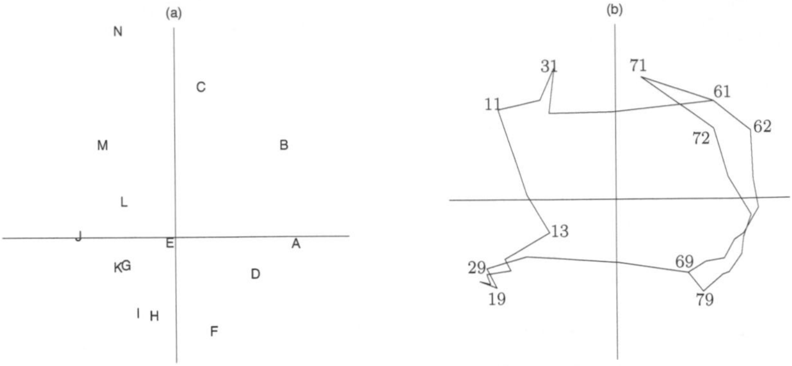

Figure 1.3 *Configurations of size classes of cars (a) and of age and income groups of purchasers (b) obtained by unconstrained CA.*

specifically interested in finding out which size classes of cars are more and less popular among which age and income groups. Correspondence analysis (CA; Greenacre 1984), also known as dual scaling (DS; Nishisato 1980), is a popular technique for analyzing such a table. It is also a special case of CPCA, as will be discussed in Section 4.6. Here, it is sufficient to understand it as a special kind of PCA designed for the analysis of contingency tables. See also Section 1.5 for analytic details.

Figure 1.3a shows the two-dimensional configuration of the 14 size classes of cars (corresponding to the columns of Table 1.5) derived from (unconstrained) CA of the data in Table 1.5. The first component (the horizontal axis) roughly corresponds with the size dimension (large cars on the right, and smaller cars toward the left). The second component (the vertical axis), on the other hand, represents luxuriousness of cars (more luxurious cars toward the top, and plain cars at the bottom).

Figure 1.3b presents a configuration of row points representing combinations of age groups and income groups. To avoid a clutter of symbols, only those lying in the periphery of the configuration are plotted. The first of two integers labeling each point represents the age group, and the second indicates the income group. For example, "21" designates the group between 25 and 34 years of age and the income above $ 75,000. The horizontal axis roughly corresponds with age and the vertical axis with income. Younger generations are located toward the left, while older people tend to be located toward the right, although beyond 75 years of age a minor reversal occurs in this tendency. Wealthy people are located toward the top, while low income groups toward the bottom. The two figures (1.3a, b) are consistent in the sense that the car size and the age of purchaser go together along the horizontal axis with older people preferring larger cars. The luxuriousness and the income level go together along the vertical axis, with wealthier people tending to afford and buy more luxurious cars.

What happens if we incorporate the information on age and income in the representation of the rows of the table, and the information on size classes in the representation of the columns of the table? Let $\mathbf{1}_9$ denote the 9-component vector of ones, $\mathbf{0}$ the vector of zeros of the same size, and $\mathbf{I}_9$ the identity matrix of order 9, and define the age and income information matrix **G** by

$$\mathbf{G} = \left[\begin{array}{cccc|c} \mathbf{1}_9 & \mathbf{0} & \cdots & \mathbf{0} & \mathbf{I}_9 \\ \mathbf{0} & \mathbf{1}_9 & \cdots & \mathbf{0} & \mathbf{I}_9 \\ \vdots & \vdots & \ddots & \vdots & \vdots \\ \mathbf{0} & \mathbf{0} & \cdots & \mathbf{1}_9 & \mathbf{I}_9 \end{array}\right]. \tag{1.2}$$

The first part of **G** represents age groups, while the second part represents income groups. (The basic construction of each part of this matrix is analogous to the **G** matrix in Section 1.1.) Similarly, let

$$\mathbf{H} = \left[\begin{array}{cccccc|ccccc} 1&0&0&0&0&0&0&1&0&0&0 \\ 1&0&0&0&0&0&1&0&0&0&0 \\ 0&0&0&0&0&0&1&0&0&0&0 \\ 0&1&0&0&0&0&0&0&0&1&0 \\ 0&1&0&0&0&0&0&0&1&0&0 \\ 0&0&1&0&0&0&0&0&0&1&0 \\ 0&0&1&0&0&0&0&0&1&0&0 \\ 0&0&0&1&0&0&0&0&0&1&0 \\ 0&0&0&1&0&0&0&0&1&0&0 \\ 0&0&0&0&1&0&0&0&0&0&0 \\ 0&0&0&0&0&1&0&0&0&1&0 \\ 0&0&0&0&0&1&0&1&0&0&0 \\ 0&0&0&0&0&1&0&0&0&0&1 \\ 0&0&0&0&0&1&1&0&0&0&0 \end{array}\right] \begin{array}{c} \mathrm{A}\\ \mathrm{B}\\ \mathrm{C}\\ \mathrm{D}\\ \mathrm{E}\\ \mathrm{F}\\ \mathrm{G}\\ \mathrm{H}\\ \mathrm{I}\\ \mathrm{J}\\ \mathrm{K}\\ \mathrm{L}\\ \mathrm{M}\\ \mathrm{N} \end{array} \tag{1.3}$$

denote the information matrix on size classes. This matrix has 14 rows corresponding to the 14 size classes, and 11 columns, the first six of which represent the size, and the last five of which represent the luxuriousness of cars. (Column 1 indicates full-sized cars, Column 2 intermediate-sized cars, Column 3 compact cars, Column 4 subcompact cars, Column 5 passenger utility cars, and Column 6 imported cars. Column 7 indicates luxurious cars, Column 8 standard cars, Column 9 specialty cars, Column 10 regular cars, and Column 11 sports cars.) Row 1, for example, has ones in columns 1 and 8, meaning that this row represents the full-sized standard cars. Note that class C has no applicable size category, and class J has no applicable luxuriousness category.

Constrained CA incorporating the above constraints are applied to the data in Table 1.5. This special kind of CA is called canonical correspondence analysis (CCA; ter Braak 1986), which, like unconstrained CA, is a special case of CPCA (Section 4.7). Figures 1.4a and b show two-dimensional representations of column and row points derived from this analysis. It may be observed that these figures are strikingly similar to Figures 1.3a and b, despite the fact that the latter were obtained without explicit reference to the external information given in (1.2) and (1.3). It is especially remarkable that there are only slight differences in the locations of size classes (compare Figures 1.3a and 1.4a). This is partly because there are 11 predictor variables for 14 size classes, and most of the variability among the size classes can be captured

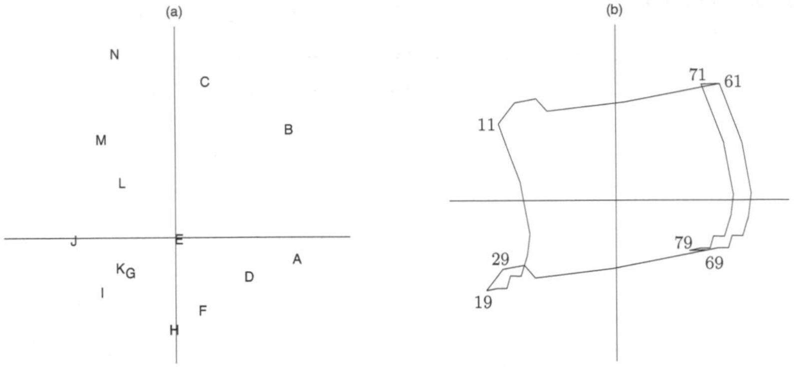

Figure 1.4 *Two-dimensional configurations of size classes of cars (a) and of age and income groups of purchasers (b) obtained by constrained CA.*

by the 11 variables. This implies that a major portion of variability in the original table can be explained by the external information. In the unconstrained CA, 90.5% of the total SS in the original data can be explained by the two components, and in the constrained CA, 86.5% of the total SS can be explained by the two components.

It may be noted that in Figures 1.4a and b, there is no one-to-one correspondence between external information and the dimensions extracted. The horizontal axes do not perfectly coincide with the size and age groups, or the vertical axes with the luxuriousness and income groups. This is because PCA extracts the most dominant components from the portion of the original data matrix that can be explained by **G** treated as a whole, and **H** treated as a whole. Therefore, the principal components are typically mixtures of both size and luxuriousness, and of the age and income groups. In order to obtain components which have a one-to-one correspondence with external information (e.g., the first component represents strictly the size and age, while the second component represents strictly the luxuriousness and income), one has to use an analysis technique called DCDD (Different constraints on different dimensions), to be described in Chapter 6 of this book.

1.4 Analysis of Tocher's Data

As in the previous section, we again analyze a contingency table in this section, demonstrating a more rigorous use of external information. This is in contrast to Section 1.3, in which external information was used either informally for interpretation, or for assessing how much of the variability (SS) in the original data can be explained by the external information. Specifically, in this section we view external information as representing certain hypotheses about a contingency table, and use chi-square tests to test these hypotheses.

A small contingency table given in Table 1.6 cross-classifies 5,387 subjects into four categories of eye color (Blue, Light, Medium, and Dark) and five categories of hair color (Fair, Red, Medium, Dark, and Black). This data set was used by Fisher

Table 1.6 *Tocher's eye color and hair color data**.

Eye Color	Hair Color				
	Fair	Red	Medium	Dark	Black
Blue	326	38	241	110	3
Light	688	116	384	188	4
Medium	343	84	909	412	26
Dark	98	48	403	681	85

*Reproduced from Fisher (1940) with permission from Wiley.

(1940) to demonstrate hypothesis testing in contingency tables. However, the method he used was rather heuristic. In what follows, we reexamine his hypotheses using a more rigorous method.

Fisher first applied unconstrained CA (Section 4.6) to this data set and derived a set of scores for rows and columns of the table (row and column representations). He then examined how consistent his hypotheses were with these scores. He observed that the scores for the Blue and Light eye colors were very close, and so the difference between them might not be significant. If so, these two categories may be combined into one category. He calculated the standard error of the difference between the two scores, and concluded that this difference was not statistically significant. He then posed another question: The scores given to the three row categories remaining after the first two categories were merged into one looked fairly linear, but was there any significant departure from linearity? Fisher again used his heuristic method to test this hypothesis and found that there was a significant departure from linearity.

These hypotheses may be rigorously tested using chi-square tests. The N (the total sample size) times the total SS in the contingency table is known to be equal to Pearson's (1900) chi-square statistic. This statistic indicates the total association between rows and columns of a contingency table. For Table 1.6, the value of chi-square is found to be 1240.0 with 12 degrees of freedom (df). This indicates a highly significant association or departure from independence between rows and columns of the table. If we incorporate external information in the analysis as row and/or column constraints, we can partition the total chi-square into the portions that can and cannot be explained by the external information. Both part chi-squares follow independent chi-square distributions with df's that add up to the df in the original chi-square statistic.

To actually perform these tests, define the following three contrast vectors for the rows of the table:

$$\mathbf{t}_1 = \begin{pmatrix} 1 \\ -1 \\ 0 \\ 0 \end{pmatrix}, \quad \mathbf{t}_2 = \begin{pmatrix} -3 \\ -3 \\ 1 \\ 5 \end{pmatrix}, \text{ and } \mathbf{t}_3 = \begin{pmatrix} 1 \\ 1 \\ -4 \\ 2 \end{pmatrix}. \tag{1.4}$$

Each of the above contrast vectors represents a specific hypothesis about the rows of the contingency table. For example, $\mathbf{t}_1$ represents the difference between the first two rows. Stating that there is no effect of $\mathbf{t}_1$ thus implies that there is no difference between them. Similarly, $\mathbf{t}_2$ represents a linear trend in row scores after the first two rows are merged into one, and $\mathbf{t}_3$ represents all remaining variation among the rows of the table. Note that all three contrast vectors are orthogonal to a constant vector (qualifying that they are contrast vectors), and they are mutually orthogonal.

We first apply CA with $\mathbf{t}_2$ and $\mathbf{t}_3$ used as external information (constraints), and take residuals from this analysis. This has the effect of eliminating the effects of $\mathbf{t}_2$ and $\mathbf{t}_3$ from $\mathbf{t}_1$, representing the "pure" effect of the difference between the first two row categories. The chi-square statistic representing this effect turns out to be 8.3 with 4 df, indicating that there is no significant difference between the first two row categories of the table at the 5% significance level. This result is consistent with Fisher's conclusion.

We then fit $\mathbf{t}_2$, and take residuals from this analysis. The residual chi-square indicates the joint effects of $\mathbf{t}_1$ and $\mathbf{t}_3$ (i.e., the difference between the first two row categories plus the nonlinear trend in scores given to the remaining three row categories) after eliminating the effect of $\mathbf{t}_2$ (the linear trend). The chi-square corresponding to the joint effects is found to be 178.0 with 8 df. By subtracting the effect of $\mathbf{t}_1$ found above from this chi-square, we obtain the "pure" effect of $\mathbf{t}_3$. The chi-square representing this effect is 169.7 (= 178.0 - 8.3) with 4 df (= 8 df - 4 df), indicating a significant departure from linearity. Fisher's conclusion is again confirmed to be correct, using a more rigorous chi-square test.

An important message here is that in the analysis of contingency tables, the total chi-square can be partitioned into nonoverlapping part chi-squares, each representing the unique effect of a contrast or a set of constraints. These part chi-squares may be used to test specific hypotheses represented by the contrasts. For theoretical details, see Section 4.7. Although in this section uses of chi-square tests have been demonstrated only for the rows of the table, it is obvious that similar decompositions are possible with respect to column categories.

1.5 A Summary of the Analyses in This Chapter

We give a brief technical summary of the analyses reported in this chapter. Readers with little or no background in PCA or CA may read Sections 4.1, 4.6, and 4.7 first, and then come back to this section.

In the analysis of Mezzich's data in Section 1.1, we denote the raw data matrix by an n by m matrix $\mathbf{Z}^*$. We rowwise standardize the data by first rowwise centering, and then rowwise normalizing the data. Let $\mathbf{Z}_r$ and $\mathbf{Z}$ denote the rowwise centered data and the rowwise standardized data, respectively. Then,

$$\mathbf{Z}_r = \mathbf{Z}^* - \mathbf{f}\mathbf{1}_m' = \mathbf{Z}^*(\mathbf{I}_m - \mathbf{1}_m\mathbf{1}_m'/m), \tag{1.5}$$

where $\mathbf{1}_m$ is the m-component vector of ones, and $\mathbf{f} = \mathbf{Z}^*\mathbf{1}_m/m$ is the row mean vector. The rowwise standardized data matrix can then be derived by

$$\mathbf{Z} = (\text{diag}(\mathbf{Z}_r\mathbf{Z}_r'/m))^{-1/2}\mathbf{Z}_r, \tag{1.6}$$

where diag($\mathbf{S}$) indicates a diagonal matrix whose diagonal elements are equal to the diagonal elements of the square matrix $\mathbf{S}$. We apply PCA to $\mathbf{Z}$, which amounts to applying singular value decomposition (SVD; Section 2.3.4) to $\mathbf{Z}$. This is written as

$$\mathbf{Z} = \mathbf{U}\mathbf{D}\mathbf{V}', \tag{1.7}$$

where we assume, as is usually the case, that singular values in the diagonals of $\mathbf{D}$ are arranged in the descending order of magnitude. Suppose that we retain only the two most dominant components. (See Section 5.1 for how to choose the number of components.) Let $\mathbf{U}_2$ and $\mathbf{V}_2$ denote the portions of $\mathbf{U}$ and $\mathbf{V}$ pertaining to the first two dominant singular values, and let $\mathbf{D}_2$ denote the leading 2 by 2 diagonal block of $\mathbf{D}$, containing the two largest singular values in the diagonals. Figure 1.1a is obtained by plotting rows of

$$\mathbf{A}_2 = \mathbf{V}_2\mathbf{D}_2/\sqrt{n}. \tag{1.8}$$

This is called the component loading matrix. Figure 1.2b, on the other hand, is obtained by plotting rows of

$$\mathbf{B}_2 = \sqrt{n}\mathbf{U}_2. \tag{1.9}$$

This is called the matrix of component scores. As can be seen in (1.8) and (1.9), $\mathbf{V}_2$ and $\mathbf{U}_2$ must be rescaled to obtain the component loading and component score matrices. However, we may sometimes call $\mathbf{V}_2$ and $\mathbf{U}_2$ the component loading and component score matrices as if the rescaling has already been done. The centroids of the four diagnostic categories are obtained by

$$\mathbf{M}_2 = (\mathbf{G}'\mathbf{G})^{-1}\mathbf{G}'\mathbf{B}_2, \tag{1.10}$$

where $\mathbf{G}$ is as defined in (1.1).

In CPCA, we denote the orthogonal projector onto Sp($\mathbf{G}$), the space spanned by column vectors of $\mathbf{G}$, by

$$\mathbf{P}_G = \mathbf{G}(\mathbf{G}'\mathbf{G})^{-1}\mathbf{G}' \tag{1.11}$$

(see Section 2.2), and obtain the SVD of the between-group data matrix defined by $\mathbf{P}_G\mathbf{Z}$. Figure 1.2a is obtained by plotting (1.8) calculated from $\mathbf{V}_2$ and $\mathbf{D}_2$ obtained by SVD($\mathbf{P}_G\mathbf{Z}$). Let

$$\mathbf{Q}_G = \mathbf{I}_n - \mathbf{P}_G \tag{1.12}$$

denote the orthogonal projector onto the space orthogonal to Sp($\mathbf{G}$). (This space is called the orthogonal complement subspace to Sp($\mathbf{G}$) and is denoted as Ker($\mathbf{G}'$).) Figure 1.2b is obtained by the SVD of the within-group data matrix defined by $\mathbf{Q}_G\mathbf{Z}$.

The total SS in Table 1.2 is calculated by SS($\mathbf{Z}$) = tr($\mathbf{Z}'\mathbf{Z}$), which is equal to $n \times m$ under the rowwise standardization convention. The componentwise SSs are calculated by d_j^2 ($j = 1,2$), where d_j is the jth diagonal element of $\mathbf{D}$ from SVD($\mathbf{Z}$). The percent SSs (%SSs) are calculated by dividing the SSs by the total SS and multiplying it by 100. The between-group and within-group SSs are calculated similarly by SS($\mathbf{P}_G\mathbf{Z}$) and SS($\mathbf{Q}_G\mathbf{Z}$), respectively. The componentwise SSs and %SSs for the between- and within-group data matrices are also calculated in similar ways. Note that %SSs for these effects are calculated out of the total SS in the original data matrix $\mathbf{Z}$.

We do not have to say much about separate PCAs of subsets of variables in Section 1.2. However, it may be worthwhile pointing out that this may be considered as a special case of CPCA. Let $\mathbf{I}_m$ denote the identity matrix, where m is the total number of variables in the data set. Suppose that we are splitting the m variables into the first m_1 variables and the remaining $m_2 = m - m_1$ variables. We split $\mathbf{I}_m$ into two parts, say $\mathbf{H}_1$ consisting of the first m_1 columns of $\mathbf{I}_m$, and $\mathbf{H}_2$ consisting of the remaining columns of $\mathbf{I}_m$. Note that $\mathbf{H}_1'\mathbf{H}_2 = \mathbf{O}$, and $\mathbf{H}_1$ and $\mathbf{H}_2$ together cover the entire row space of the data matrix $\mathbf{Z}$. The matrices $\mathbf{H}_j$ $(j = 1, 2)$ are called selection matrices. We then have the following decomposition of the data matrix $\mathbf{Z}$ according to the External Analysis of CPCA:

$$\begin{aligned} \mathbf{Z} &= \mathbf{Z}\mathbf{P}_{H_1} + \mathbf{Z}\mathbf{P}_{H_2}, \\ &= [\mathbf{Z}_1, \mathbf{O}] + [\mathbf{O}, \mathbf{Z}_2], \end{aligned} \tag{1.13}$$

where the $\mathbf{P}_{H_j}$ $(j = 1, 2)$ are the orthogonal projectors onto $\text{Sp}(\mathbf{H}_j)$, and the $\mathbf{Z}_j$ are submatrices (row blocks) of the relevant portions of $\mathbf{Z}$. The separate PCAs of the two terms on the right-hand side constitute the Internal Analysis of CPCA, and provide results essentially equivalent to separate PCAs of $\mathbf{Z}_1$ and $\mathbf{Z}_2$. Note that the two terms on the right-hand side of (1.13) are not columnwise orthogonal. However, they are trace-orthogonal (i.e., $\text{tr}(\mathbf{P}_{H_1}\mathbf{Z}'\mathbf{Z}\mathbf{P}_{H_2}) = 0$), so that the total SS in $\mathbf{Z}$ is still partitioned into the sum of part SSs. Separate PCAs of more than two subsets of variables can be done similarly.

Another way of performing essentially the same analysis is to form a block diagonal matrix of the form

$$\mathbf{D}_Z = \begin{bmatrix} \mathbf{Z}_1 & \mathbf{O} \\ \mathbf{O} & \mathbf{Z}_2 \end{bmatrix}, \tag{1.14}$$

and obtain the SVD of $\mathbf{D}_Z$. This provides separate SVDs of $\mathbf{Z}_1$ and $\mathbf{Z}_2$ after reordering singular values and vectors appropriately.

In correspondence analysis (CA) of Greenacre's data in Section 1.3, we first calculate

$$\mathbf{Z} = \mathbf{D}_R^{-1}(\mathbf{F} - \mathbf{D}_R\mathbf{1}_R\mathbf{1}_C'\mathbf{D}_C/n)\mathbf{D}_C^{-1}, \tag{1.15}$$

where $\mathbf{F}$ is the $R \times C$ contingency table given in Table 1.5, $\mathbf{D}_R$ and $\mathbf{D}_C$ are diagonal matrices of row and column totals of $\mathbf{F}$, $\mathbf{1}_R$ and $\mathbf{1}_C$ are R- and C-component vectors of ones, and n is the total sample size (i.e., $n = \mathbf{1}_R'\mathbf{F}\mathbf{1}_C = \mathbf{1}_R'\mathbf{D}_R\mathbf{1}_R = \mathbf{1}_C'\mathbf{D}_C\mathbf{1}_C$). The matrices $\mathbf{D}_R$ and $\mathbf{D}_C$ are called metric matrices. The matrix $\mathbf{Z}$ has the following properties:

$$\mathbf{1}_R'\mathbf{D}_R\mathbf{Z} = \mathbf{0}', \quad \text{and} \quad \mathbf{Z}\mathbf{D}_C\mathbf{1}_C = \mathbf{0}. \tag{1.16}$$

This operation has the effect of *a priori* removing the trivial solution in CA. (See Section 4.6.) The CA of $\mathbf{Z}$ obtains the GSVD of $\mathbf{Z}$ with the metric matrices $\mathbf{D}_R$ and $\mathbf{D}_C$ (Section 2.3.6), which is often written as $\text{GSVD}(\mathbf{Z})_{D_R, D_C}$. Let this GSVD be denoted as

$$\mathbf{Z} = \mathbf{U}\mathbf{D}\mathbf{V}'. \tag{1.17}$$

Then Figures 1.3a and b are obtained by plotting the rows of $\mathbf{A}_2 = \sqrt{n}\mathbf{V}_2$ and of $\mathbf{B}_2 = \sqrt{n}\mathbf{U}_2\mathbf{D}_2$, respectively. Note the difference in scaling convention from that in

PCA. Note also that the columns of $\mathbf{A}_2$ have unit variances with respect to the metric matrix $\mathbf{D}_R$, while the columns of $\mathbf{B}_2$ have variances equal to $\mathbf{D}_2^2$ (squares of generalized singular values). We say that $\mathbf{A}_2$ is in standard coordinates, and $\mathbf{B}_2$ in principal coordinates (Greenacre 1993). The $\text{GSVD}(\mathbf{Z})_{D_R,D_C}$ is obtained by first obtaining $\text{SVD}(\mathbf{D}_R^{1/2}\mathbf{Z}\mathbf{D}_C^{1/2})$. Let this SVD be denoted as

$$\mathbf{D}_R^{1/2}\mathbf{Z}\mathbf{D}_C^{1/2} = \mathbf{U}^*\mathbf{D}^*\mathbf{V}^{*\prime}. \tag{1.18}$$

Then, $\text{GSVD}(\mathbf{Z})_{D_R,D_C}$ is obtained by $\mathbf{U} = \mathbf{D}_R^{-1/2}\mathbf{U}^*$, $\mathbf{V} = \mathbf{D}_C^{-1/2}\mathbf{V}^*$, and $\mathbf{D} = \mathbf{D}^*$ (Section 2.3.6).

In constrained CA (Section 4.7), we first define projectors with the metric matrices $\mathbf{D}_R$ and $\mathbf{D}_C$ by

$$\mathbf{P}_{G/D_R} = \mathbf{G}(\mathbf{G}'\mathbf{D}_R\mathbf{G})^{-1}\mathbf{G}'\mathbf{D}_R, \tag{1.19}$$

and

$$\mathbf{P}_{H/D_C} = \mathbf{H}(\mathbf{H}'\mathbf{D}_C\mathbf{H})^{-1}\mathbf{H}'\mathbf{D}_C, \tag{1.20}$$

and then obtain $\text{GSVD}(\mathbf{P}_{G/D_R}\mathbf{Z}\mathbf{P}'_{H/D_C})_{D_R,D_C}$.

The total SS in CA is defined as

$$\text{SS}(\mathbf{Z})_{D_R,D_C} = \text{tr}(\mathbf{Z}'\mathbf{D}_R\mathbf{Z}\mathbf{D}_C) = \text{SS}(\mathbf{D}_R^{1/2}\mathbf{Z}\mathbf{D}_C^{1/2}). \tag{1.21}$$

The SS in constrained CA is analogously defined as

$$\text{SS}(\mathbf{P}_{G/D_R}\mathbf{Z}\mathbf{P}'_{H/D_C})_{D_R,D_C} = \text{SS}(\mathbf{D}_R^{1/2}\mathbf{P}_{G/D_R}\mathbf{Z}\mathbf{P}'_{H/D_C}\mathbf{D}_C^{1/2}). \tag{1.22}$$

The componentwise SS is equal to the square of the (generalized) singular value corresponding to the component. The computation of SSs in CA is somewhat different from that in PCA, mainly because in the former the nonidentity metric matrices $\mathbf{D}_R$ and $\mathbf{D}_C$ are used.

In the analysis of Tocher's data in Section 1.4, Pearson's chi-square is defined by

$$\chi_P^2 = \sum_i^R \sum_j^C \left(\frac{f_{ij} - f_{i.}f_{.j}/n}{\sqrt{f_{i.}f_{.j}/n}} \right)^2, \tag{1.23}$$

where f_{ij} is the (i, j)th element of $\mathbf{F}$, and $f_{i.}$ and $f_{.j}$ are the ith and jth diagonal elements of $\mathbf{D}_R$ and $\mathbf{D}_C$, respectively. This statistic asymptotically follows the chi-square distribution with $(R-1)(C-1)$ degrees of freedom (df) under the hypothesis that there is no association between rows and columns of a contingency table. The following relation holds:

$$\chi_P^2 = n \cdot \text{SS}(\mathbf{Z})_{D_R,D_C}. \tag{1.24}$$

For calculating a part χ^2, say, under the row constraint $\mathbf{t}_1$, set $\mathbf{G} = \mathbf{t}_1$ in (1.19). Alternatively, we may set $\mathbf{G} = \mathbf{Q}_{1/D_R}\mathbf{t}_1$. In this case, we may also set $\mathbf{Z} = \mathbf{D}_R^{-1}\mathbf{F}\mathbf{D}_C^{-1}$.

The MATLAB codes for all the computations in this chapter are available from http://takane.brinkster.net/Yoshio/.

Chapter 2

Mathematical Foundation

In this chapter we introduce some of the mathematical (mostly linear algebraic) concepts useful in subsequent chapters. We do not assume advanced knowledge of linear algebra, but the reader will find this chapter considerably easier to follow with some basic knowledge of vector and matrix operations (e.g., addition, subtraction, multiplication, etc.). Those who lack the background knowledge in this area are encouraged to read Green and Carroll (1976; Carroll et al. 1997), Searle (1966), or introductory chapters in Harville (1997) and Yanai et al. (2011). Green and Carroll provide a good starting point, as it gives a very nontechnical introduction to the subject, followed by Searle, which is slightly more technical, but provides a comprehensive account of general topics in linear algebra. Harville and Yanai et al. also cover fairly advanced and specialized topics.

An overview of this chapter is as follows: Section 2.1 presents a fairly large body of basic results in linear algebra. This section should serve as a refresher course for advanced readers. For others, it should serve as a guide to what level of knowledge in linear algebra is expected in subsequent sections. Sections 2.2 and 2.3 provide in-depth accounts of two important mathematical ingredients in CPCA, projection matrices for External Analysis and singular value decomposition (SVD) for Internal Analysis. With a few exceptions, no proofs are given of the results presented. Those looking for mathematical rigor in these topics may consult Rao and Mitra (1971), Rao (1973), Yanai et al. (2011), or specific papers cited for the given results. There may also be topics for which statistical motivations are not immediately obvious. These topics may be safely skipped on first reading, but it would be beneficial to return for a closer look when reading later chapters.

2.1 Preliminaries

In this section we overview some of the basic notions in linear algebra. As the concepts and definitions are quite basic, readers already familiar with the topic may wish to proceed directly to Section 2.2.

2.1.1 Vectors

Sets of n real numbers $a_1, \cdots, a_n$ and $b_1, \cdots, b_n$ arranged in the following manner are called m-component column vectors:

$$\mathbf{a} = \begin{pmatrix} a_1 \\ \vdots \\ a_n \end{pmatrix}, \text{ and } \mathbf{b} = \begin{pmatrix} b_1 \\ \vdots \\ b_n \end{pmatrix}. \tag{2.1}$$

The real numbers in **a** and **b** are called elements or components in the vectors. One-component vectors are called scalars. When the elements of vectors are arranged horizontally, i.e.,

$$\mathbf{a}' = (a_1, \cdots, a_n), \text{ and } \mathbf{b}' = (b_1, \cdots, b_n), \tag{2.2}$$

they are called row vectors. The symbol $'$ (read as prime) on the vectors indicates a transposing operation that turns column vectors into row vectors and row vectors into column vectors ($(\mathbf{a}')' = \mathbf{a}$). We use $\mathbf{1}_n$ and $\mathbf{0}_n$ to indicate n-component vectors of ones and zeroes.

We define an inner product between two vectors of the same size by

$$\langle \mathbf{a}, \mathbf{b} \rangle = \sum_{i=1}^{n} a_i b_i = \mathbf{a}'\mathbf{b} = \mathbf{b}'\mathbf{a}, \tag{2.3}$$

and the length of a vector by

$$||\mathbf{a}|| = \sqrt{\langle \mathbf{a}, \mathbf{a} \rangle}. \tag{2.4}$$

The length of **a** is also called a norm of **a**, and its square is sometimes denoted as $\mathrm{SS}(\mathbf{a})$. The following properties hold regarding the inner product and the norm of vectors:

$$\begin{aligned}
&\text{(i)} \ \ ||\mathbf{a}+\mathbf{b}||^2 = ||\mathbf{a}||^2 + ||\mathbf{b}||^2 + 2\langle \mathbf{a}, \mathbf{b} \rangle. \\
&\text{(ii)} \ \ \langle c\mathbf{a}, \mathbf{b} \rangle = \langle \mathbf{a}, c\mathbf{b} \rangle = c\langle \mathbf{a}, \mathbf{b} \rangle, \text{ where } c \text{ is a scalar.} \\
&\text{(iii)} \ \ ||\mathbf{a}|| = 0 \Leftrightarrow \mathbf{a} = \mathbf{0}. \\
&\text{(iv)} \ \ \langle \mathbf{a}, \mathbf{b} \rangle^2 \le ||\mathbf{a}||^2 ||\mathbf{b}||^2 \ \text{(Schwarz' inequality).} \\
&\text{(v)} \ \ ||\mathbf{a}+\mathbf{b}|| \le ||\mathbf{a}|| + ||\mathbf{b}|| \ \text{(the triangle inequality).} \\
&\text{(vi)} \ \ \cos\theta = \frac{\langle \mathbf{a}, \mathbf{b} \rangle}{||\mathbf{a}|| \cdot ||\mathbf{b}||}, \text{ where } \theta \text{ indicates the angle between } \mathbf{a} \text{ and } \mathbf{b}.
\end{aligned} \tag{2.5}$$

In (iii) above, "$\Leftrightarrow$" indicates an equivalence or "if and only if" relationship.

2.1.2 Matrices

The m n-component vectors $\mathbf{a}_1, \cdots, \mathbf{a}_m$ arranged side by side form an $n \times m$ matrix

$$\mathbf{A} = [\mathbf{a}_1, \cdots, \mathbf{a}_m] = \begin{bmatrix} a_{11} & \cdots & a_{1m} \\ \vdots & \ddots & \vdots \\ a_{n1} & \cdots & a_{nm} \end{bmatrix} = [a_{ij}]. \tag{2.6}$$

The tall nm-component vector obtained by vertically stacking the m n-component vectors of $\mathbf{A}$ end to end is denoted by vec($\mathbf{A}$), i.e.,

$$\text{vec}(\mathbf{A}) = (\mathbf{a}_1', \cdots \mathbf{a}_m')'. \tag{2.7}$$

The matrix obtained by interchanging the rows and columns of $\mathbf{A}$ is called the transpose of $\mathbf{A}$ and is denoted by

$$\mathbf{A}' = \begin{bmatrix} a_{11} & \cdots & a_{n1} \\ \vdots & \ddots & \vdots \\ a_{1m} & \cdots & a_{nm} \end{bmatrix} = [\tilde{\mathbf{a}}_1, \cdots, \tilde{\mathbf{a}}_n]. \tag{2.8}$$

The vector $\tilde{\mathbf{a}}_j$ in the jth column of $\mathbf{A}'$ is the transpose of the jth row vector of $\mathbf{A}$. The following property holds for the transpose of a product of two matrices:

$$(\mathbf{AB})' = \mathbf{B}'\mathbf{A}'. \tag{2.9}$$

An $n \times m$ matrix $\mathbf{A}$ is called a square matrix when $n = m$. (The matrix $\mathbf{A}$ is said to be rectangular if $n \neq m$.) A square matrix whose off-diagonal elements are all zero is called a diagonal matrix. The diagonal matrix whose diagonal elements are equal to those of $\mathbf{A}$ is denoted by diag($\mathbf{A}$) = diag($a_{11}, \cdots, a_{nn}$). The diagonal matrix with all diagonal elements equal to unity is called an identity matrix and is denoted by $\mathbf{I}$ (or by $\mathbf{I}_n$, where n is the order of the matrix). A square matrix $\mathbf{A}$ whose transpose is identical to $\mathbf{A}$, that is,

$$\mathbf{A}' = \mathbf{A} \tag{2.10}$$

is called a symmetric matrix. A square matrix $\mathbf{A}$ that satisfies

$$\mathbf{A}'\mathbf{A} = \mathbf{AA}' = \mathbf{I}_n \tag{2.11}$$

is called a (fully) orthogonal matrix.

The sum of the diagonal elements of a square matrix $\mathbf{A}$ is called the trace of $\mathbf{A}$ and is denoted by

$$\text{tr}(\mathbf{A}) = \sum_{i=1}^{n} a_{ii}. \tag{2.12}$$

Let $\mathbf{C}$ be an $n \times m$ matrix, and define $||\mathbf{C}||^2 = \text{tr}(\mathbf{C}'\mathbf{C}) = \text{tr}(\mathbf{CC}') \equiv \text{SS}(\mathbf{C})$. The $||\mathbf{C}||$ is called the Frobenius norm or the SS (sum of squares) norm of $\mathbf{C}$. Some statisticians, e.g., Rao (1980), call this the Euclidean norm. However, many numerical linear algebraists, e.g., Lawson and Hanson (1974), use the term "Euclidean norm" differently. To avoid confusion, we consistently use the term "SS norm" in this book. The SS norm is sometimes called the L_2 norm.

The following properties hold regarding the trace and the SS norm of a matrix:

(i) $\text{tr}(\mathbf{AB}) = \text{tr}(\mathbf{BA})$.

(ii) $||\mathbf{C}||^2 = \sum_{i=1}^{n}\sum_{j=1}^{m} c_{ij}^2$.

(iii) $||\mathbf{C}||^2 = 0 \Leftrightarrow \mathbf{C} = \mathbf{O}$, where $\mathbf{O}$ indicates a zero matrix. (2.13)

(iv) $(\text{tr}(\mathbf{A}'\mathbf{B}))^2 \leq ||\mathbf{A}||^2||\mathbf{B}||^2$ (Schwarz' inequality).

(v) $||\mathbf{A}+\mathbf{B}|| \leq ||\mathbf{A}|| + ||\mathbf{B}||$ (the triangle inequality).

In (i) above, $\mathbf{AB}$ and $\mathbf{BA}$ may be square matrices of different orders. The $\text{tr}(\mathbf{A}'\mathbf{B})$ is sometimes written as $\langle \mathbf{A}, \mathbf{B} \rangle$ using the inner product notation. Note that $\text{tr}(\mathbf{A}'\mathbf{B}) = \langle \text{vec}(\mathbf{A}), \text{vec}(\mathbf{B}) \rangle$, $||\mathbf{A}|| = ||\text{vec}(\mathbf{A})||$, and $||\mathbf{B}|| = ||\text{vec}(\mathbf{B})||$.

There are a number of ways to define a norm of a matrix other than the SS norm. An orthogonally invariant norm is a norm that is invariant under arbitrary orthogonal transformations. In other words, the equality

$$||\mathbf{A}|| = ||\mathbf{UAV}'|| \tag{2.14}$$

holds for any orthogonal matrices $\mathbf{U}$ and $\mathbf{V}$ (Mirsky 1960; Rao 1980). It should be clear that the SS norm is a special case of the orthogonally invariant norm. Most of the solutions to the minimization problems discussed in this book hold more generally for the orthogonally invariant norm, although in most cases we restrict our attention to the SS norm.

Let $\mathbf{A}$ be an $n \times m$ matrix, and let $\mathbf{b}$ be an m-component vector. The n-component vector obtained by $\mathbf{Ab} = \sum_{j=1}^{m} b_j \mathbf{a}_j$, where b_j is the jth element of $\mathbf{b}$, and $\mathbf{a}_j$ is the jth column vector of $\mathbf{A}$, is called a linear combination of the column vectors of $\mathbf{A}$. The column vectors of $\mathbf{A}$ are said to be linearly independent if $\mathbf{Ab} = \mathbf{0}$ implies $\mathbf{b} = \mathbf{0}$. Otherwise they are said to be linearly dependent. The maximum number of linearly independent vectors (column or row) in $\mathbf{A}$ is called the rank of $\mathbf{A}$ and is denoted as $\text{rank}(\mathbf{A})$. The matrix $\mathbf{A}$ is said to be columnwise nonsingular if $\text{rank}(\mathbf{A}) = m$, and rowwise nonsingular if $\text{rank}(\mathbf{A}) = n$. It is called singular if $\text{rank}(\mathbf{A}) < \min(n,m)$. A square matrix $\mathbf{A}$ is said to be nonsingular if $\text{rank}(\mathbf{A}) = n = m$, and is singular otherwise.

The following properties hold regarding the rank of a matrix:

(i) $\text{rank}(\mathbf{A}) = \text{rank}(\mathbf{A}')$.

(ii) $\text{rank}(\mathbf{A}'\mathbf{A}) = \text{rank}(\mathbf{A})$.

(iii) $\text{rank}(\mathbf{AB}) \leq \min(\text{rank}(\mathbf{A}), \text{rank}(\mathbf{B}))$. (2.15)

(iv) $\text{rank}(\mathbf{A}+\mathbf{B}) \leq \text{rank}([\mathbf{A},\mathbf{B}]) \leq \text{rank}(\mathbf{A}) + \text{rank}(\mathbf{B})$.

In (iv) above, $\mathbf{A}$ and $\mathbf{B}$ are assumed to be of the same order. See Marsaglia and Styan (1974) for a large collection of other interesting rank formulae.

2.1.3 Vector space

The set of all n-component vectors is called an n-dimensional vector space and is denoted by E^n. A subset of E^n, denoted by U, that satisfies the following two properties, is called a subspace of E^n:

(i) If $\mathbf{x} \in U$, then $a\mathbf{x} \in U$, where a is an arbitrary scalar.

(ii) If $\mathbf{x}, \mathbf{y} \in U$, then $\mathbf{x}+\mathbf{y} \in U$.

Let U and V be two subspaces of E^n. Then

$$U \cap V = \{\mathbf{x}|\mathbf{x} \in U, \mathbf{x} \in V\}$$

is called the intersection space (or the product space) between U and V. The subspace defined by

$$U + V = \{\mathbf{x}+\mathbf{y}|\mathbf{x} \in U, \mathbf{y} \in V\},$$

on the other hand, is called the sum space of U and V. If

$$U \cap V = \{\mathbf{0}\},$$

U and V are said to be mutually disjoint. In such a case, the sum of U and V is called the direct sum and is denoted by $U \oplus V$. When

$$U \oplus V = E^n$$

(that is, when U and V are mutually disjoint and jointly cover the entire n-dimensional vector space E^m), U and V are said to be complementary subspaces of each other. For a given subspace U, its complementary subspace is not uniquely determined. When the inner product is zero between all vectors $\mathbf{x} \in U$ and $\mathbf{y} \in V$, U and V are said to be mutually orthogonal, denoted by $U \perp V$. In such a case, the direct sum of U and V is called the orthogonal direct sum and is denoted by $U \dot{\oplus} V$. When

$$U \dot{\oplus} V = E^n,$$

U and V are said to be orthocomplement subspaces of each other. For a given subspace, its orthocomplement subspace is uniquely determined.

Let

$$U = \text{Sp}(\mathbf{A}) = \{\mathbf{A}\mathbf{x}|\mathbf{x} \in E^m\} \tag{2.16}$$

represent the subspace spanned by the column vectors of an $n \times m$ matrix $\mathbf{A}$ ($n \geq m$). The maximum number of linearly independent vectors contained in a subspace is called the dimensionality of the subspace and is denoted by $\dim(U)$. This coincides with the rank of $\mathbf{A}$, that is,

$$\dim(U) = \text{rank}(\mathbf{A}). \tag{2.17}$$

The subspace of maximum dimensionality consisting of all vectors orthogonal to all column vectors of $\mathbf{A}$ is called the orthocomplement subspace of $\mathbf{A}$ and is denoted by $V = \text{Ker}(\mathbf{A}')$. The dimensionality of this subspace is equal to

$$\dim(V) = \dim(\text{Ker}(\mathbf{A}')) = n - \text{rank}(\mathbf{A}). \tag{2.18}$$

The subspaces $\text{Sp}(\mathbf{A})$ and $\text{Ker}(\mathbf{A}')$ constitute orthocomplement subspaces of each other, that is,

$$\text{Sp}(\mathbf{A}) \dot{\oplus} \text{Ker}(\mathbf{A}') = E^n \quad (\dim(E^n) = n). \tag{2.19}$$

The subspaces $\text{Sp}(\mathbf{A}')$ and $\text{Ker}(\mathbf{A})$ also constitute orthocomplement subspaces of each other, for which it holds that

$$\text{Sp}(\mathbf{A}') \dot{\oplus} \text{Ker}(\mathbf{A}) = E^m \quad (\dim(E^m) = m). \tag{2.20}$$

The following properties hold concerning Sp and Ker:

(i) $\mathrm{Sp}(\mathbf{AB}) \subset \mathrm{Sp}(\mathbf{A})$.

(ii) $\mathrm{Ker}(\mathbf{B}) \subset \mathrm{Ker}(\mathbf{AB})$. (2.21)

A set of linearly independent vectors spanning U are called basis vectors of U. If those vectors all have unit length and are mutually orthogonal, they are said to be orthonormal ($\mathbf{A}'\mathbf{A} = \mathbf{I}$) basis vectors of U. In general, basis vectors of a subspace are not uniquely determined.

Let U be as defined in (2.16), and let $V = \mathrm{Sp}(\mathbf{B})$. The disjointness of the subspaces U and V is equivalent to

$$\mathrm{rank}([\mathbf{A}, \mathbf{B}]) = \mathrm{rank}(\mathbf{A}) + \mathrm{rank}(\mathbf{B}), \tag{2.22}$$

and the orthogonality of U and V is equivalent to $\mathbf{A}'\mathbf{B} = \mathbf{O}$. That U and V cover the entire space of E^n is equivalent to $\mathrm{rank}([\mathbf{A}, \mathbf{B}]) = n$.

2.1.4 *Kronecker products*

Let $\mathbf{A} = [a_{ij}]$ and $\mathbf{B}$ be $n \times m$ and $p \times q$ matrices, respectively. The Kronecker product of $\mathbf{A}$ and $\mathbf{B}$ is defined by

$$\mathbf{A} \otimes \mathbf{B} = [a_{ij}\mathbf{B}] = \begin{bmatrix} a_{11}\mathbf{B} & \cdots & a_{1m}\mathbf{B} \\ \vdots & \ddots & \vdots \\ a_{n1}\mathbf{B} & \cdots & a_{nm}\mathbf{B} \end{bmatrix}. \tag{2.23}$$

The following properties hold regarding the Kronecker product and the vec operation introduced in (2.7):

(i) $(a\mathbf{A}) \otimes (b\mathbf{B}) = ab(\mathbf{A} \otimes \mathbf{B})$.

(ii) $(\mathbf{A} \otimes \mathbf{B})' = \mathbf{A}' \otimes \mathbf{B}'$.

(iii) $(\mathbf{A} \otimes \mathbf{B})(\mathbf{C} \otimes \mathbf{D}) = (\mathbf{AC}) \otimes (\mathbf{BD})$.

(iv) $\mathrm{rank}(\mathbf{A} \otimes \mathbf{B}) = \mathrm{rank}(\mathbf{A})\mathrm{rank}(\mathbf{B})$.

(v) $\mathrm{tr}(\mathbf{A} \otimes \mathbf{B}) = \mathrm{tr}(\mathbf{A})\mathrm{tr}(\mathbf{B})$, where $\mathbf{A}$ and $\mathbf{B}$ are both square. (2.24)

(vi) $\mathrm{tr}(\mathbf{A}'\mathbf{B}) = \mathrm{vec}(\mathbf{A})'\mathrm{vec}(\mathbf{B})$.

(vii) $\mathrm{vec}(\mathbf{ABC}) = (\mathbf{C}' \otimes \mathbf{A})\mathrm{vec}(\mathbf{B})$.

(viii) $\mathrm{vec}(\mathbf{AB}) = (\mathbf{I} \otimes \mathbf{A})\mathrm{vec}(\mathbf{B}) = (\mathbf{B}' \otimes \mathbf{I})\mathrm{vec}(\mathbf{A})$.

(ix) $\mathrm{tr}(\mathbf{A}'\mathbf{BCD}') = \mathrm{vec}(\mathbf{A})'(\mathbf{D} \otimes \mathbf{B})\mathrm{vec}(\mathbf{C})$.

See Harville (1997) for proofs and for more results.

2.1.5 *Inverse matrices*

For a square matrix $\mathbf{A}$ of order n, another matrix $\mathbf{B}$ of the same size satisfying

$$\mathbf{AB} = \mathbf{BA} = \mathbf{I}_n, \tag{2.25}$$

if it exists, is called the inverse (matrix) of $\mathbf{A}$ and is denoted by $\mathbf{A}^{-1}$. The matrix $\mathbf{A}$ must be nonsingular (i.e., it has to have full rank) for its inverse to exist. The following properties hold for $\mathbf{A}^{-1}$:

(i) $(\mathbf{AB})^{-1} = \mathbf{B}^{-1}\mathbf{A}^{-1}$, where $\mathbf{B}$ is another nonsingular matrix of order n.

(ii) $(\mathbf{A} \otimes \mathbf{B})^{-1} = \mathbf{A}^{-1} \otimes \mathbf{B}^{-1}$. (2.26)

Let $\mathbf{A}$ and $\mathbf{D}$ be nonsingular matrices of order n and m ($m \leq n$), respectively, and let $\mathbf{B}$ be an $n \times m$ matrix. Then,

$$(\mathbf{A} + \mathbf{BDB}')^{-1} = \mathbf{A}^{-1} - \mathbf{A}^{-1}\mathbf{B}(\mathbf{B}'\mathbf{A}^{-1}\mathbf{B} + \mathbf{D}^{-1})^{-1}\mathbf{B}'\mathbf{A}^{-1}. \tag{2.27}$$

Let $\mathbf{A}$ be a nonsingular matrix of order n, and let $\mathbf{b}$ be an n-component vector. Then,

$$(\mathbf{A} + \mathbf{bb}')^{-1} = \mathbf{A}^{-1} - \mathbf{A}^{-1}\mathbf{b}(\mathbf{b}'\mathbf{A}^{-1}\mathbf{b} + 1)^{-1}\mathbf{b}'\mathbf{A}^{-1}. \tag{2.28}$$

This is a special case of (2.27), in which $m = 1$ and $\mathbf{D} = 1$.

Let

$$\mathbf{M} = \begin{bmatrix} \mathbf{A} & \mathbf{B} \\ \mathbf{B}' & \mathbf{D} \end{bmatrix} \tag{2.29}$$

be a nonsingular symmetric matrix. Then

$$\mathbf{M}^{-1} = \begin{bmatrix} \mathbf{A}^{-1} + \mathbf{A}^{-1}\mathbf{BFB}'\mathbf{A}^{-1} & -\mathbf{A}^{-1}\mathbf{BF} \\ -\mathbf{FB}'\mathbf{A}^{-1} & \mathbf{F} \end{bmatrix}, \tag{2.30}$$

where $\mathbf{F} = (\mathbf{D} - \mathbf{B}'\mathbf{A}^{-1}\mathbf{B})^{-1}$, or alternatively

$$\mathbf{M}^{-1} = \begin{bmatrix} \mathbf{E} & -\mathbf{EBD}^{-1} \\ -\mathbf{D}^{-1}\mathbf{B}'\mathbf{E} & \mathbf{D}^{-1} + \mathbf{D}^{-1}\mathbf{B}'\mathbf{EBD}^{-1} \end{bmatrix}, \tag{2.31}$$

where $\mathbf{E} = (\mathbf{A} - \mathbf{BD}^{-1}\mathbf{B}')^{-1}$.

When $\mathbf{M}$ is singular, a regular inverse of $\mathbf{M}$ does not exist. However, under some conditions, the regular inverses in the formula above can be replaced by their respective symmetric generalized inverses (Section 2.2.4). This condition stipulates that $\mathrm{Sp}(\mathbf{A}) \supset \mathrm{Sp}(\mathbf{B})$ (or $\mathbf{AA}^{-}\mathbf{B} = \mathbf{B}$), which is also equivalent to $\mathbf{M}$ being an *nnd* matrix (the next section).

2.1.6 Signature of square matrices

A square matrix $\mathbf{A}$ for which $\mathbf{x}'\mathbf{Ax} > 0$ for any nonzero vector $\mathbf{x}$ is called a positive-definite (*pd*) matrix. A square matrix $\mathbf{A}$ for which $\mathbf{x}'\mathbf{Ax} \geq 0$ and $\mathbf{x}'\mathbf{Ax} = 0$ for at least one nonzero vector $\mathbf{x}$ is called a positive-semidefinite (*psd*) matrix. Negative-definite (*nd*) and negative-semidefinite (*nsd*) matrices are analogously defined. Matrices which are either *pd* or *psd* are called nonnegative-definite (*nnd*) matrices. Nonpositive-definite (*npd*) matrices are analogously defined. Matrices which are neither *nnd* nor *npd* are called indefinite matrices.

A *pd* matrix $\mathbf{A}$ is often denoted by $\mathbf{A} > \mathbf{O}$, and an *nnd* matrix by $\mathbf{A} \geq \mathbf{O}$. We often write $\mathbf{A} > \mathbf{B}$ when $\mathbf{A} - \mathbf{B} > \mathbf{O}$, and $\mathbf{A} \geq \mathbf{B}$ when $\mathbf{A} - \mathbf{B} \geq \mathbf{O}$. Let $\mathbf{A}$ denote a symmetric *nnd* matrix of order n and of rank r. Then there exists a columnwise nonsingular matrix of order $n \times r$ such that

$$\mathbf{A} = \mathbf{C}\mathbf{C}', \tag{2.32}$$

where $\mathbf{C}$ is called a square root factor of $\mathbf{A}$.

Let the leading diagonal block of $\mathbf{M}^{-1}$ in (2.30) be denoted as $\mathbf{A}^{11}$. Then $\mathbf{A}^{11} \geq \mathbf{A}^{-1}$ since the second term in $\mathbf{A}^{11}$ is *nnd* because it can be factored in the form of $\mathbf{C}\mathbf{C}'$. In general, we have

$$\mathbf{A} \geq \mathbf{B} \Leftrightarrow \mathbf{B}^{-1} \geq \mathbf{A}^{-1}, \tag{2.33}$$

where $\mathbf{A}$ and $\mathbf{B}$ are both nonsingular matrices.

2.1.7 *Metric matrices*

Let $\mathbf{K}$ denote a symmetric *pd* matrix of order n. When the inner product between two vectors $\mathbf{x}$ and $\mathbf{y}$ in E^n is defined by

$$\langle \mathbf{x}, \mathbf{y} \rangle_K = \langle \mathbf{x}, \mathbf{K}\mathbf{y} \rangle = \mathbf{x}'\mathbf{K}\mathbf{y} = \langle \mathbf{K}\mathbf{x}, \mathbf{y} \rangle, \tag{2.34}$$

the matrix $\mathbf{K}$ is called a metric matrix. Using this definition, the (squared) norm and the orthogonality of vectors and matrices can also be generalized as follows.

(i) The squared norm of a vector:

$$\langle \mathbf{x}, \mathbf{x} \rangle = \mathbf{x}'\mathbf{x} \rightarrow \langle \mathbf{x}, \mathbf{x} \rangle_K = \mathbf{x}'\mathbf{K}\mathbf{x} = \mathrm{SS}(\mathbf{x})_K, \tag{2.35}$$

where "$\rightarrow$" indicates that the left-hand side is generalized into the right-hand side.

(ii) The orthogonality of two vectors:

$$\langle \mathbf{x}, \mathbf{y} \rangle = \mathbf{x}'\mathbf{y} = 0 \rightarrow \langle \mathbf{x}, \mathbf{y} \rangle_K = \mathbf{x}'\mathbf{K}\mathbf{y} = 0. \tag{2.36}$$

Let $\mathbf{L}$ denote a symmetric *pd* matrix of order m. Then

(iii) The squared norm of an $n \times m$ matrix $\mathbf{A}$:

$$\mathrm{SS}(\mathbf{A}) = \mathrm{tr}(\mathbf{A}'\mathbf{A}) = \mathrm{tr}(\mathbf{A}\mathbf{A}') \rightarrow$$
$$\mathrm{SS}(\mathbf{A})_{K,L} = \mathrm{tr}(\mathbf{A}'\mathbf{K}\mathbf{A}\mathbf{L}) = \mathrm{tr}(\mathbf{A}\mathbf{L}\mathbf{A}'\mathbf{K}). \tag{2.37}$$

(iv) The columnwise orthogonality of $\mathbf{A}$ ($n \times m$) and $\mathbf{B}$ ($n \times p$) denoted as $\mathrm{Sp}(\mathbf{A}) \perp \mathrm{Sp}(\mathbf{B})$:

$$\mathbf{A}'\mathbf{B} = \mathbf{O} \rightarrow \mathbf{A}'\mathbf{K}\mathbf{B} = \mathbf{O}. \tag{2.38}$$

(v) The rowwise orthogonality of $\mathbf{A}$ ($n \times m$) and $\mathbf{B}$ ($p \times m$) denoted as $\mathrm{Sp}(\mathbf{A}') \perp \mathrm{Sp}(\mathbf{B}')$:

$$\mathbf{A}\mathbf{B}' = \mathbf{O} \rightarrow \mathbf{A}\mathbf{L}\mathbf{B}' = \mathbf{O}. \tag{2.39}$$

(vi) The trace-orthogonality of two $n \times m$ matrices $\mathbf{A}$ and $\mathbf{B}$:

$$\mathrm{tr}(\mathbf{A}'\mathbf{B}) = \mathrm{tr}(\mathbf{B}\mathbf{A}') = 0 \rightarrow \mathrm{tr}(\mathbf{A}'\mathbf{K}\mathbf{B}\mathbf{L}) = \mathrm{tr}(\mathbf{B}\mathbf{L}\mathbf{A}'\mathbf{K}) = 0. \tag{2.40}$$

If $\mathbf{A}$ and $\mathbf{B}$ are either columnwise or rowwise orthogonal, they are also trace- orthogonal.

Let $\mathbf{R}_K$ and $\mathbf{R}_L$ be square root factors of $\mathbf{K}$ and $\mathbf{L}$, namely,

$$\mathbf{K} = \mathbf{R}_K\mathbf{R}_K', \tag{2.41}$$

and

$$\mathbf{L} = \mathbf{R}_L\mathbf{R}_L'. \tag{2.42}$$

A square root factor of a *pd* matrix can be found by the Cholesky decomposition (see Section 2.1.10), in which case it is a lower triangular matrix. It may also be computed by the eigendecomposition (Section 2.3.3), in which case it can be symmetric. The symmetric square root of $\mathbf{K}$ may be denoted as $\mathbf{K}^{1/2}$. A symmetric square root factor is rarely needed, however. Using $\mathbf{R}_K$ and $\mathbf{R}_L$ defined above, $\mathrm{SS}(\mathbf{A})_{K,L}$ defined in (2.37) can be rewritten as

$$\mathrm{SS}(\mathbf{A})_{K,L} = \mathrm{SS}(\mathbf{R}_K'\mathbf{A}\mathbf{R}_L) = \mathrm{SS}(\mathbf{A}^*), \tag{2.43}$$

where $\mathbf{A}^* = \mathbf{R}_K'\mathbf{A}\mathbf{R}_L$, indicating how the (squared) SS norm with the metric matrices $\mathbf{K}$ and $\mathbf{L}$ can be restated in terms of the (squared) SS norm under the identity metric matrices. Similarly, $\langle\mathbf{A},\mathbf{B}\rangle_{K,L} = \mathrm{tr}(\mathbf{A}'\mathbf{K}\mathbf{B}\mathbf{L})$ can be rewritten as

$$\mathrm{tr}(\mathbf{A}'\mathbf{K}\mathbf{B}\mathbf{L}) = \mathrm{tr}(\mathbf{A}^{*'}\mathbf{B}^*), \tag{2.44}$$

where $\mathbf{A}^* = \mathbf{R}_K'\mathbf{A}\mathbf{R}_L$ and $\mathbf{B}^* = \mathbf{R}_K'\mathbf{B}\mathbf{R}_L$, indicating that the inner product between $\mathbf{A}$ and $\mathbf{B}$ under the nonidentity metric matrices can also be restated in terms of the inner product under the identity metric matrices.

Let $||\mathbf{A}||$ represent a norm (not necessarily the SS norm) of $\mathbf{A}$, and let $\mathbf{U}^*$ and $\mathbf{V}^*$ be matrices that satisfy $\mathbf{U}^{*'}\mathbf{K}\mathbf{U}^* = \mathbf{K}$ and $\mathbf{V}^{*'}\mathbf{L}\mathbf{V}^* = \mathbf{L}$. The norm is said to be *KL*-invariant if it is invariant under transformation by $\mathbf{U}^*$ and $\mathbf{V}^{*'}$ (Rao 1980); i.e., it satisfies the equality

$$||\mathbf{A}|| = ||\mathbf{U}^*\mathbf{A}\mathbf{V}^{*'}||. \tag{2.45}$$

The $\mathrm{SS}(\mathbf{A})_{K,L}$ is a *KL*-invariant norm because

$$\begin{aligned}\mathrm{SS}(\mathbf{A})_{K,L} &= \mathrm{SS}(\mathbf{R}_K'\mathbf{A}\mathbf{R}_L) = \mathrm{SS}(\mathbf{U}\mathbf{R}_K'\mathbf{A}\mathbf{R}_L\mathbf{V}')\\ &= \mathrm{SS}(\mathbf{R}_K'(\mathbf{R}_K')^{-1}\mathbf{U}\mathbf{R}_K'\mathbf{A}\mathbf{R}_L\mathbf{V}'\mathbf{R}_L^{-1}\mathbf{R}_L)\\ &= \mathrm{SS}((\mathbf{R}_K')^{-1}\mathbf{U}\mathbf{R}_K'\mathbf{A}\mathbf{R}_L\mathbf{V}'\mathbf{R}_L^{-1})_{K,L}\\ &= \mathrm{SS}(\mathbf{U}^*\mathbf{A}\mathbf{V}^{*'})_{K,L},\end{aligned} \tag{2.46}$$

where $\mathbf{U}$ and $\mathbf{V}$ are the orthogonal matrices analogous to those used in (2.14), and $\mathbf{U}^* = (\mathbf{R}_K')^{-1}\mathbf{U}\mathbf{R}_K'$ and $\mathbf{V}^* = (\mathbf{R}_L')^{-1}\mathbf{V}'\mathbf{R}_L'$. It can be readily verified that these $\mathbf{U}^*$ and $\mathbf{V}^*$ satisfy $\mathbf{U}^{*'}\mathbf{K}\mathbf{U}^* = \mathbf{K}$ and $\mathbf{V}^{*'}\mathbf{L}\mathbf{V}^* = \mathbf{L}$. In this book we only use $\mathrm{SS}(\mathbf{A})_{K,L}$, but most of the optimality conditions that hold under the SS norm with the metric matrices $\mathbf{K}$ and $\mathbf{L}$ hold more generally under the *KL*-invariant norm.

Similarly to the SS norm in (2.43), the KL-invariant norm $||\mathbf{A}||_{K,L}$ can be transformed into an orthogonally invariant norm by $||\mathbf{R}_K'\mathbf{A}\mathbf{R}_L||$. This can be shown by

$$\begin{aligned}||\mathbf{R}_K'\mathbf{A}\mathbf{R}_L|| &= ||\mathbf{A}||_{K,L} = ||\mathbf{U}^*\mathbf{A}\mathbf{V}^{*'}||_{K,L} \\ &= ||\mathbf{U}^*(\mathbf{R}_K')^{-1}\mathbf{R}_K'\mathbf{A}\mathbf{R}_L\mathbf{R}_L^{-1}\mathbf{V}^{*'}||_{K,L} \\ &= ||\mathbf{R}_K'\mathbf{U}^*(\mathbf{R}_K')^{-1}\mathbf{R}_K'\mathbf{A}\mathbf{R}_L\mathbf{R}_L^{-1}\mathbf{V}^{*'}\mathbf{R}_L|| \\ &= ||\mathbf{U}\mathbf{R}_K'\mathbf{A}\mathbf{R}_L\mathbf{V}'||, \end{aligned} \tag{2.47}$$

where $\mathbf{U} = \mathbf{R}_K'\mathbf{U}^*(\mathbf{R}_K')^{-1}$ and $\mathbf{V} = \mathbf{R}_L'\mathbf{V}^*(\mathbf{R}_L')^{-1}$. It can be readily verified that these $\mathbf{U}$ and $\mathbf{V}$ satisfy $\mathbf{U}'\mathbf{U} = \mathbf{I}_n$ and $\mathbf{V}'\mathbf{V} = \mathbf{I}_m$. Equation (2.47) generalizes (2.43).

In most practical data analysis situations, $\mathbf{K}$ and $\mathbf{L}$ are nonsingular, but occasionally they are singular. A norm under the *psd* metric matrices $\mathbf{K}$ and $\mathbf{L}$ is often called a seminorm (Mitra and Rao 1974). In this case, no regular inverse matrices exist for $\mathbf{R}_K'$ and $\mathbf{R}_L'$. In most cases, however, they can be replaced by the unique Moore-Penrose inverses (see Section 2.2.4). The same holds for inverse matrices of $\mathbf{K}$ and $\mathbf{L}$. Note that

$$(\mathbf{R}_K')^+ = \mathbf{R}_K(\mathbf{R}_K'\mathbf{R}_K)^{-1}, \tag{2.48}$$

and

$$\mathbf{K}^+ = (\mathbf{R}_K\mathbf{R}_K')^+ = (\mathbf{R}_K')^+\mathbf{R}_K^+ = \mathbf{R}_K(\mathbf{R}_K'\mathbf{R}_K)^{-2}\mathbf{R}_K', \tag{2.49}$$

where $^+$ indicates a Moore-Penrose inverse, and $(\mathbf{R}_K'\mathbf{R}_K)^{-2} = ((\mathbf{R}_K'\mathbf{R}_K)^{-1})^2$. The Moore-Penrose inverse of $\mathbf{R}_L^+$ and $\mathbf{L}^+$ can be similarly found.

Only a handful of textbooks discuss metric matrices systematically. Cailliez and Pagés (1976) is probably the most comprehensive, although unfortunately it is in French. Being closely related to the weighted least squares (WLS) estimation in linear models, metric matrices play an important role in generalizations of g-inverses (Section 2.2.5), orthogonal projectors (Sections 2.2.3 and 2.2.7), and singular value decomposition (Sections 2.3.4 and 2.3.6).

2.1.8 *Determinants*

The determinant of a square matrix $\mathbf{A}$ is a scalar function of $\mathbf{A}$ and is denoted by $\det(\mathbf{A})$ or $|\mathbf{A}|$. It plays a crucial role in specifying a multivariate normal density function. Without going into its precise definition, we provide several important properties of the determinant:

(i) $|\mathbf{AB}| = |\mathbf{A}|\cdot|\mathbf{B}|$, where $\mathbf{A}$ and $\mathbf{B}$ are both square matrices.

(ii) $|\mathbf{A}'| = |\mathbf{A}|$.

(iii) $|\mathbf{A}| \neq 0 \Leftrightarrow \mathbf{A}$ is nonsingular $\Leftrightarrow \mathbf{A}^{-1}$ exists. $\Leftrightarrow \mathbf{x} = \mathbf{0}$ is the only solution to $\mathbf{Ax} = \mathbf{0}$.

(iv) $|\mathbf{A}^{-1}| = 1/|\mathbf{A}|$. (2.50)

(v) $\det\left(\begin{bmatrix}\mathbf{A} & \mathbf{B}\\ \mathbf{C} & \mathbf{D}\end{bmatrix}\right) = |\mathbf{A}|\cdot|\mathbf{D}-\mathbf{C}\mathbf{A}^{-1}\mathbf{B}| = |\mathbf{D}|\cdot|\mathbf{A}-\mathbf{B}\mathbf{D}^{-1}\mathbf{C}|$.

(vi) $|\mathbf{A} \otimes \mathbf{B}| = |\mathbf{A}|^n |\mathbf{B}|^m$, where **A** and **B** are square matrices of orders n and m, respectively.

(vii) $|\mathbf{A}| \leq a_{11}a_{22}\cdots a_{nn}$, where **A** is a square matrix of order n.

The equality holds in (vii) if **A** is triangular. These properties of determinants are important for manipulations of a multivariate normal (MVN) density function that involves the determinant of the variance-covariance matrix among variables. See Harville (1997) for other interesting properties of determinants.

2.1.9 Vector and matrix derivatives

In multivariate analysis we often encounter the problem of finding a minimum or maximum of a scalar function of vectors and matrices. Least squares and maximum likelihood criteria are two representative examples of such functions. A minimum or maximum of such a function is often located where the first derivatives of the function vanish. Let $f(\mathbf{w})$ denote a scalar function of the p-component vector **w**. Then the derivatives of $f(\mathbf{w})$ with respect to (the elements of) **w** are defined as

$$f_d(\mathbf{w}) = (\partial f(\mathbf{w})/\partial w_1, \cdots, \partial f(\mathbf{w})/\partial w_p)'. \tag{2.51}$$

Let $f(\mathbf{W})$ be a scalar function of the $p \times q$ matrix **W**. Then $f_d(\mathbf{W}) = \partial f(\mathbf{W})/\partial \mathbf{W}$ is a $p \times q$ matrix whose ijth element is equal to $\partial f(\mathbf{W})/\partial w_{ij}$. Formulae of the first derivatives of a number of key functions with respect to their vector or matrix arguments are given below (Magnus and Neudecker 1988):

$$\begin{array}{ll}
\text{(i)}\ f(\mathbf{w}) = \mathbf{a}'\mathbf{w} & f_d(\mathbf{w}) = \mathbf{a} \\
\text{(ii)}\ f(\mathbf{w}) = \mathbf{w}'\mathbf{A}\mathbf{w} & f_d(\mathbf{w}) = (\mathbf{A}+\mathbf{A}')\mathbf{w}\ (= 2\mathbf{A}\mathbf{w},\ \text{if } \mathbf{A}' = \mathbf{A}) \\
\text{(iii)}\ f(\mathbf{W}) = \mathrm{tr}(\mathbf{W}'\mathbf{A}) & f_d(\mathbf{W}) = \mathbf{A} \\
\text{(iv)}\ f(\mathbf{W}) = \mathrm{tr}(\mathbf{A}\mathbf{W}) & f_d(\mathbf{W}) = \mathbf{A}' \\
\text{(v)}\ f(\mathbf{W}) = \mathrm{tr}(\mathbf{W}'\mathbf{A}\mathbf{W}) & f_d(\mathbf{W}) = (\mathbf{A}+\mathbf{A}')\mathbf{W}\ (= 2\mathbf{A}\mathbf{W},\ \text{if } \mathbf{A}' = \mathbf{A}) \\
\text{(vi)}\ f(\mathbf{W}) = \mathrm{tr}(\mathbf{W}'\mathbf{A}\mathbf{W}\mathbf{B}) & f_d(\mathbf{W}) = \mathbf{A}\mathbf{W}\mathbf{B} + \mathbf{A}'\mathbf{W}\mathbf{B}' \\
\text{(vii)}\ f(\mathbf{W}) = \log(|\mathbf{W}|) & f_d(\mathbf{W}) = \mathbf{W}^{-1}|\mathbf{W}|
\end{array} \tag{2.52}$$

In (vii) above, **W** is assumed to be nonsymmetric. (The expression is a bit more complicated if **W** is symmetric. See Harville (1997).)

Let $f(\mathbf{w})$ and $g(\mathbf{w})$ be two scalar functions of **w**. Then the following relations hold as in the case in which **x** is a scalar:

$$\frac{\partial(f(\mathbf{w})+g(\mathbf{w}))}{\partial \mathbf{w}} = f_d(\mathbf{w}) + g_d(\mathbf{w}), \tag{2.53}$$

$$\frac{\partial(f(\mathbf{w})g(\mathbf{w}))}{\partial \mathbf{w}} = f_d(\mathbf{w})g(\mathbf{w}) + f(\mathbf{w})g_d(\mathbf{w}), \tag{2.54}$$

and

$$\frac{\partial(f(\mathbf{w})/g(\mathbf{w}))}{\partial\mathbf{w}} = \frac{f_d(\mathbf{w})g(\mathbf{w}) - f(\mathbf{w})g_d(\mathbf{w})}{g(\mathbf{w})^2}. \tag{2.55}$$

These formulae remain essentially the same even if f and g are functions of a matrix $\mathbf{W}$.

When $f(\mathbf{w})$ is a composite function such that $\mathbf{w} = \mathbf{g}(\mathbf{y})$, where $\mathbf{y}$ is a q-component vector, the derivatives of $f(\mathbf{w}) = f(\mathbf{g}(\mathbf{y}))$ with respect to $\mathbf{y}$ can be found by the chain rule:

$$\frac{\partial f(\mathbf{g}(\mathbf{y}))}{\partial\mathbf{y}} = \frac{\partial\mathbf{g}(\mathbf{y})}{\partial\mathbf{y}}\frac{\partial f(\mathbf{w})}{\partial\mathbf{w}}. \tag{2.56}$$

Note that $\mathbf{w} = \mathbf{g}(\mathbf{y})$ is a vector function of another vector, and its derivatives with respect to $\mathbf{y}$ constitute a matrix of the form:

$$\frac{\partial\mathbf{w}}{\partial\mathbf{y}} = \frac{\partial\mathbf{g}(\mathbf{y})}{\partial\mathbf{y}} = \begin{bmatrix} \frac{\partial g_1(\mathbf{y})}{\partial y_1} & \cdots & \frac{\partial \mathbf{g}_p(\mathbf{y})}{\partial y_1} \\ \vdots & \ddots & \vdots \\ \frac{\partial \mathbf{g}_1(\mathbf{y})}{\partial y_q} & \cdots & \frac{\partial g_p(\mathbf{y})}{\partial y_q} \end{bmatrix}, \tag{2.57}$$

where $w_i = g_i(\mathbf{y})$ is the ith element of $\mathbf{w} = \mathbf{g}(\mathbf{y})$.

Consider, as an example, the following least squares problem. Let

$$\mathbf{z} = \mathbf{A}\mathbf{b} + \mathbf{e} \tag{2.58}$$

represent a linear regression model, where $\mathbf{z}$, $\mathbf{A}$, $\mathbf{b}$, and $\mathbf{e}$ represent a vector of observations on a criterion variable, a matrix of predictor variables, a vector of regression coefficients, and a vector of disturbance terms, respectively. We wish to find an estimate of $\mathbf{b}$ that minimizes

$$f(\mathbf{b}) = \mathrm{SS}(\mathbf{z} - \mathbf{A}\mathbf{b}) = \mathrm{tr}(\mathbf{z}'\mathbf{z}) - 2\mathrm{tr}(\mathbf{z}'\mathbf{A}\mathbf{b}) + \mathrm{tr}(\mathbf{b}'\mathbf{A}'\mathbf{A}\mathbf{b}) \tag{2.59}$$

with respect to $\mathbf{b}$. By differentiating f with respect to $\mathbf{b}$, we find

$$\frac{\partial f(\mathbf{b})}{\partial\mathbf{b}} = -2\mathbf{A}'\mathbf{z} + 2\mathbf{A}'\mathbf{A}\mathbf{b}. \tag{2.60}$$

By setting (2.60) to zero, we find

$$\hat{\mathbf{b}} = (\mathbf{A}'\mathbf{A})^{-1}\mathbf{A}'\mathbf{z}, \tag{2.61}$$

where $\mathbf{A}'\mathbf{A}$ is assumed nonsingular. Note that the above results are easily extensible to the situation in which we have multivariate dependent variables $\mathbf{Z}$ and a matrix of regression coefficients $\mathbf{B}$.

The derivation above used (2.53) and (2.52)-(i) and (ii). Exactly the same result can be found using the chain rule (2.56) with $\mathbf{g}(\mathbf{b}) = \mathbf{z} - \mathbf{A}\mathbf{b}$ and $f(\mathbf{g}(\mathbf{b})) = \mathbf{g}(\mathbf{b})'\mathbf{g}(\mathbf{b})$:

$$\frac{\partial f(\mathbf{b})}{\partial\mathbf{b}} = -\mathbf{A}' \cdot 2(\mathbf{z} - \mathbf{A}\mathbf{b}), \tag{2.62}$$

where $\partial \mathbf{g}(\mathbf{b})/\partial \mathbf{b} = \mathbf{A}'$ according to (2.57). This is identical to (2.60), as should be the case.

Another common situation in multivariate analysis involves maximizing the ratio of two quadratic forms. A quadratic function of vector $\mathbf{w}$ of the form $\mathbf{w}'\mathbf{S}\mathbf{w}$ where $\mathbf{S}$ is any symmetric matrix is called a quadratic form. Let

$$f(\mathbf{w}) = \lambda = \frac{\mathbf{w}'\mathbf{S}\mathbf{w}}{\mathbf{w}'\mathbf{w}}, \tag{2.63}$$

and consider maximizing f with respect to $\mathbf{w}$. By differentiating f with respect to $\mathbf{w}$ using (2.55), we find

$$\frac{\partial f(\mathbf{w})}{\partial \mathbf{w}} = \frac{2(\mathbf{S}\mathbf{w}(\mathbf{w}'\mathbf{w}) - (\mathbf{w}'\mathbf{S}\mathbf{w})\mathbf{w})}{(\mathbf{w}'\mathbf{w})^2} = \frac{2(\mathbf{S}\mathbf{w} - \lambda\mathbf{w})}{\mathbf{w}'\mathbf{w}}. \tag{2.64}$$

Setting (2.64) to zero gives the stationary equation to be solved:

$$\mathbf{S}\mathbf{w} = \lambda\mathbf{w}. \tag{2.65}$$

This is called an eigenequation (Section 2.3.3). Note that a solution to (2.65) is not unique. If $\mathbf{w}$ is a solution to (2.65), $a\mathbf{w}$ is also a solution to (2.65) for any scalar a. It is customary to choose $\mathbf{w}$ with unit length.

The problem of maximizing the ratio of quadratic forms described above can equivalently be formulated as a constrained maximization problem. Consider maximizing $f(\mathbf{w}) = \mathbf{w}'\mathbf{S}\mathbf{w}$ with respect to $\mathbf{w}$ under the constraint that $\mathbf{w}'\mathbf{w} = 1$. The Lagrange multiplier method can be used to deal with this kind of constrained maximization problems. In the Lagrange multiplier method we define another function $h(\mathbf{w}, \lambda)$ by

$$h(\mathbf{w}, \lambda) = f(\mathbf{w}) - \lambda(\mathbf{w}'\mathbf{w} - 1), \tag{2.66}$$

where λ is called a Lagrange multiplier. Note that this function takes the same value as the original function $f(\mathbf{w})$ when the constraint is satisfied (i.e., $\mathbf{w}'\mathbf{w} = 1$). We maximize $h(\mathbf{w}, \lambda)$ with respect to both $\mathbf{w}$ and λ. By differentiating $\mathbf{h}$ with respect to $\mathbf{w}$ and λ, and setting the results equal to zero, we find

$$\mathbf{S}\mathbf{w} = \lambda\mathbf{w}, \tag{2.67}$$

and

$$\mathbf{w}'\mathbf{w} = 1. \tag{2.68}$$

Equation (2.67) is identical to (2.65). The additional equation (2.68) requires $\mathbf{w}$ to have unit length (satisfying the constraint) thereby determining $\mathbf{w}$ uniquely (up to reflection).

The maximization problem described above can be generalized to that of maximizing $\mathbf{w}'\mathbf{S}\mathbf{w}/\mathbf{w}'\mathbf{T}\mathbf{w}$ (or equivalently the maximization of $\mathbf{w}'\mathbf{S}\mathbf{w}$ subject to $\mathbf{w}'\mathbf{T}\mathbf{w} = 1$), where $\mathbf{T}$ is an arbitrary *nnd* matrix. This leads to a generalized eigenproblem to be discussed in Section 2.3.7. As before, these problems can easily be generalized to multidimensional cases, in which we maximize $\text{tr}(\mathbf{W}'\mathbf{S}\mathbf{W})$ subject to the orthonormalization constraint that $\mathbf{W}'\mathbf{T}\mathbf{W} = \mathbf{I}$.

2.1.10 *Cholesky decomposition*

Let $\mathbf{S}$ denote a *pd* matrix of order m. The factorization of $\mathbf{S}$ into the product of a lower triangular matrix $\mathbf{R}$ and its transpose, namely

$$\mathbf{S} = \mathbf{R}\mathbf{R}', \tag{2.69}$$

is called the Cholesky decomposition (factorization) of $\mathbf{S}$. The upper triangular matrix $\mathbf{R}'$ can be calculated as follows: Let s_{ij} and r_{ij} denote the ijth elements of $\mathbf{S}$ and $\mathbf{R}'$. Then,

$$\begin{aligned}
r_{11} &= (s_{11})^{1/2}, \\
r_{1j} &= s_{1j}/r_{11} \quad (j = 2, \cdots, m), \\
r_{jj} &= (s_{jj} - \sum_{k=1}^{j-1} r_{kj}^2)^{1/2} \quad (j = 2, \cdots, m), \\
r_{ij} &= (s_{ij} - \sum_{k=1}^{j-1} r_{ki} r_{kj})/r_{jj} \quad (j > i,\ i = 2, \cdots, m).
\end{aligned} \tag{2.70}$$

The matrix $\mathbf{R}'$ obtained by the Schmidt orthogonalization (the next section) or by the compact QR decomposition (Section 2.1.12) of a columnwise nonsingular matrix $\mathbf{A}$ is identical to $\mathbf{R}'$ obtained by the Cholesky decomposition of $\mathbf{S} = \mathbf{A}'\mathbf{A}$. When $\mathbf{S}$ is singular, $\mathbf{R}'$ becomes upper trapezoidal (see (2.73)). The Cholesky decomposition of a *psd* $\mathbf{S}$ is called the incomplete Cholesky decomposition.

The Cholesky decomposition is useful for calculating a square root factor of a metric matrix. It is also useful to solve the normal equation $\mathbf{A}'\mathbf{A}\mathbf{b} = \mathbf{A}'\mathbf{z}$ obtained in linear regression analysis (see (2.60) and (2.62)). More recently, however, the method based on the QR decomposition is becoming popular in solving linear regression problems. (See Section 2.1.12.)

2.1.11 *Schmidt orthogonalization method*

The column vectors, $\mathbf{a}_1, \cdots \mathbf{a}_m$, of an $n \times m$ matrix $\mathbf{A}$ can be orthonormalized as follows: Assume for the moment that $\mathbf{A}$ is columnwise nonsingular (i.e., $\text{rank}(\mathbf{A}) = m$), and let $\text{norm}(\mathbf{u})$ denote the normalization of $\mathbf{u}$ (i.e., $\text{norm}(\mathbf{u}) = \mathbf{u}/||\mathbf{u}||$, where $||\mathbf{u}||$ is the SS norm of $\mathbf{u}$). Then,

$$\begin{aligned}
\mathbf{f}_1 &= \text{norm}(\mathbf{a}_1), \\
\mathbf{f}_2 &= \text{norm}(\mathbf{a}_2 - \langle \mathbf{f}_1, \mathbf{a}_2 \rangle \mathbf{a}_1), \\
\mathbf{f}_3 &= \text{norm}(\mathbf{a}_3 - \sum_{i=1}^{2} \langle \mathbf{f}_i, \mathbf{a}_3 \rangle \mathbf{a}_i), \\
&\vdots \\
\mathbf{f}_m &= \text{norm}(\mathbf{a}_m - \sum_{i=1}^{m-1} \langle \mathbf{f}_i, \mathbf{a}_m \rangle \mathbf{a}_i).
\end{aligned} \tag{2.71}$$

Here, $\langle \mathbf{x}, \mathbf{y} \rangle = \mathbf{x}'\mathbf{y}$ indicates an inner product between $\mathbf{x}$ and $\mathbf{y}$ under the identity metric. If we set $\mathbf{F} = [\mathbf{f}_1, \cdots, \mathbf{f}_m]$, then $\mathbf{F}$ is columnwise orthogonal (i.e., $\mathbf{F}'\mathbf{F} = \mathbf{I}_m$). If we set $r_{ij} = \langle \mathbf{f}_i, \mathbf{a}_j \rangle$ for $i \leq j$, and define

$$\mathbf{R}' = \begin{bmatrix} r_{11} & r_{12} & \cdots & r_{1m} \\ 0 & r_{22} & \cdots & r_{2m} \\ \vdots & \vdots & \ddots & \vdots \\ 0 & 0 & \cdots & r_{mm} \end{bmatrix}, \tag{2.72}$$

we obtain $\mathbf{F}'\mathbf{A} = \mathbf{R}'$, and $\mathbf{FF}'\mathbf{A} = \mathbf{A} = \mathbf{FR}'$. It is clear that $\mathrm{Sp}(\mathbf{F}) = \mathrm{Sp}(\mathbf{A})$ from the way the columns of $\mathbf{F}$ are constructed in (2.71), and so $\mathbf{FF}'\mathbf{A} = \mathbf{A}$. The matrix $\mathbf{R}'$ is upper triangular.

If we replace the ordinary inner product above by the inner product under $\mathbf{K}$, we find K-orthogonal $\mathbf{F}^*$ (instead of the I-orthogonal $\mathbf{F}$) such that $\mathbf{F}^{*\prime}\mathbf{KF}^* = \mathbf{I}_m$. If $\mathrm{rank}(\mathbf{A}) = p < m$, we can find only up to $\mathbf{f}_p$ in (2.71), and $\mathbf{R}'$ is $p \times m$ upper trapezoidal of the form,

$$\mathbf{R}' = \begin{bmatrix} r_{11} & r_{12} & \cdots & r_{1p} & \cdots & r_{1m} \\ 0 & r_{22} & \cdots & r_{2p} & \cdots & r_{2m} \\ \vdots & \vdots & \ddots & \vdots & \ddots & \vdots \\ 0 & 0 & \cdots & r_{pp} & \cdots & r_{mm} \end{bmatrix}. \tag{2.73}$$

2.1.12 *QR decomposition*

Let $\mathbf{A}$ denote an $n \times m$ $(n \geq m)$ matrix of rank p. Then, as shown in the previous subsection, $\mathbf{A}$ can be expressed as

$$\mathbf{A} = \mathbf{FR}', \tag{2.74}$$

where $\mathbf{F}$ is an $n \times p$ columnwise orthogonal matrix (i.e., $\mathbf{F}'\mathbf{F} = \mathbf{I}_p$), and $\mathbf{R}'$ is an $p \times m$ upper trapezoidal matrix. This decomposition is often called a compact (or incomplete) QR decomposition of $\mathbf{A}$. When $\mathrm{rank}(\mathbf{A}) = m$, $\mathbf{R}'$ is upper triangular. If we define $\mathbf{F}_1 = [\mathbf{F}, \mathbf{F}_0]$, where $\mathbf{F}_0$ is a matrix of arbitrary orthonormal basis vectors spanning $\mathrm{Ker}(\mathbf{A}') = \mathrm{Ker}(\mathbf{F}')$, then $\mathbf{F}_1$ is square and (fully) orthogonal (i.e., $\mathbf{F}_1'\mathbf{F}_1 = \mathbf{F}_1\mathbf{F}_1' = \mathbf{I}_n$), and $\mathbf{A}$ can also be expressed as

$$\mathbf{A} = \mathbf{F}_1 \begin{bmatrix} \mathbf{R}' \\ \mathbf{O} \end{bmatrix}. \tag{2.75}$$

This is called a complete QR decomposition of $\mathbf{A}$. The $\mathbf{F}_0$ part of $\mathbf{F}_1$ is often unnecessary in practical data analysis, However, it can be useful in mathematical proofs.

One way to find the compact QR decomposition is to use the Schmidt orthogonalization method (the previous subsection). Then $\mathbf{F}_0$ cannot be obtained. To find the complete QR decomposition, Householder's (1958) reflection transformation is often used. This transformation is defined by

$$\mathbf{Q} = \mathbf{I}_n - 2\mathbf{uu}', \tag{2.76}$$

where $\mathbf{u}$ is an arbitrary m-component vector of unit length (i.e., $||\mathbf{u}|| = 1$). The transformation matrix $\mathbf{Q}$ has the following properties: $\mathbf{Q}' = \mathbf{Q}$, and $\mathbf{Q}^2 = \mathbf{Q}'\mathbf{Q} = \mathbf{Q}\mathbf{Q}' = \mathbf{I}_n$. That is, $\mathbf{Q} = \mathbf{Q}^{-1}$. Such a matrix is called an involutory matrix.

Let $\mathbf{a}_1$ represent the first column of $\mathbf{A}$. Define

$$\mathbf{u}_1^* = (a_{11} - ||\mathbf{a}_1||, a_{21}, \cdots, a_{n1})', \tag{2.77}$$

where a_{i1} is the ith element of $\mathbf{a}_1$, and

$$\mathbf{u}_1 = \mathbf{u}_1^* / ||\mathbf{u}_1^*||. \tag{2.78}$$

(When $\mathbf{u}_1^* = \mathbf{0}$, set $\mathbf{u}_1 = \mathbf{0}$.) Define $\mathbf{Q}_1 = \mathbf{I}_n - 2\mathbf{u}_1\mathbf{u}_1'$. Then, by premultiplying $\mathbf{A}$ by $\mathbf{Q}_1$, we find

$$\mathbf{A}_1 \equiv \mathbf{Q}_1\mathbf{A} = \left[\begin{array}{c|c} r_{11} & \\ 0 & \\ \vdots & \mathbf{A}_{(1)} \\ 0 & \end{array}\right], \tag{2.79}$$

where $r_{11} = ||\mathbf{a}_1||$. Let $\mathbf{a}_2$ represent the subvector obtained by eliminating the first element from the first column vector of $\mathbf{A}_{(1)}$, and define $\mathbf{u}_2$ similarly to $\mathbf{u}_1$ above. Define

$$\mathbf{Q}_2 = \begin{bmatrix} 1 & \mathbf{0}' \\ \mathbf{0} & \mathbf{I}_{n-1} - 2\mathbf{u}_2\mathbf{u}_2' \end{bmatrix}. \tag{2.80}$$

By premultiplying $\mathbf{A}_1$ by $\mathbf{Q}_2$, we find

$$\mathbf{A}_2 \equiv \mathbf{Q}_2\mathbf{A}_1 = \left[\begin{array}{cc|c} r_{11} & r_{12} & \\ 0 & r_{22} & \\ 0 & 0 & \mathbf{A}_{(2)} \\ \vdots & \vdots & \\ 0 & 0 & \end{array}\right], \tag{2.81}$$

where $r_{22} = ||\mathbf{a}_2||$. After repeating similar operations p times, we find

$$\mathbf{Q}_p\mathbf{Q}_{p-1}\cdots\mathbf{Q}_1\mathbf{A} = \begin{bmatrix} \mathbf{R}' \\ \mathbf{O} \end{bmatrix}. \tag{2.82}$$

We then set $\mathbf{F}_1 = (\mathbf{Q}_p\mathbf{Q}_{p-1}\cdots\mathbf{Q}_1)' = \mathbf{Q}_1'\cdots\mathbf{Q}_{p-1}'\mathbf{Q}_p'$. (See Golub and van Loan (1996) for more detail.)

When $\mathbf{A}$ is singular, $\mathbf{R}'$ becomes upper trapezoidal, and $\hat{\mathbf{b}}$ is not uniquely determined. There is an algorithm, however, that obtains $\hat{\mathbf{b}}$ that satisfies $\mathbf{F}'\mathbf{z} = \mathbf{R}'\hat{\mathbf{b}}$ and that has the minimum norm $||\hat{\mathbf{b}}||$. See Lawson and Hanson (1974) for more details.

Note 2.1 The QR decomposition provides a numerically stable solution for the linear least squares (LS) problem (Lawson and Hanson 1974). As before (Section 2.1.9), let

$$f(\mathbf{b}) = \mathrm{SS}(\mathbf{z} - \mathbf{A}\mathbf{b}) \tag{2.83}$$

represent the LS criterion in linear regression analysis, and let $\mathbf{A} = \mathbf{F}\mathbf{R}'$ be the compact QR decomposition of the $n \times m$ matrix of predictor variables $\mathbf{A}$. Then

$$\begin{aligned} f(\mathbf{b}) &= \mathrm{SS}\{(\mathbf{F}\mathbf{F}'\mathbf{z} - \mathbf{F}\mathbf{R}'\mathbf{b}) + (\mathbf{z} - \mathbf{F}\mathbf{F}'\mathbf{z})\} \\ &= \mathrm{SS}(\mathbf{F}(\mathbf{F}'\mathbf{z} - \mathbf{R}'\mathbf{b})) + \mathrm{SS}(\mathbf{z} - \mathbf{F}\mathbf{F}'\mathbf{z}) \\ &\qquad - 2\mathrm{tr}(\mathbf{F}\mathbf{F}'\mathbf{z} - \mathbf{F}\mathbf{R}'\mathbf{b})'(\mathbf{z} - \mathbf{F}\mathbf{F}'\mathbf{z}) \qquad (2.84) \\ &= \mathrm{SS}(\mathbf{F}'\mathbf{z} - \mathbf{R}'\mathbf{b}) + \mathrm{SS}(\mathbf{z} - \mathbf{F}\mathbf{F}'\mathbf{z}), \qquad (2.85) \end{aligned}$$

since the third term in (2.84) vanishes, and $\mathrm{SS}(\mathbf{F}(\mathbf{F}'\mathbf{z} - \mathbf{R}'\mathbf{b})) = \mathrm{SS}(\mathbf{F}'\mathbf{z} - \mathbf{R}'\mathbf{b})_{F'F} = \mathrm{SS}(\mathbf{F}'\mathbf{z} - \mathbf{R}'\mathbf{b})$. Assume, for simplicity, that $\mathbf{A}$ is columnwise nonsingular ($\mathrm{rank}(\mathbf{A}) = m$). Then $\hat{\mathbf{b}}$ that satisfies $\mathbf{F}'\mathbf{z} = \mathbf{R}'\hat{\mathbf{b}}$ exactly can be uniquely determined. Let $z_1^*, \cdots, z_{m-1}^*, z_m^*$ denote the elements of $\mathbf{F}'\mathbf{z}$, and let $\hat{b}_1, \cdots, \hat{b}_{m-1}, \hat{b}_m$ denote the elements of $\hat{\mathbf{b}}$. Then because $\mathbf{R}'$ is upper triangular, the problem reduces to solving the simultaneous equations of the form

$$\begin{pmatrix} z_1^* \\ \vdots \\ z_{m-1}^* \\ z_m^* \end{pmatrix} = \begin{bmatrix} r_{11} & \cdots & r_{1,m-1} & r_{1m} \\ \vdots & \ddots & \vdots & \vdots \\ 0 & \cdots & r_{m-1,m-1} & r_{m-1,m} \\ 0 & \cdots & 0 & r_{mm} \end{bmatrix} \begin{pmatrix} \hat{b}_1 \\ \vdots \\ \hat{b}_{m-1} \\ \hat{b}_m \end{pmatrix}. \qquad (2.86)$$

These equations are solved from the bottom to the top. First, $\hat{b}_m$ is determined by $\hat{b}_m = z_m^*/r_{mm}$ from $z_m^* = r_{mm}\hat{b}_m$, and then $\hat{b}_{m-1} = (z_{m-1}^* - z_m^*)/r_{m-1,m-1}$ from $z_{m-1}^* = r_{m-1,m-1}\hat{b}_{m-1} + r_{mm}\hat{b}_m = r_{m-1,m-1}\hat{b}_{m-1} + z_m^*$, and so on. This process is called back substitutions. Note that the vector in the second term of (2.85)

$$(\mathbf{I} - \mathbf{F}\mathbf{F}')\mathbf{z} = \mathbf{F}_0\mathbf{F}_0'\mathbf{z} \qquad (2.87)$$

represents the residual vector. Here, $\mathbf{F}\mathbf{F}' = \mathbf{A}(\mathbf{A}'\mathbf{A})^{-1}\mathbf{A}'$ is the orthogonal projector onto $\mathrm{Sp}(\mathbf{A})$, and $\mathbf{F}_0\mathbf{F}_0' = \mathbf{I} - \mathbf{A}(\mathbf{A}'\mathbf{A})^{-1}\mathbf{A}'$ is the orthogonal projector onto $\mathrm{Ker}(\mathbf{A}')$, to be discussed in Section 2.2.3. See, in particular, (2.107), which is a slight generalization of $\mathbf{A}(\mathbf{A}'\mathbf{A})^{-1}\mathbf{A}'$.

2.2 Projection Matrices

In this section, we present an in-depth account of projection matrices. Regression analysis amounts to projection of a criterion variable or variables onto the space spanned by predictor variables. However, much of the exposition in this section is given at a fairly abstract level until Sections 2.2.9 and 2.2.10. Those who would like statistical motivations may find it helpful to read these sections first, and then come back to the beginning of this section.

2.2.1 Definition

Let U and V be subspaces in E^n such that $U \oplus V = E^n$ (i.e., U and V are disjoint and jointly cover the entire space of E^n). Then, for an arbitrary $\mathbf{x} \in E^n$, there exist $\mathbf{y}$ and $\mathbf{z}$ such that $\mathbf{x} = \mathbf{y} + \mathbf{z}$, where $\mathbf{y} \in U$, and $\mathbf{z} \in V$ are uniquely determined. Let $\mathbf{P}$ and $\mathbf{Q}$ denote square matrices of order n such that $\mathbf{y} = \mathbf{P}\mathbf{x}$ and $\mathbf{z} = \mathbf{Q}\mathbf{x}$. Then

$$\mathbf{x} = \mathbf{P}\mathbf{x} + \mathbf{Q}\mathbf{x}. \qquad (2.88)$$

We call $\mathbf{P}$ the projection matrix (projector) onto U along V, and $\mathbf{Q}$ the projection matrix onto V along U. For (2.88) to hold for any $\mathbf{x}$, it must hold that

$$\mathbf{P}+\mathbf{Q}=\mathbf{I}_n \quad (\mathbf{Q}=\mathbf{I}_n-\mathbf{P}). \tag{2.89}$$

By substituting $\mathbf{y}=\mathbf{Px}$ for $\mathbf{y}$ in $\mathbf{Py}=\mathbf{y}$ and $\mathbf{Qy}=\mathbf{0}$, we find $\mathbf{P}^2\mathbf{x}=\mathbf{Px}$, and $\mathbf{QPx}=\mathbf{0}$. For these relations to hold for arbitrary $\mathbf{x}$, it must be that

$$\mathbf{P}^2=\mathbf{P}, \text{ and } \mathbf{QP}=\mathbf{O}. \tag{2.90}$$

Similarly,

$$\mathbf{Q}^2=\mathbf{Q}, \text{ and } \mathbf{PQ}=\mathbf{O}. \tag{2.91}$$

These properties ((2.90) and (2.91)) characterize essential features of projection matrices. In fact, $\mathbf{P}^2=\mathbf{P}$ (idempotency) gives the necessary and sufficient (*ns*) condition for $\mathbf{P}$ to be a projection matrix. In this case $\mathbf{P}$ is the projection matrix onto $\text{Sp}(\mathbf{P})$ along $\text{Ker}(\mathbf{P})$. The matrix $\mathbf{Q}=\mathbf{I}_n-\mathbf{P}$ is also idempotent (i.e., $\mathbf{Q}^2=\mathbf{Q}$), so it is also a projector. It is the projector onto $\text{Sp}(\mathbf{Q})=\text{Ker}(\mathbf{P})$ along $\text{Ker}(\mathbf{Q})=\text{Sp}(\mathbf{P})$. It also follows that $(\mathbf{P}')^2=\mathbf{P}'$, so $\mathbf{P}'$ is also a projector. It is the projector onto $\text{Sp}(\mathbf{P}')$ along $\text{Ker}(\mathbf{P}')$. Similarly, $\mathbf{Q}'$ is the projector onto $\text{Sp}(\mathbf{Q}')=\text{Ker}(\mathbf{P}')$ along $\text{Ker}(\mathbf{Q}')=\text{Sp}(\mathbf{P}')$.

If $\mathbf{P}$ is a projector, its eigenvalues (see Section 2.3.3) are either 1 or 0. Hence, we have

$$\text{rank}(\mathbf{P})=\text{tr}(\mathbf{P})= \text{ the number of unit eigenvalues.} \tag{2.92}$$

2.2.2 *Sum, difference, and product of projectors*

The following three theorems by Rao and Mitra (1971) concerning the sum, difference, and product of two projectors play important roles in additive decompositions of projection matrices (to be discussed in Section 2.2.11):

(i) Let $\mathbf{P}_1$ and $\mathbf{P}_2$ be projectors of order n, and let $\mathbf{P}=\mathbf{P}_1+\mathbf{P}_2$. The *ns* (necessary and sufficient) condition for $\mathbf{P}$ to also be a projection matrix is given by

$$\mathbf{P}_1\mathbf{P}_2=\mathbf{P}_2\mathbf{P}_1=\mathbf{O}. \tag{2.93}$$

In this case $\mathbf{P}$ is the projector onto $\text{Sp}(\mathbf{P}_1)\oplus\text{Sp}(\mathbf{P}_2)$ along $\text{Ker}(\mathbf{P}_1)\cap\text{Ker}(\mathbf{P}_2)=\text{Ker}(\mathbf{P}_1+\mathbf{P}_2)$ (Rao and Mitra 1971, Theorem 5.1.2).

Let $\mathbf{Q}_1=\mathbf{I}-\mathbf{P}_1$ and $\mathbf{Q}_2=\mathbf{I}-\mathbf{P}_2$. Then, the condition in (2.93) is equivalent to the following two conditions: $\mathbf{P}_1\mathbf{Q}_2=\mathbf{P}_1=\mathbf{Q}_2\mathbf{P}_1$ and $\mathbf{P}_2\mathbf{Q}_1=\mathbf{P}_2=\mathbf{Q}_1\mathbf{P}_2$. The former is equivalent to $\text{Sp}(\mathbf{P}_1)\subset\text{Sp}(\mathbf{Q}_2)=\text{Ker}(\mathbf{P}_2)$, and the latter to $\text{Sp}(\mathbf{P}_2)\subset\text{Sp}(\mathbf{Q}_1)=\text{Ker}(\mathbf{P}_1)$. We then have $(\text{Ker}(\mathbf{P}_1)\cap\text{Ker}(\mathbf{P}_2))\oplus\text{Sp}(\mathbf{P}_1)\oplus\text{Sp}(\mathbf{P}_2)=\mathbf{E}^n$.

(ii) Let $\mathbf{P}_1$ and $\mathbf{P}_2$ be as introduced above. The *ns* condition for their difference $\mathbf{P}=\mathbf{P}_1-\mathbf{P}_2$ to also be a projection matrix is

$$\mathbf{P}_1\mathbf{P}_2=\mathbf{P}_2\mathbf{P}_1=\mathbf{P}_2. \tag{2.94}$$

In this case, $\mathbf{P}$ is the projector onto $\text{Ker}(\mathbf{P}_1)\cap\text{Sp}(\mathbf{P}_2)$ along $\text{Sp}(\mathbf{P}_1)\oplus\text{Ker}(\mathbf{P}_2)$ (Rao and Mitra 1971, Theorem 5.1.3). Clearly, $\text{Sp}(\mathbf{P}_2) \subset \text{Sp}(\mathbf{P}_1)$ and $\text{Ker}(\mathbf{P}_1) \subset \text{Ker}(\mathbf{P}_2)$, and $\text{Sp}(\mathbf{P}_1) \oplus \text{Ker}(\mathbf{P}_2) \oplus (\text{Ker}(\mathbf{P}_1) \cap \text{Sp}(\mathbf{P}_2)) = E^n$.

(iii) Let $\mathbf{P}_1$ and $\mathbf{P}_2$ be as introduced in (i). A sufficient condition for their product $\mathbf{P} = \mathbf{P}_1\mathbf{P}_2$ to also be a projection matrix is

$$\mathbf{P}_1\mathbf{P}_2 = \mathbf{P}_2\mathbf{P}_1. \tag{2.95}$$

In this case $\mathbf{P}$ is the projector onto $\text{Sp}(\mathbf{P}_1) \cap \text{Sp}(\mathbf{P}_2)$ along $\text{Ker}(\mathbf{P}_1) + \text{Ker}(\mathbf{P}_2)$ (Rao and Mitra 1971, Theorem 5.1.4). Note that $\text{Ker}(\mathbf{P}_1)$ and $\text{Ker}(\mathbf{P}_2)$ are not necessarily disjoint, and that $\text{Ker}(\mathbf{P}_1) + \text{Ker}(\mathbf{P}_2) \oplus (\text{Sp}(\mathbf{P}_1) \cap \text{Sp}(\mathbf{P}_2)) = E^n$.

Note that whereas (2.93) and (2.94) are *ns* conditions, (2.95) is only a sufficient condition. If both $\mathbf{P}_1$ and $\mathbf{P}_2$ are orthogonal projectors (Section 2.2.3), however, the commutativity of $\mathbf{P}_1$ and $\mathbf{P}_2$ is indeed *ns* for the product to also be an orthogonal projector. For proofs of the above theorems, see Rao and Mitra (1971) or Yanai et al. (2011, Theorems 2.7, 2.8 and 2.9).

Note 2.2 Theorem (iii) above has been elaborated recently, which we summarize below.

(iv) The following conditions are equivalent:

(1) $(\mathbf{P}_1\mathbf{P}_2)^2 = \mathbf{P}_1\mathbf{P}_2$ ($\mathbf{P}_1\mathbf{P}_2$ is a projector.)
(2) $\text{Sp}(\mathbf{P}_1\mathbf{P}_2) \subset \text{Sp}(\mathbf{P}_2) \oplus (\text{Ker}(\mathbf{P}_1) \cap \text{Ker}(\mathbf{P}_2))$ (Brown and Page (1970)).
(3) $\text{Sp}(\mathbf{P}_2) \subset \text{Sp}(\mathbf{P}_1) \oplus (\text{Ker}(\mathbf{P}_1) \cap \text{Sp}(\mathbf{P}_2)) \oplus (\text{Ker}(\mathbf{P}_1) \cap \text{Ker}(\mathbf{P}_2))$ (Werner 1992).
(4) $\text{Sp}(\mathbf{P}_1\mathbf{P}_2) = \text{Sp}(\mathbf{P}_1) \cap (\text{Sp}(\mathbf{P}_2) \oplus (\text{Ker}(\mathbf{P}_1) \cap \text{Ker}(\mathbf{P}_2))$ (Groß and Trenkler 1998).
(5) $\text{Sp}(\mathbf{I} - \mathbf{P}_1\mathbf{P}_2) = \text{Ker}(\mathbf{P}_1\mathbf{P}_2) = \text{Ker}(\mathbf{P}_2) \oplus (\text{Ker}(\mathbf{P}_1) \cap \text{Sp}(\mathbf{P}_2))$.

The second equality in (5) always holds, while the first equality imposes a condition to be satisfied. For proofs, see Groß and Trenkler (1998), and Takane and Yanai (1999).

Similar equivalent conditions can be established for $\mathbf{Q}_1\mathbf{P}_2$, $\mathbf{P}_1\mathbf{Q}_2$, and $\mathbf{Q}_1\mathbf{Q}_2$ to be idempotent, where $\mathbf{Q}_1 = \mathbf{I} - \mathbf{P}_1$ and $\mathbf{Q}_2 = \mathbf{I} - \mathbf{P}_2$. For $\text{Sp}(\mathbf{P}_1\mathbf{P}_2) = \text{Sp}(\mathbf{P}_1) \cap \text{Sp}(\mathbf{P}_2)$ to hold, both $\mathbf{P}_1\mathbf{P}_2$ and $\mathbf{Q}_1\mathbf{P}_2$ must be idempotent, and for $\text{Ker}(\mathbf{P}_1\mathbf{P}_2) = \text{Ker}(\mathbf{P}_1) + \text{Ker}(\mathbf{P}_2)$ to hold, both $\mathbf{P}_1\mathbf{P}_2$ and $\mathbf{P}_1\mathbf{Q}_2$ must be idempotent. For $\mathbf{P}_1\mathbf{P}_2$ to be the projector onto $\text{Sp}(\mathbf{P}_1) \cap \text{Sp}(\mathbf{P}_2)$ along $\text{Ker}(\mathbf{P}_1) + \text{Ker}(\mathbf{P}_2)$, both of these conditions must hold, that is, $\mathbf{P}_1\mathbf{P}_2$, $\mathbf{Q}_1\mathbf{P}_2$, and $\mathbf{P}_1\mathbf{Q}_2$ must all be idempotent. Finally, $\mathbf{P}_1\mathbf{P}_2 = \mathbf{P}_2\mathbf{P}_1$ holds if and only if $\mathbf{P}_1\mathbf{P}_2$, $\mathbf{Q}_1\mathbf{P}_2$, $\mathbf{P}_1\mathbf{Q}_2$, and $\mathbf{Q}_1\mathbf{Q}_2$ are all idempotent. See Takane and Yanai (1999) for more details.

2.2.3 *Orthogonal projectors*

When $V = U^\perp$ in the definition of a projector, $\mathbf{P}$ is called an orthogonal projector. Let $\mathbf{x}, \mathbf{y} \in E^n$, and let $\mathbf{K}$ denote a metric matrix associated with E^n. Then we have $\langle \mathbf{Px}, (\mathbf{I} - \mathbf{P})\mathbf{y} \rangle_K = \langle \mathbf{Px}, \mathbf{y} \rangle_K - \langle \mathbf{Px}, \mathbf{Py} \rangle_K = 0$. Hence, we find $\langle \mathbf{KPx}, \mathbf{y} \rangle = \langle \mathbf{Px}, \mathbf{y} \rangle_K = \langle \mathbf{Px}, \mathbf{Py} \rangle_K = \langle \mathbf{KPx}, \mathbf{Py} \rangle = \langle \mathbf{P}'\mathbf{KPx}, \mathbf{y} \rangle$, that is,

$$\mathbf{KP} = \mathbf{P}'\mathbf{KP} \tag{2.96}$$

must hold. Clearly, $\mathbf{P}'\mathbf{KP}$ is symmetric, and thus (2.96) can also be written as

$$(\mathbf{KP})' = \mathbf{P}'\mathbf{K} = \mathbf{KP}. \tag{2.97}$$

When $\mathbf{P}^2 = \mathbf{P}$ and (2.96) hold, $\mathbf{P}$ is called a K-orthogonal projector. When $\mathbf{K} = \mathbf{I}$, that is, when

$$\mathbf{P}' = \mathbf{P}, \tag{2.98}$$

$\mathbf{P}$ is called an I-orthogonal projector or simply an orthogonal projector. Let $\mathbf{P}_U$ denote a K-orthogonal projector onto the subspace U, and let $\mathbf{x} \in E^n$ and $\mathbf{y} \in U$. Define $f(\mathbf{y}) = \mathrm{SS}(\mathbf{x} - \mathbf{y})_K$. Then $\mathbf{y}$ that minimizes $f(\mathbf{y})$ is given by $\mathbf{P}_U\mathbf{x}$. This shows that orthogonal projectors play important roles in least squares estimation problems (Section 2.2.9).

Clearly, it holds in general that

$$\mathrm{SS}(\mathbf{P}_U\mathbf{x})_K \leq \mathrm{SS}(\mathbf{x})_K, \tag{2.99}$$

where $\mathbf{P}_U$ is the K-orthogonal projector onto U, because

$$\mathbf{P}_U'\mathbf{K}\mathbf{Q}_U = \mathbf{O}, \tag{2.100}$$

where $\mathbf{Q}_U = \mathbf{I} - \mathbf{P}_U$, and thus $\mathrm{SS}(\mathbf{x})_K = \mathrm{SS}(\mathbf{P}_U\mathbf{x})_K + \mathrm{SS}(\mathbf{Q}_U\mathbf{x})_K \geq \mathrm{SS}(\mathbf{P}_U\mathbf{x})_K$. In general, $\mathbf{K} - \mathbf{P}'\mathbf{K}\mathbf{P} = \mathbf{K}(\mathbf{I} - \mathbf{P}) = (\mathbf{I} - \mathbf{P}')\mathbf{K}$ is *nnd*.

2.2.4 Generalized inverse matrices

So far, no explicit representations of projection matrices have been given. To do so requires some knowledge of generalized inverse matrices.

Let $\mathbf{A}$ be an $n \times m$ matrix. An $m \times n$ matrix $\mathbf{A}^-$ satisfying

$$\mathbf{A}\mathbf{A}^-\mathbf{A} = \mathbf{A} \tag{2.101}$$

is called a generalized inverse matrix (g-inverse matrix) of $\mathbf{A}$. The matrix $\mathbf{A}^-$ has the following properties:

$$(\mathbf{A}\mathbf{A}^-)^2 = \mathbf{A}\mathbf{A}^-, \tag{2.102}$$

where $\mathrm{rank}(\mathbf{A}\mathbf{A}^-) = \mathrm{rank}(\mathbf{A})$, and

$$(\mathbf{A}^-\mathbf{A})^2 = \mathbf{A}^-\mathbf{A}, \tag{2.103}$$

where $\mathrm{rank}(\mathbf{A}^-\mathbf{A}) = \mathrm{rank}(\mathbf{A})$. These identities indicate that $\mathbf{A}\mathbf{A}^-$ and $\mathbf{A}^-\mathbf{A}$ are both idempotent and are therefore projectors.

For a rectangular matrix $\mathbf{A}$ ($n \neq m$) or for a singular square matrix $\mathbf{A}$, a g-inverse of $\mathbf{A}$ is not uniquely determined. Let $\{\mathbf{A}^-\}$ denote the set of all g-inverses of $\mathbf{A}$, and let $\{(\mathbf{A}')^-\}$ denote the set of all g-inverses of $\mathbf{A}'$. Then

$$(\mathbf{A}'\mathbf{A})^-\mathbf{A}' \in \{\mathbf{A}^-\}, \tag{2.104}$$

that is, $(\mathbf{A}'\mathbf{A})^-\mathbf{A}'$ is a g-inverse of $\mathbf{A}$. Also,

$$\mathbf{A}(\mathbf{A}'\mathbf{A})^- \in \{(\mathbf{A}')^-\}, \tag{2.105}$$

indicating that $\mathbf{A}(\mathbf{A}'\mathbf{A})^-$ is a g-inverse of $\mathbf{A}'$. In general,

$$\{(\mathbf{A}^-)'\} = \{(\mathbf{A}')^-\}. \tag{2.106}$$

That is, the set of transposes of g-inverses of $\mathbf{A}$ is equal to the set of g-inverses of the transpose of $\mathbf{A}$. The relation in (2.104) implies that

$$\mathbf{P}_A = \mathbf{A}(\mathbf{A}'\mathbf{A})^-\mathbf{A}' \tag{2.107}$$

is idempotent, i.e.,

$$\mathbf{P}_A^2 = \mathbf{P}_A, \tag{2.108}$$

and that

$$\mathbf{P}_A\mathbf{A} = \mathbf{A}. \tag{2.109}$$

These two identities indicate that $\mathbf{P}_A$ is a projection matrix onto $\text{Sp}(\mathbf{A})$. Similarly, (2.105) implies that

$$\mathbf{A}'\mathbf{P}_A = \mathbf{A}', \tag{2.110}$$

indicating that $\mathbf{P}_A$ is a projector along $\text{Ker}(\mathbf{A}')$. The identities (2.109) and (2.110) together indicate that $\mathbf{P}_A$ is in fact the orthogonal projector onto $\text{Sp}(\mathbf{A})$ (along $\text{Ker}(\mathbf{A}')$). That $\mathbf{P}_A$ is an orthogonal projector implies that it is symmetric despite the fact that $(\mathbf{A}'\mathbf{A})^-$ may not be symmetric. It is also the case that $\mathbf{P}_A$ is invariant over the choice of $(\mathbf{A}'\mathbf{A})^-$. These two properties are shown in the following note.

Note 2.3 The invariance of $\mathbf{P}_A$ over the choice of $(\mathbf{A}'\mathbf{A})^-$ can be shown as follows: Let $(\mathbf{A}'\mathbf{A})^-_{(1)}$ and $(\mathbf{A}'\mathbf{A})^-_{(2)}$ be two arbitrary g-inverses of $\mathbf{A}'\mathbf{A}$. Then, $\mathbf{A}(\mathbf{A}'\mathbf{A})^-_{(1)}\mathbf{A}' = (\mathbf{A}(\mathbf{A}'\mathbf{A})^-_{(1)}\mathbf{A}')^2 = \mathbf{A}(\mathbf{A}'\mathbf{A})^-_{(1)}\mathbf{A}'\mathbf{A}(\mathbf{A}'\mathbf{A})^-_{(2)}\mathbf{A}' = (\mathbf{A}(\mathbf{A}'\mathbf{A})^-_{(2)}\mathbf{A}')^2 = \mathbf{A}(\mathbf{A}'\mathbf{A})^-_{(2)}\mathbf{A}'$. The invariance of $\mathbf{P}_A$ over the choice of $(\mathbf{A}'\mathbf{A})^-$ more directly follows from Lemma 2.2.4(iii) of Rao and Mitra (1971), which states that $\mathbf{B}\mathbf{A}^-\mathbf{C}$ is invariant over the choice of $\mathbf{A}^-$ if $\text{Sp}(\mathbf{B}') \supset \text{Sp}(\mathbf{A}')$ and $\text{Sp}(\mathbf{C}) \supset \text{Sp}(\mathbf{A})$. The symmetry of $\mathbf{P}_A$ irrespective of the choice of $(\mathbf{A}'\mathbf{A})^-$, on the other hand, can be shown as follows: Note first that if $(\mathbf{A}'\mathbf{A})^-$ is a g-inverse of $\mathbf{A}'\mathbf{A}$, so is its transpose. The average of the two, which is symmetric, is also a g-inverse of $\mathbf{A}'\mathbf{A}$. The matrix $\mathbf{P}_A$ must be symmetric for this choice of a g-inverse of $\mathbf{A}'\mathbf{A}$, which implies that it is symmetric for any choice of $(\mathbf{A}'\mathbf{A})^-$ because it is invariant over any choice of $(\mathbf{A}'\mathbf{A})^-$.

As has been noted already, a g-inverse of $\mathbf{A}$ is generally not uniquely determined. Additional constraints may be imposed to narrow down its range. We discuss some of the representative constraints below:

$$\mathbf{A}^-\mathbf{A}\mathbf{A}^- = \mathbf{A}^- \quad \text{(reflexivity)}, \tag{2.111}$$

$$(\mathbf{A}\mathbf{A}^-)' = \mathbf{A}\mathbf{A}^- \quad \text{(least squares)}, \tag{2.112}$$

and

$$(\mathbf{A}^-\mathbf{A})' = \mathbf{A}^-\mathbf{A} \quad \text{(minimum norm)}. \tag{2.113}$$

These three conditions together with (2.101) are often called the Penrose (1955) conditions. Condition (2.111) requires that $\mathbf{A}$ is a g-inverse of $\mathbf{A}^-$, and (2.112) and

(2.113) require that $\mathbf{AA}^-$ and $\mathbf{A}^-\mathbf{A}$ are symmetric, so that they are I-orthogonal projectors. A variety of g-inverse matrices are defined depending on which subsets of the Penrose conditions are satisfied:

(i) $\mathbf{A}_r^-$: A reflexive g-inverse that satisfies (2.101) and (2.111).

(ii) $\mathbf{A}_\ell^-$: A least squares g-inverse that satisfies (2.101) and (2.112).

(iii) $\mathbf{A}_m^-$: A minimum norm g-inverse that satisfies (2.101) and (2.113).

(iv) $\mathbf{A}_{\ell r}^-$: A least squares reflexive g-inverse that satisfies (2.101), (2.111), and (2.112).

(v) $\mathbf{A}_{mr}^-$: A minimum norm reflexive g-inverse that satisfies (2.101), (2.111), and (2.113).

(vi) $\mathbf{A}^+$: The Moore-Penrose g-inverse that satisfies all of the Penrose conditions, i.e., (2.101), (2.111), (2.112), and (2.113).

A reflexive g-inverse has the property that

$$\text{rank}(\mathbf{A}_r^-) = \text{rank}(\mathbf{A}). \tag{2.114}$$

We generally have

$$(\mathbf{A}_\ell^-)' \in \{(\mathbf{A}')_m^-\}, \text{ and } (\mathbf{A}_m^-)' \in \{(\mathbf{A}')_\ell^-\}. \tag{2.115}$$

The Moore-Penrose g-inverse (vi) is uniquely determined, while the other g-inverses given above are usually nonunique.

Note 2.4 The uniqueness of the Moore-Penrose g-inverse can be shown as follows (Kalman 1976): Suppose that the Moore-Penrose g-inverse is nonunique, and let $\mathbf{X}$ and $\mathbf{Y}$ be two Moore-Penrose g-inverses of $\mathbf{A}$. We show that $\mathbf{X} = \mathbf{Y}$. We have $\mathbf{X} = \mathbf{XAX} = \mathbf{A}'\mathbf{X}'\mathbf{X} = \mathbf{A}'\mathbf{Y}'\mathbf{A}'\mathbf{X}'\mathbf{X} = \mathbf{YAXAX} = \mathbf{YAXX}'\mathbf{A}' = \mathbf{YX}'\mathbf{A}'\mathbf{X}'\mathbf{A}'\mathbf{Y}'\mathbf{A}' = \mathbf{YX}'\mathbf{A}'\mathbf{Y}'\mathbf{A}' = \mathbf{YAXAY} = \mathbf{YAY} = \mathbf{Y}$.

The g-inverses from (ii) to (vi) above can be explicitly written as:

$$\mathbf{A}_\ell^- = (\mathbf{A}'\mathbf{A})^-\mathbf{A}' + (\mathbf{I}_m - (\mathbf{A}'\mathbf{A})^-\mathbf{A}'\mathbf{A})\mathbf{Z}, \tag{2.116}$$

where $\mathbf{Z}$ is an arbitrary $m \times n$ matrix.

$$\mathbf{A}_{\ell r}^- = (\mathbf{A}'\mathbf{A})^-\mathbf{A}', \tag{2.117}$$

which is in fact the $\mathbf{A}^-$ introduced in (2.104).

$$\mathbf{A}_m^- = \mathbf{A}'(\mathbf{AA}')^- + \mathbf{Z}(\mathbf{I}_n - \mathbf{AA}'(\mathbf{AA}')^-), \tag{2.118}$$

where $\mathbf{Z}$ is an arbitrary $m \times n$ matrix. (Don't confuse the subscript m for the minimum norm g-inverse, and the same symbol indicating the size of a matrix.)

$$\mathbf{A}_{mr}^- = \mathbf{A}'(\mathbf{AA}')^-. \tag{2.119}$$

$$\mathbf{A}^+ = \mathbf{A}_m^-\mathbf{A}\mathbf{A}_\ell^- = \mathbf{A}'(\mathbf{AA}')^-\mathbf{A}(\mathbf{A}'\mathbf{A})^-\mathbf{A}'. \tag{2.120}$$

Let $\mathbf{A} = \mathbf{FR}'$ represent the compact QR decomposition of $\mathbf{A}$. Then

$$\mathbf{A}^+ = \mathbf{R}(\mathbf{R}'\mathbf{R})^{-1}\mathbf{F}'. \tag{2.121}$$

We also have

$$(\mathbf{A}'\mathbf{A})^+ = \mathbf{A}^+(\mathbf{A}')^+ = (\mathbf{RR}')^+ = \mathbf{R}(\mathbf{R}'\mathbf{R})^{-2}\mathbf{R}' \tag{2.122}$$

where $(\mathbf{R}'\mathbf{R})^{-2} = ((\mathbf{R}'\mathbf{R})^{-1})^2$ and $(\mathbf{A}')^+ = (\mathbf{A}^+)'$, and

$$(\mathbf{AA}')^+ = \mathbf{P}_A(\mathbf{AA}')^-\mathbf{P}_A = \mathbf{A}(\mathbf{A}'\mathbf{A})^{+2}\mathbf{A}', \tag{2.123}$$

where $(\mathbf{AA}')^{+2} = ((\mathbf{AA}')^+)^2$. As will be shown (Section 2.3), the g-inverses mentioned above can also be expressed using the complete singular value decomposition (Section 2.3.5).

2.2.5 *Metric matrices and g-inverses*

Generalized inverses can be further generalized by incorporating nonidentity metric matrices. Let $\mathbf{K}$ $(n \times n)$ and $\mathbf{L}$ $(m \times m)$ be *nnd* matrices associated with the row and column sides, respectively, of an $n \times m$ matrix $\mathbf{A}$. We assume that they satisfy the following rank conditions:

$$\text{rank}(\mathbf{KA}) = \text{rank}(\mathbf{A}), \tag{2.124}$$

and

$$\text{rank}(\mathbf{AL}) = \text{rank}(\mathbf{A}). \tag{2.125}$$

If $\mathbf{K}$ and $\mathbf{L}$ are *pd*, these conditions are automatically satisfied. We also assume that

$$\text{Sp}(\mathbf{L}) \supset \text{Sp}(\mathbf{A}'), \tag{2.126}$$

which implies that $\mathbf{LL}^-\mathbf{A}' = \mathbf{A}'$. Together with (2.125), this implies that $\text{rank}(\mathbf{AL}^-) = \text{rank}(\mathbf{A})$. This condition is also automatically satisfied if $\mathbf{L}$ is *pd*.

We may generalize the least squares (2.112) and minimum norm (2.113) conditions in the Penrose conditions as follows:

$$(\mathbf{KAA}^-)' = \mathbf{KAA}^- \quad (K\text{-least squares}), \tag{2.127}$$

and

$$(\mathbf{LA}^-\mathbf{A})' = \mathbf{LA}^-\mathbf{A} \quad (L\text{-minimum norm}). \tag{2.128}$$

A g-inverse of $\mathbf{A}$ that satisfies (2.127) in addition to (2.101) and (2.111) is called a least squares reflexive g-inverse in the metric $\mathbf{K}$, and is denoted by $\mathbf{A}^-_{\ell(K)r}$. Similarly, an $\mathbf{A}^-$ that satisfies (2.128), (2.101), and (2.111) is called a minimum norm reflexive g-inverse of $\mathbf{A}$ in the metric $\mathbf{L}$, and is denoted by $\mathbf{A}^-_{m(L)r}$. A g-inverse of $\mathbf{A}$ that satisfies (2.127), (2.128), (2.101), and (2.111), is called the Moore-Penrose g-inverse of $\mathbf{A}$ in the metrics $\mathbf{K}$ and $\mathbf{L}$, and is denoted by $\mathbf{A}^+_{(KL)}$.

These generalized versions of g-inverses can be explicitly written as:

$$\mathbf{A}^-_{\ell(K)r} = (\mathbf{A}'\mathbf{KA})^-\mathbf{A}'\mathbf{K}, \tag{2.129}$$

$$\mathbf{A}^-_{m(L)r} = \mathbf{L}^-\mathbf{A}'(\mathbf{A}\mathbf{L}^-\mathbf{A}')^-, \tag{2.130}$$

and

$$\mathbf{A}^+_{(KL)} = \mathbf{L}^-\mathbf{A}'(\mathbf{A}\mathbf{L}^-\mathbf{A}')^-\mathbf{A}(\mathbf{A}'\mathbf{K}\mathbf{A})^-\mathbf{A}'\mathbf{K}. \tag{2.131}$$

In general, we have

$$(\mathbf{A}^-_{\ell(K)r})' \in \{(\mathbf{A}')^-_{m(K^{-1})r}\}, \text{ and } (\mathbf{A}^-_{m(L)r})' \in \{(\mathbf{A}')^-_{\ell(L^{-1})r}\}. \tag{2.132}$$

Let $\mathbf{A} = \mathbf{F}\mathbf{R}'$ be the compact QR decomposition of $\mathbf{A}$. Then, $\mathbf{A}^+_{(KL)}$ can also be expressed as

$$\mathbf{A}^+_{(KL)} = \mathbf{L}^-\mathbf{R}(\mathbf{R}'\mathbf{L}^-\mathbf{R})^{-1}(\mathbf{F}'\mathbf{K}\mathbf{F})^{-1}\mathbf{F}'\mathbf{K}. \tag{2.133}$$

Similarly to $\mathbf{A}^+$, there is an expression of $\mathbf{A}^+_{(KL)}$ in terms of the complete generalized singular value decomposition of $\mathbf{A}$ with the metric matrices $\mathbf{K}$ and $\mathbf{L}$ (Section 2.3.6).

For given $\mathbf{K}$ and $\mathbf{L}$, conditions (2.127) and (2.128) impose restrictions on the form of $\mathbf{A}^-$. On the other hand, we can always find a $\mathbf{K}$ and $\mathbf{L}$ that satisfy (2.127) and (2.128), respectively, for any arbitrary $\mathbf{A}^-$. The following (slightly generalized version) of a theorem by Rao and Mitra (1971, Lemma 2.7.1) provides a theoretical basis for this assertion. Let $\mathbf{P}$ be an arbitrary idempotent matrix ($\mathbf{P}^2 = \mathbf{P}$). Then there exists a symmetric *nnd* matrix $\mathbf{S}$ such that

$$\mathbf{P}'\mathbf{S} = \mathbf{S}\mathbf{P}. \tag{2.134}$$

Such an $\mathbf{S}$ is not uniquely determined, but can be systematically generated by

$$\mathbf{S} = \mathbf{P}'\Delta_1\mathbf{P} + \mathbf{Q}'\Delta_2\mathbf{Q}, \tag{2.135}$$

where $\mathbf{Q} = \mathbf{I} - \mathbf{P}$, and Δ_1 and Δ_2 are arbitrary *nnd* matrices. This implies that $\mathbf{A}\mathbf{A}^-$ and $\mathbf{A}^-\mathbf{A}$ are always S-symmetric, and are consequently S-orthogonal projectors. Such an $\mathbf{S}$ can be found by setting $\mathbf{P}$ equal to $\mathbf{A}\mathbf{A}^-$ or $\mathbf{A}^-\mathbf{A}$ and using (2.135). It is interesting to note that with $\mathbf{S}$ defined as above, we have $\mathbf{P}'\mathbf{S}\mathbf{P} = \mathbf{S}\mathbf{P} = \mathbf{P}'\mathbf{S}$, which implies that

$$\mathbf{P}'\mathbf{S}\mathbf{Q} = \mathbf{O}. \tag{2.136}$$

That is, for an arbitrary projector $\mathbf{P}$, $\mathbf{P}$ and $\mathbf{Q}$ are always S-orthogonal (i.e., orthogonal with respect to $\mathbf{S}$). Also, note the close parallel between (2.134) and (2.97), and between (2.136) and (2.100).

2.2.6 *Constrained g-inverse matrices*

Let $\mathbf{M}$ and $\mathbf{N}$ be given $n \times p$ and $m \times q$ matrices, respectively. Then g-inverses of $\mathbf{A}$ that satisfy the following conditions are called constrained g-inverses of $\mathbf{A}$ (Rao and Mitra 1971, Lemma 2.25, Sections 4.4.1 and 4.11; see also Takane et al. (2007)):

(i) There exists an $\mathbf{X}$ such that

$$\mathbf{A}^- = \mathbf{X}\mathbf{M}', \tag{2.137}$$

and

$$\mathbf{M}'\mathbf{A}\mathbf{A}^- = \mathbf{M}'. \tag{2.138}$$

That is, $(\mathbf{A}\mathbf{A}^-)'$ is a projector onto $\text{Sp}(\mathbf{M})$ (or equivalently $\mathbf{A}\mathbf{A}^-$ is a projector along $\text{Ker}(\mathbf{M}')$).

(ii) There exists a $\mathbf{Y}$ such that

$$\mathbf{A}^- = \mathbf{N}\mathbf{Y}, \tag{2.139}$$

and

$$\mathbf{A}^-\mathbf{A}\mathbf{N} = \mathbf{N}. \tag{2.140}$$

That is, $\mathbf{A}^-\mathbf{A}$ is a projector onto $\text{Sp}(\mathbf{N})$.

(iii) An $\mathbf{A}^-$ that satisfies both (i) and (ii) above.

How can such an $\mathbf{A}^-$ be explicitly represented? Let $\mathbf{A}^-_{M'}$ denote a g-inverse of $\mathbf{A}$ that satisfies (i) above. We assume that

$$\text{rank}(\mathbf{M}'\mathbf{A}) = \text{rank}(\mathbf{A}), \tag{2.141}$$

and

$$\text{rank}(\mathbf{M}'\mathbf{A}) = \text{rank}(\mathbf{M}). \tag{2.142}$$

From (2.137) and (2.138) we have $\mathbf{M}'\mathbf{A}\mathbf{X}\mathbf{M}' = \mathbf{M}'$. Under (2.141), this and $\mathbf{M}'\mathbf{A}\mathbf{X}\mathbf{M}'\mathbf{A} = \mathbf{M}'\mathbf{A}$ have the same solution, that is, $\mathbf{X} = (\mathbf{M}'\mathbf{A})^-$, and hence we find

$$\mathbf{A}^-_{M'} = (\mathbf{M}'\mathbf{A})^-\mathbf{M}'. \tag{2.143}$$

Conversely, if (2.141) holds, we have $\mathbf{A} = \mathbf{W}\mathbf{M}'\mathbf{A}$ for some $\mathbf{W}$. Hence, $\mathbf{A}\mathbf{A}^-_{M'}\mathbf{A} = \mathbf{A}(\mathbf{M}'\mathbf{A})^-\mathbf{M}'\mathbf{A} = \mathbf{W}\mathbf{M}'\mathbf{A}(\mathbf{M}'\mathbf{A})^-\mathbf{M}'\mathbf{A} = \mathbf{W}\mathbf{M}'\mathbf{A} = \mathbf{A}$, confirming that $\mathbf{A}^-_{M'}$ is a g-inverse of $\mathbf{A}$. If, on the other hand, (2.142) also holds, we have $\mathbf{M}' = \mathbf{M}'\mathbf{A}\mathbf{V}$ for some $\mathbf{V}$. Hence, we have $\mathbf{A}^-_{M'}\mathbf{A}\mathbf{A}^-_{M'} = (\mathbf{M}'\mathbf{A})^-\mathbf{M}'\mathbf{A}(\mathbf{M}'\mathbf{A})^-\mathbf{M}' = (\mathbf{M}'\mathbf{A})^-\mathbf{M}'\mathbf{A}(\mathbf{M}'\mathbf{A})^-\mathbf{M}'\mathbf{A}\mathbf{V} = (\mathbf{M}'\mathbf{A})^-\mathbf{M}'$, indicating $\mathbf{A}^-_{M'}$ is a reflexive g-inverse of $\mathbf{A}$.

Similarly, under the condition that

$$\text{rank}(\mathbf{A}\mathbf{N}) = \text{rank}(\mathbf{A}) = \text{rank}(\mathbf{N}), \tag{2.144}$$

a reflexive g-inverse of $\mathbf{A}$ that satisfies (ii), denoted as $\mathbf{A}^-_N$, can be expressed as

$$\mathbf{A}^-_N = \mathbf{N}(\mathbf{A}\mathbf{N})^-. \tag{2.145}$$

In addition, under the condition that

$$\text{rank}(\mathbf{M}'\mathbf{A}\mathbf{N}) = \text{rank}(\mathbf{A}) = \text{rank}(\mathbf{M}) = \text{rank}(\mathbf{N}), \tag{2.146}$$

a reflexive g-inverse of $\mathbf{A}$ that satisfies (iii), denoted as $\mathbf{A}^-_{M',N}$, can be expressed as

$$\mathbf{A}^-_{M',N} = \mathbf{N}(\mathbf{M}'\mathbf{A}\mathbf{N})^-\mathbf{M}' = \mathbf{N}(\mathbf{A}\mathbf{N})^-\mathbf{A}(\mathbf{M}'\mathbf{A})^-\mathbf{M}' \tag{2.147}$$

(Yanai 1990, Theorem 3.3). The second equality in (2.147) holds because $(\mathbf{AN})^-\mathbf{A}(\mathbf{M}'\mathbf{A})^-$ is a g-inverse of $\mathbf{M}'\mathbf{AN}$. It may also be observed that $\mathbf{AN}(\mathbf{M}'\mathbf{AN})^-$ is a g-inverse of $\mathbf{M}'$ under $\text{rank}(\mathbf{M}'\mathbf{AN}) = \text{rank}(\mathbf{M})$, and it is a reflexive g-inverse of $\mathbf{M}'$ if $\text{rank}(\mathbf{M}'\mathbf{AN}) = \text{rank}(\mathbf{AN})$ also holds. Similarly, $(\mathbf{M}'\mathbf{AN})^-\mathbf{M}'\mathbf{A}$ is a g-inverse of $\mathbf{N}$ if $\text{rank}(\mathbf{M}'\mathbf{AN}) = \text{rank}(\mathbf{N})$, and it is a reflexive g-inverse of $\mathbf{N}$ if $\text{rank}(\mathbf{M}'\mathbf{AN}) = \text{rank}(\mathbf{M}'\mathbf{A})$ also holds.

In the previous section we discussed a way to find symmetric *nnd* matrices with respect to which $\mathbf{AA}^-$ and $\mathbf{A}^-\mathbf{A}$ are symmetric. Since $\mathbf{A}_{M'}^-$, $\mathbf{A}_N^-$, and $\mathbf{A}_{M',N}^-$ are all special cases of $\mathbf{A}^-$, essentially the same can be done with these g-inverses. This also means that under (2.141) and (2.142), (2.144), and (2.146), $\mathbf{A}_{M'}^-$, $\mathbf{A}_N^-$, and $\mathbf{A}_{M',N}^-$ are all special cases of $\mathbf{A}_{(KL)}^+$ introduced in (2.131).

2.2.7 *A variety of projectors*

It should be clear by now that by inserting a variety of g-inverses of $\mathbf{A}$ in $\mathbf{AA}^-$ and $\mathbf{A}^-\mathbf{A}$, we find a variety of projectors. If we choose $\mathbf{A}_\ell^-$ for $\mathbf{A}^-$ in $\mathbf{AA}^-$, we find $\mathbf{P}_A$ introduced in (2.107). This is the orthogonal projector onto $\text{Sp}(\mathbf{A})$. If we choose $\mathbf{A}_m^-$ for $\mathbf{A}^-$ in $\mathbf{A}^-\mathbf{A}$, we find

$$\mathbf{P}_{A'} = \mathbf{A}'(\mathbf{AA}')^-\mathbf{A}, \tag{2.148}$$

which is the orthogonal projector onto $\text{Sp}(\mathbf{A}')$. Similarly, if we choose $\mathbf{A}_{\ell(K)}^-$ for $\mathbf{A}^-$ in $\mathbf{AA}^-$, we find

$$\mathbf{P}_{A/K} = \mathbf{A}(\mathbf{A}'\mathbf{KA})^-\mathbf{A}'\mathbf{K}, \tag{2.149}$$

which is the K-orthogonal projector onto $\text{Sp}(\mathbf{A})$ under (2.124) (Tian and Takane 2009). If we choose $\mathbf{A}_{m(L)}^-$ for $\mathbf{A}^-$ in $\mathbf{A}^-\mathbf{A}$, we find

$$\mathbf{P}'_{A'/L^-} = \mathbf{L}^-\mathbf{A}'(\mathbf{AL}^-\mathbf{A}')^-\mathbf{A} = \mathbf{P}_{L^-A'/L}. \tag{2.150}$$

Likewise,

$$\mathbf{P}_{A:M^\perp} = \mathbf{AA}_{M'}^- = \mathbf{A}(\mathbf{M}'\mathbf{A})^-\mathbf{M}' \tag{2.151}$$

is the oblique projector onto $\text{Sp}(\mathbf{A})$ along $\text{Ker}(\mathbf{M}') = \text{Sp}(\mathbf{M}^\perp)$ under (2.141) and (2.142) (The symbol $\mathbf{M}^\perp$ indicates a matrix whose column vectors span $\text{Ker}(\mathbf{M}')$.), and

$$\mathbf{P}_{N:(A')^\perp} = \mathbf{A}_N^-\mathbf{A} = \mathbf{N}(\mathbf{AN})^-\mathbf{A} \tag{2.152}$$

is the oblique projector onto $\text{Sp}(\mathbf{N})$ along $\text{Ker}(\mathbf{A}) = \text{Sp}((\mathbf{A}')^\perp)$ under (2.144). In all cases we can take a complementary projector by $\mathbf{Q} = \mathbf{I} - \mathbf{P}$, where $\mathbf{P}$ can be any of the projectors introduced above. In general, if $\mathbf{P}$ is the projector onto $\text{Sp}(\mathbf{P}) = \text{Ker}(\mathbf{Q})$ along $\text{Ker}(\mathbf{P}) = \text{Sp}(\mathbf{Q})$, $\mathbf{Q}$ is the projector onto $\text{Sp}(\mathbf{Q}) = \text{Ker}(\mathbf{P})$ along $\text{Ker}(\mathbf{Q}) = \text{Sp}(\mathbf{P})$.

The K-orthogonal projectors can easily be interpreted as oblique projectors. For example, $\mathbf{P}_{A/K}$ in (2.149) is equal to $\mathbf{P}_{A:(KA)^\perp}$, which is the projector onto $\text{Sp}(\mathbf{A})$ along $\text{Ker}(\mathbf{A}'\mathbf{K})$. Similarly, $\mathbf{P}'_{A/K} = \mathbf{P}_{KA:A^\perp}$, $\mathbf{Q}_{A/K} = \mathbf{P}_{(KA)^\perp:A} = \mathbf{Q}_{A:(KA)^\perp}$, and $\mathbf{Q}'_{A/K} = \mathbf{P}_{A^\perp:KA} = \mathbf{Q}_{KA:A^\perp}$.

Oblique projectors can also be easily turned into K-orthogonal projectors for some $\mathbf{K}$. Under (2.141), we have $\text{Sp}(\mathbf{M}) = \text{Sp}(\mathbf{P}_M\mathbf{A})$, and consequently, $\text{Ker}(\mathbf{M}') = \text{Ker}(\mathbf{A}'\mathbf{P}_M)$, where $\mathbf{P}_M$ is the orthogonal projector onto $\text{Sp}(\mathbf{M})$. This implies that $\mathbf{P}_{A:M'}$ can be rewritten as

$$\mathbf{P}_{A/P_M} = \mathbf{A}(\mathbf{A}'\mathbf{P}_M\mathbf{A})^-\mathbf{A}'\mathbf{P}_M. \tag{2.153}$$

The metric matrix $\mathbf{P}_M$ is typically singular. However, it can always be made nonsingular by adding $\mathbf{Q}_A$ or more generally $\tilde{\mathbf{A}}\Delta\tilde{\mathbf{A}}'$, where, as before, columns of $\tilde{\mathbf{A}}$ span the space orthogonal to $\text{Sp}(\mathbf{A})$, and Δ is an arbitrary *pd* matrix. Similarly, we find $\mathbf{P}'_{A:M^\perp} = \mathbf{P}_{M:A^\perp} = \mathbf{P}_{M/P_A}$, $\mathbf{Q}_{A:M^\perp} = \mathbf{P}_{M^\perp:A} = \mathbf{Q}_{A/P_M}$, and $\mathbf{Q}'_{A:M^\perp} = \mathbf{P}_{A^\perp:M} = \mathbf{Q}_{M/P_A}$.

It may be pointed out that (2.141) implies $\text{rank}(\mathbf{M}'\mathbf{A}) = \text{rank}(\mathbf{P}_M\mathbf{A}) = \text{rank}(\mathbf{A})$, but not vice versa. The latter is required for $\mathbf{P}_{A/P_M}$ to be the projector onto $\text{Sp}(\mathbf{A})$ along $\text{Ker}(\mathbf{A}'\mathbf{P}_M)$. This means that $\text{rank}(\mathbf{M})$ can be larger than $\text{rank}(\mathbf{A})$ in (2.153), and so $\mathbf{P}_{A/P_M}$ is in fact slightly more general than $\mathbf{P}_{A:M'}$. This means that the equalities in the preceding paragraph hold strictly only when (2.141) holds.

The following theorem shows which projector will "win" when two projectors onto the same subspace but with different metric matrices are successively applied (ter Braak and de Jong 1998):

(i) $\mathbf{P}_{A/K}\mathbf{P}_{A/L} = \mathbf{P}_{A/L}$.

(ii) $\mathbf{P}'_{A/K}\mathbf{P}'_{A/L} = \mathbf{P}'_{A/K}$.

(iii) $\mathbf{Q}_{A/K}\mathbf{Q}_{A/L} = \mathbf{Q}_{A/K}$. (2.154)

(iv) $\mathbf{Q}'_{A/K}\mathbf{Q}'_{A/L} = \mathbf{Q}'_{A/L}$. (2.155)

These relations can be easily verified directly.

The following result on the orthogonal projector defined by the Kronecker product of two matrices may also be of interest:

$$\mathbf{P}_{A\otimes B} = \mathbf{P}_A \otimes \mathbf{P}_B \tag{2.156}$$

This can be shown in a straightforward manner.

$$\begin{aligned}\mathbf{P}_{A\otimes B} &= (\mathbf{A}\otimes\mathbf{B})((\mathbf{A}\otimes\mathbf{B})'(\mathbf{A}\otimes\mathbf{B}))^-(\mathbf{A}\otimes\mathbf{B})' \\ &= (\mathbf{A}\otimes\mathbf{B})(\mathbf{A}'\mathbf{A}\otimes\mathbf{B}'\mathbf{B})^-(\mathbf{A}\otimes\mathbf{B})' \\ &= (\mathbf{A}\otimes\mathbf{B})((\mathbf{A}'\mathbf{A})^-\otimes(\mathbf{B}'\mathbf{B})^-)(\mathbf{A}\otimes\mathbf{B})' \\ &= \mathbf{A}(\mathbf{A}'\mathbf{A})^-\mathbf{A}'\otimes\mathbf{B}(\mathbf{B}'\mathbf{B})^-\mathbf{B}' = \mathbf{P}_A\otimes\mathbf{P}_B.\end{aligned} \tag{2.157}$$

This formula can be readily extended to projectors with nonidentity metric matrices.

2.2.8 Khatri's lemma

Let $\mathbf{K}$ and $\mathbf{L}$ denote *nnd* matrices of order m, and let $\mathbf{B}$ and $\mathbf{C}$ be matrices such that:

(i) $\text{rank}(\mathbf{KB}) = \text{rank}(\mathbf{B})$,

(ii) $\text{rank}(\mathbf{LC}) = \text{rank}(\mathbf{C})$, (2.158)

(iii) $\text{Ker}(\mathbf{B}'\mathbf{K}) = \text{Sp}(\mathbf{LC})$.

Then,

$$\mathbf{B}(\mathbf{B}'\mathbf{K}\mathbf{B})^{-}\mathbf{B}'\mathbf{K}+\mathbf{L}\mathbf{C}(\mathbf{C}'\mathbf{L}\mathbf{C})^{-}\mathbf{C}'=\mathbf{I}. \tag{2.159}$$

For a proof, see Yanai and Takane (1992). The first two conditions above are automatically satisfied if $\mathbf{K}$ and $\mathbf{L}$ are *pd*. In this case we find, by setting $\mathbf{L}=\mathbf{K}^{-1}$ in (2.159), that

$$\mathbf{B}(\mathbf{B}'\mathbf{K}\mathbf{B})^{-}\mathbf{B}'\mathbf{K}=\mathbf{I}_m-\mathbf{K}^{-1}\mathbf{C}(\mathbf{C}'\mathbf{K}^{-1}\mathbf{C})^{-}\mathbf{C}', \tag{2.160}$$

or

$$\mathbf{P}_{B/K}=\mathbf{Q}'_{C/K^{-1}}=\mathbf{Q}_{K^{-1}C/K}. \tag{2.161}$$

Equation (2.160) is called Khatri's (1966) lemma. This lemma is one of the most useful lemmas in various extensions of CPCA, particularly in the estimation of growth curve models (Sections 4.12 and 4.13; also see Takane and Zhou (2012)). Specifically, it allows us to rewrite K-orthogonal projectors in the form of $\mathbf{Q}=\mathbf{I}-\mathbf{P}$ into equivalent $\mathbf{P}$ projectors. For example, $\mathbf{Q}_{A/K}=\mathbf{P}'_{A^{\perp}/K^{-1}}=\mathbf{P}_{K^{-1}A^{\perp}/K}$. Note that Khatri's lemma requires $\mathbf{K}$ to be *pd*. To rewrite $\mathbf{Q}_{A/P_M}$ introduced in Section 2.2.7 into the $\mathbf{P}$ form, we first replace the singular metric matrix $\mathbf{P}_M$ by a nonsingular matrix. This can be done easily by adding $\mathbf{Q}_A$, or more generally $\tilde{\mathbf{A}}\Delta\tilde{\mathbf{A}}'$, where $\tilde{\mathbf{A}}$ spans the orthocomplement subspace of $\mathrm{Sp}(\mathbf{A})$, and Δ is an arbitrary *pd* matrix. Clearly, (2.161) reduces to $\mathbf{P}_B=\mathbf{Q}_C$ when $\mathbf{K}=\mathbf{L}=\mathbf{I}_m$. It may be pointed out that Khatri's lemma can also be obtained by setting $\mathbf{A}$ to be a square nonsingular matrix of order m, $\mathbf{X}=\mathbf{K}^{1/2}\mathbf{B}$, and $\mathbf{Y}=\mathbf{K}^{-1/2}\mathbf{C}$ in (a) of (2.197).

Khatri's lemma (2.160) has been extended to a rectangular $\mathbf{K}$. See Takane and Hunter (2011, Appendix A) in connection with the Wedderburn–Guttman decomposition (Section 4.16).

2.2.9 *Theory of linear estimation*

So far projection matrices have been discussed largely without statistical motivations. This section provides some statistical motivation and background. As has been alluded to several times already, projection matrices are closely related to problems of linear estimation in regression analysis.

In Section 2.1.9 we briefly discussed the simplest case of linear LS estimation. We follow this example a little further. Recall that $\mathbf{z}=\mathbf{A}\mathbf{b}+\mathbf{e}$ represents a regression model, where $\mathbf{z}$, $\mathbf{A}$, $\mathbf{b}$, and $\mathbf{e}$ denote, respectively, the criterion vector, the matrix of predictor variables, the vector of regression coefficients, and the vector of disturbance terms. The ordinary LS estimate (OLSE) of $\mathbf{b}$ that minimizes

$$f(\mathbf{b})=\mathrm{SS}(\mathbf{z}-\mathbf{A}\mathbf{b})=(\mathbf{z}-\mathbf{A}\mathbf{b})'(\mathbf{z}-\mathbf{A}\mathbf{b}) \tag{2.162}$$

is given by

$$\hat{\mathbf{b}}_{OLSE}=(\mathbf{A}'\mathbf{A})^{-1}\mathbf{A}'\mathbf{z}, \tag{2.163}$$

assuming that $\mathbf{A}'\mathbf{A}$ is nonsingular. Assume further that $\mathbf{e}$ is a random vector such that

$$\mathrm{Ex}[\mathbf{e}]=\mathbf{0}, \tag{2.164}$$

where Ex takes expected values of its arguments, and that

$$\text{Var}[\mathbf{e}] = \sigma^2 \mathbf{I}, \tag{2.165}$$

where Var indicates a variance-covariance operator. The latter indicates that the elements of $\mathbf{e}$ are uncorrelated and have equal variances. We then have

$$\text{Ex}[\hat{\mathbf{b}}_{OLSE}] = (\mathbf{A}'\mathbf{A})^{-1}\mathbf{A}'\text{Ex}[\mathbf{z}] = (\mathbf{A}'\mathbf{A})^{-1}\mathbf{A}'\mathbf{A}\mathbf{b} = \mathbf{b}, \tag{2.166}$$

showing that $\hat{\mathbf{b}}_{OLSE}$ is an unbiased estimator of $\mathbf{b}$, and

$$\begin{aligned} \text{Var}[\hat{\mathbf{b}}_{OLSE}] &= \text{Ex}[(\hat{\mathbf{b}}_{OLSE} - \text{Ex}[\hat{\mathbf{b}}_{OLSE}])(\hat{\mathbf{b}}_{OLSE} - \text{Ex}[\hat{\mathbf{b}}_{OLSE}])'] \\ &= (\mathbf{A}'\mathbf{A})^{-1}\mathbf{A}'\text{Ex}[(\mathbf{z} - \mathbf{A}\mathbf{b})(\mathbf{z} - \mathbf{A}\mathbf{b})']\mathbf{A}(\mathbf{A}'\mathbf{A})^{-1} \\ &= (\mathbf{A}'\mathbf{A})^{-1}\mathbf{A}'(\sigma^2\mathbf{I})\mathbf{A}(\mathbf{A}'\mathbf{A})^{-1} = \sigma^2(\mathbf{A}'\mathbf{A})^{-1}. \end{aligned} \tag{2.167}$$

It is well known (e.g., Puntanen et al. 2011) that under (2.164) and (2.165), the OLSE of $\mathbf{b}$ is the best among all linear unbiased estimators of the form $\mathbf{L}'\mathbf{z}$ in the sense that it has the smallest variance. Such an estimator is called the BLUE (the best linear unbiased estimator) of $\mathbf{b}$.

In the discussion above, it is assumed that $\mathbf{A}'\mathbf{A}$ is nonsingular. What happens if it is singular? In this case, $\hat{\mathbf{b}}_{OLSE} = (\mathbf{A}'\mathbf{A})^{-}\mathbf{A}'\mathbf{z}$ that minimizes (2.162) is not uniquely determined. So we consider linear functions of $\mathbf{b}$ for which invariant estimators exist. A linear function of $\mathbf{b}$ of the form $\mathbf{c}'\mathbf{b}$ is said to be estimable if there exists a vector $\mathbf{t}$ such that $\text{Ex}(\mathbf{t}'\mathbf{z}) = \mathbf{c}'\mathbf{b}$ (i.e., $\mathbf{t}'\mathbf{z}$ is an unbiased estimator of $\mathbf{c}'\mathbf{b}$). The $\mathbf{c}$ that satisfies this condition is necessarily of the form $\mathbf{A}'\mathbf{t}$, since $\text{Ex}[\mathbf{t}'\mathbf{z}] = \mathbf{t}'\mathbf{A}\mathbf{b} = \mathbf{c}'\mathbf{b}$ has to hold for any arbitrary $\mathbf{b}$. It follows that $\mathbf{c}'\hat{\mathbf{b}} = \mathbf{t}'\mathbf{P}_A\mathbf{z}$ is invariant for any choice of $(\mathbf{A}'\mathbf{A})^{-}$. It has been shown (Rao 1962) that $\mathbf{c}'\hat{\mathbf{b}}_{OLSE}$ is the BLUE for $\mathbf{c}'\mathbf{b}$ among all possible linear unbiased estimator $\mathbf{t}'\mathbf{z}$. (See also Searle (1971, Section 5.4).) Obviously, the prediction vector $\mathbf{A}\mathbf{b}$ is estimable, and its OLSE $\hat{\mathbf{z}} = \mathbf{A}\hat{\mathbf{b}}_{OLSE} = \mathbf{P}_A\mathbf{z}$ is the BLUE of $\mathbf{A}\mathbf{b}$ under (2.164) and (2.165). The estimate of the vector of disturbance terms (often called the residual vector), on the other hand, is obtained by $\hat{\mathbf{e}} = \mathbf{z} - \mathbf{A}\hat{\mathbf{b}}_{OLSE} = \mathbf{z} - \mathbf{P}_A\mathbf{z} = \mathbf{Q}_A\mathbf{z}$, where $\mathbf{Q}_A = \mathbf{I} - \mathbf{P}_A$. Note that the diagonal elements of $\mathbf{P}_A$ are often called leverages and are useful in diagnosing influential observations (outliers) in regression analysis (e.g., Belsley et al. 1980; Cook and Weisberg 1982). See also Section 5.3.

The $\hat{\mathbf{b}}_{OLSE}$ given in (2.163) is not necessarily the BLUE of $\mathbf{b}$ if (2.165) is not true. Suppose more generally that

$$\text{Var}[\mathbf{e}] = \Sigma, \tag{2.168}$$

where Σ is an arbitrary *pd* matrix. It is well known (e.g., Puntanen et al. 2011) that under (2.168) the BLUE of $\mathbf{b}$ is given by

$$\hat{\mathbf{b}}_{BLUE} = (\mathbf{A}'\Sigma^{-1}\mathbf{A})^{-1}\mathbf{A}'\Sigma^{-1}\mathbf{z}. \tag{2.169}$$

Under (2.168), $\text{Var}[\hat{\mathbf{b}}_{OLSE}] = (\mathbf{A}'\mathbf{A})^{-1}\mathbf{A}'\Sigma\mathbf{A}(\mathbf{A}'\mathbf{A})^{-1}$, whereas $\text{Var}[\hat{\mathbf{b}}_{BLUE}] = (\mathbf{A}'\Sigma\mathbf{A})^{-1}$. It has been shown that $\text{Var}[\hat{\mathbf{b}}_{OLSE}] \geq \text{Var}[\hat{\mathbf{b}}_{BLUE}]$ (e.g., Rao 1967, Lemma 2c). The $\hat{\mathbf{b}}_{BLUE}$ given above is also called the generalized LS estimator (GLSE) of $\mathbf{b}$, which is a special case of the weighted LS estimator (WLSE), where the weight matrix is given by $\mathbf{K} = \Sigma^{-1}$. The WLSE of $\mathbf{b}$ that minimizes

$$f(\mathbf{b}) = \mathrm{SS}(\mathbf{z} - \mathbf{Ab})_K = (\mathbf{z} - \mathbf{Ab})'\mathbf{K}(\mathbf{z} - \mathbf{Ab}), \tag{2.170}$$

is given by

$$\hat{\mathbf{b}}_{WLSE} = (\mathbf{A}'\mathbf{KA})^{-1}\mathbf{A}'\mathbf{Kz}, \tag{2.171}$$

assuming that $\mathbf{A}'\mathbf{KA}$ is invertible. In case it is not, we again consider estimable functions of $\mathbf{b}$, e.g., the prediction vector whose WLSE $\hat{\mathbf{z}} = \mathbf{P}_{A/K}\mathbf{z}$ is invariant over any choice of $(\mathbf{A}'\mathbf{KA})^-$. Recall, however, that we assumed that $\mathrm{rank}(\mathbf{KA}) = \mathrm{rank}(\mathbf{A})$ in (2.124) to make $\mathbf{P}_{A/K}$ the orthogonal projector onto $\mathrm{Sp}(\mathbf{A})$ in the metric of $\mathbf{K}$. The residual vector for the WLSE is given by $\hat{\mathbf{e}} = \mathbf{Q}_{A/K}\mathbf{z}$.

Note 2.5 Note that $\hat{\mathbf{b}}_{GLSE} = \hat{\mathbf{b}}_{BLUE}$ is true only when the true Σ is known. That is, it may not hold if Σ has to be estimated from the data (Rao 1967). Interestingly, $\hat{\mathbf{b}}_{OLSE} = \hat{\mathbf{b}}_{BLUE}$ holds under a condition weaker than (2.165). There are a number of alternative ways of stating this condition (e.g., Puntanen and Styan 1989). Some representative ones are: (1) $\mathbf{A}'\Sigma\tilde{\mathbf{A}} = \mathbf{O}$, where $\tilde{\mathbf{A}}$ is such that $\mathrm{Sp}(\tilde{\mathbf{A}}) = \mathrm{Ker}(\mathbf{A}')$; (2) $\mathrm{Sp}(\Sigma\mathbf{A}) \subset \mathrm{Sp}(\mathbf{A})$; (3) $\Sigma = \mathbf{A}\Delta_1\mathbf{A}' + \tilde{\mathbf{A}}\Delta_2\tilde{\mathbf{A}}'$, where Δ's are arbitrary *nnd* matrices; (4) $\Sigma = \mathbf{A}\Delta_1\mathbf{A}' + \tilde{\mathbf{A}}\Delta_2\tilde{\mathbf{A}}' + \sigma^2\mathbf{I}$, where Δ's are arbitrary *nnd* matrices, and σ^2 is an arbitrary nonnegative scalar; (5) $\mathbf{P}_A\Sigma = \Sigma\mathbf{P}_A$. The above conditions are all equivalent.

Now consider another class of linear estimators of the form

$$\hat{\mathbf{b}}_{IV} = (\mathbf{M}'\mathbf{A})^{-1}\mathbf{M}'\mathbf{z}, \tag{2.172}$$

which is often called an instrumental variable (IV) estimator of $\mathbf{b}$, where $\mathbf{M}$ represents the set of instrumental variables. The IV estimator is unbiased under (2.164), but has the variance of

$$\mathrm{Var}[\hat{\mathbf{b}}_{IV}] = \sigma^2(\mathbf{M}'\mathbf{A})^{-1}\mathbf{M}'\mathbf{M}(\mathbf{A}'\mathbf{M})^{-1} = \sigma^2(\mathbf{A}'\mathbf{P}_M\mathbf{A})^{-1} \tag{2.173}$$

under (2.165), which is generally larger than the variance of $\hat{\mathbf{b}}_{OLSE}$ given in (2.167). The IV estimators are thus not at all attractive under the "normal" circumstances. In the analysis of economic time series data, however, predictor variables $\mathbf{A}$ may be correlated with disturbance terms $\mathbf{e}$, in which case $\hat{\mathbf{b}}_{OLSE}$ is not consistent, and an alternative estimator such as the IV estimator is called for. The IV variables should be highly correlated with $\mathbf{A}$ (to keep the variances of the IV estimators to an acceptable level) but should be uncorrelated with $\mathbf{e}$. See books in econometrics (e.g, Johnston 1984) for possible choices of IV. The IV estimation leads to the estimate of the prediction vector of $\hat{\mathbf{z}} = \mathbf{P}_{A:M^\perp}\mathbf{z}$ and the residual vector of $\hat{\mathbf{e}} = \mathbf{Q}_{A:M^\perp}\mathbf{z}$. Recall that we needed to assume that $\mathrm{rank}(\mathbf{M}'\mathbf{A}) = \mathrm{rank}(\mathbf{A}) = \mathrm{rank}(\mathbf{M})$ in (2.141) and (2.142) to make $\mathbf{P}_{A:M^\perp}$ the projector onto $\mathrm{Sp}(\mathbf{A})$ along $\mathrm{Ker}(\mathbf{M}') = \mathrm{Sp}(\mathbf{M}^\perp)$. As has been noted in Section 2.2.7, there is an alternative representation of the IV estimator given by $\hat{\mathbf{b}}_{IV} = (\mathbf{A}'\mathbf{P}_M\mathbf{A})^-\mathbf{A}'\mathbf{P}_M\mathbf{z}$. This version of the IV estimator is specifically called the two-stage LS (2SLS) estimator (Amemiya 1985). The predictor variables $\mathbf{A}$ are first regressed onto $\mathbf{M}$ to find $\mathbf{P}_M\mathbf{A}$, which are then used as the predictor variables in the second stage (for predicting $\mathbf{z}$).

2.2.10 Constrained LS and Seber's trick

There are a couple of interesting variations on the topic of linear estimation to which we now turn. In regression analysis, we may wish to test a hypothesis about the regression coefficient $\mathbf{b}$. Such a hypothesis may generally be expressed as

$$\mathbf{C}'\mathbf{b} = \mathbf{0}. \tag{2.174}$$

To test the hypothesis, we must find an estimate of $\mathbf{b}$ under the hypothesis. As has been shown in Section 2.1.9, Lagrange's multiplier method may be used for this kind of constrained minimization problem. Define

$$h(\mathbf{b}, \lambda) = \mathrm{SS}(\mathbf{z} - \mathbf{A}\mathbf{b}) + 2\lambda'\mathbf{C}'\mathbf{b}, \tag{2.175}$$

where λ indicates the vector of Lagrangean multipliers. Taking the derivatives of $h(\mathbf{b}, \lambda)$ with respect to $\mathbf{b}$ and λ, we find:

$$\frac{\partial h(\mathbf{b}, \lambda)}{\partial \mathbf{b}} = -2\mathbf{A}'(\mathbf{z} - \mathbf{A}\mathbf{b}) + 2\mathbf{C}\lambda, \tag{2.176}$$

and

$$\frac{\partial h(\mathbf{b}, \lambda)}{\partial \lambda} = \mathbf{C}'\mathbf{b}. \tag{2.177}$$

Setting (2.176) equal to zero, we find

$$\hat{\mathbf{b}}_{CLSE} = (\mathbf{A}'\mathbf{A})^{-1}(\mathbf{A}'\mathbf{z} - \mathbf{C}\lambda) = \hat{\mathbf{b}}_{OLSE} - (\mathbf{A}'\mathbf{A})^{-1}\mathbf{C}\lambda. \tag{2.178}$$

Premultiplying (2.178) by $\mathbf{C}'$, we find

$$\mathbf{C}'\hat{\mathbf{b}}_{CLSE} = \mathbf{C}'\hat{\mathbf{b}}_{OLSE} - \mathbf{C}'(\mathbf{A}'\mathbf{A})^{-1}\mathbf{C}\lambda = \mathbf{0}, \tag{2.179}$$

or

$$\lambda = (\mathbf{C}'(\mathbf{A}'\mathbf{A})^{-1}\mathbf{C})^{-1}\mathbf{C}'\hat{\mathbf{b}}_{OLSE}. \tag{2.180}$$

Substituting this for λ in (2.178), we find

$$\hat{\mathbf{b}}_{CLSE} = \hat{\mathbf{b}}_{OLSE} - (\mathbf{A}'\mathbf{A})^{-1}\mathbf{C}(\mathbf{C}'(\mathbf{A}'\mathbf{A})^{-1}\mathbf{C})^{-1}\mathbf{C}'\hat{\mathbf{b}}_{OLSE}. \tag{2.181}$$

The $\hat{\mathbf{b}}_{CLSE}$ is called the constrained OLSE of $\mathbf{b}$ under (2.174). The CLSE of the prediction vector is now found by

$$\mathbf{A}\hat{\mathbf{b}}_{CLSE} = (\mathbf{P}_A - \mathbf{P}_{A(A'A)^{-1}C})\mathbf{z}. \tag{2.182}$$

The regular inverse of $\mathbf{A}'\mathbf{A}$ in this formula can be replaced by a g-inverse if $\mathrm{Sp}(\mathbf{C}) \subset \mathrm{Sp}(\mathbf{A}')$.

The hypothesis given in (2.174) can also be expressed in an alternative form:

$$\mathbf{b} = \mathbf{B}\mathbf{b}_c^* \tag{2.183}$$

for some $\mathbf{b}_c^*$, where $\mathbf{B}$ is such that $\mathrm{Sp}(\mathbf{B}) = \mathrm{Ker}(\mathbf{C}')$. The hypothesis (2.174) specifies the space orthogonal to (i.e., the null space of) the space in which $\mathbf{b}$ lies, whereas (2.183) specifies basis vectors of the subspace in which $\mathbf{b}$ lies. The former is called the null space method of specifying the constraint, while the latter is called a reparameterization method because $\mathbf{b}$ is reexpressed as a linear function of $\mathbf{b}_c^*$. We may substitute (2.183) for $\mathbf{b}$ in the regression model, and find an unconstrained LS estimate of $\mathbf{b}_c^*$. We then have

$$\hat{\mathbf{b}}_{CLSE} = \mathbf{B}\hat{\mathbf{b}}_c^* = \mathbf{B}(\mathbf{B}'\mathbf{A}'\mathbf{A}\mathbf{B})^-\mathbf{B}'\mathbf{A}'\mathbf{z}, \tag{2.184}$$

from which the estimate of the prediction vector can be found by

$$\mathbf{A}\hat{\mathbf{b}}_{CLSE} = \mathbf{P}_{AB}\mathbf{z}. \tag{2.185}$$

By comparing (2.182) and (2.185), we have

$$\mathbf{P}_A = \mathbf{P}_{AB} + \mathbf{P}_{A(A'A)^-C}, \tag{2.186}$$

assuming that $\text{Sp}(\mathbf{C}) \subset \text{Sp}(\mathbf{A}')$. This is one of the most fundamental decompositions of $\mathbf{P}_A$ to be discussed in the next section. The decomposition above can easily be extended to the WLSE.

The second variation we discuss concerns the situations in which the matrix of predictor variables is split into two blocks, say $\mathbf{A} = [\mathbf{X}, \mathbf{Y}]$, where we assume $\mathbf{X}$ and $\mathbf{Y}$ are disjoint. It may be that $\mathbf{X}$ represents the set of predictor variables of interest, while $\mathbf{Y}$ represents the set of nuisance variables (covariates). We have

$$\mathbf{z} = \mathbf{A}\mathbf{b} + \mathbf{e} = \mathbf{X}\mathbf{b}_1 + \mathbf{Y}\mathbf{b}_2 + \mathbf{e}, \tag{2.187}$$

where $\mathbf{b} = (\mathbf{b}_1', \mathbf{b}_2')'$. The OLSE of $\mathbf{b}$ may directly be found by $\hat{\mathbf{b}}_{OLSE} = (\mathbf{A}'\mathbf{A})^-\mathbf{A}'\mathbf{z}$, but alternatively, OLSE of $\mathbf{b}_1$ and $\mathbf{b}_2$ may be found separately as follows: We first orthogonalize the first two terms in (2.187) by

$$\begin{aligned} \mathbf{z} &= \mathbf{Q}_Y\mathbf{X}\mathbf{b}_1 + \mathbf{Y}(\mathbf{b}_2 + (\mathbf{Y}'\mathbf{Y})^-\mathbf{Y}'\mathbf{X}\mathbf{b}_1) + \mathbf{e} \\ &= \mathbf{Q}_Y\mathbf{X}\mathbf{b}_1 + \mathbf{Y}\mathbf{b}_2^* + \mathbf{e}, \end{aligned} \tag{2.188}$$

where $\mathbf{Q}_Y = \mathbf{I} - \mathbf{Y}(\mathbf{Y}'\mathbf{Y})^-\mathbf{Y}'$, and

$$\mathbf{b}_2^* = \mathbf{b}_2 + (\mathbf{Y}'\mathbf{Y})^-\mathbf{Y}'\mathbf{X}\mathbf{b}_1. \tag{2.189}$$

The OLSE of $\mathbf{b}_1$ and $\mathbf{b}_2^*$ can be easily found, because of the orthogonality of the two structural terms in (2.188), by

$$\hat{\mathbf{b}}_1 = (\mathbf{X}'\mathbf{Q}_Y\mathbf{X})^-\mathbf{X}'\mathbf{Q}_Y\mathbf{z}, \tag{2.190}$$

and

$$\hat{\mathbf{b}}_2^* = (\mathbf{Y}'\mathbf{Y})^-\mathbf{Y}'\mathbf{z}, \tag{2.191}$$

from which the OLSE of $\mathbf{b}_2$ can be easily found by

$$\hat{\mathbf{b}}_2 = (\mathbf{Y}'\mathbf{Y})^-\mathbf{Y}'(\mathbf{z} - \mathbf{X}\hat{\mathbf{b}}_1). \tag{2.192}$$

The formula for OLSE of $\mathbf{b}_1$ and $\mathbf{b}_2$ in (2.190) and (2.192) are sometimes called Seber's trick (Seber 1984, pp. 465–466). The OLSE of the prediction vector pertaining to each of the two structural terms in (2.187) is given by

$$\mathbf{X}\hat{\mathbf{b}}_1 = \mathbf{P}_{X/Q_Y}\mathbf{z}, \tag{2.193}$$

and

$$\begin{aligned}\mathbf{Y}\hat{\mathbf{b}}_2 &= \mathbf{P}_Y(\mathbf{z} - \mathbf{P}_{X/Q_Y}\mathbf{z}) \\ &= \mathbf{P}_Y\mathbf{Q}_{X/Q_Y}\mathbf{z} = \mathbf{P}_{Y/Q_X}\mathbf{z}.\end{aligned} \tag{2.194}$$

The last equality in (2.194) holds because the roles of $\mathbf{b}_1$ and $\mathbf{b}_2$ are algebraically symmetric in the regression model. This gives the following identity:

$$\mathbf{P}_{Y/Q_X} = \mathbf{P}_Y\mathbf{Q}_{X/Q_Y}, \tag{2.195}$$

and a decomposition of $\mathbf{P}_A$:

$$\mathbf{P}_A = \mathbf{P}_{X/Q_Y} + \mathbf{P}_{Y/Q_X}, \tag{2.196}$$

since $\mathbf{P}_A\mathbf{z} = \mathbf{A}\hat{\mathbf{b}}_{OLSE} = \mathbf{X}\hat{\mathbf{b}}_1 + \mathbf{Y}\hat{\mathbf{b}}_2 = (\mathbf{P}_{X/Q_Y} + \mathbf{P}_{Y/Q_X})\mathbf{z}$. Again it is straightforward to extend the decomposition to the WLSE.

2.2.11 *Decompositions of projectors*

Two decompositions (2.186) and (2.196) of $\mathbf{P}_A$ were derived in the previous subsection. In this section, we systematically investigate various decompositions of projectors $\mathbf{P}_A$, $\mathbf{P}_{A/K}$, and $\mathbf{P}_{A:M^\perp}$ when $\mathbf{A}$ is a row block matrix $\mathbf{A} = [\mathbf{X}, \mathbf{Y}]$. Depending on the relationships between $\mathbf{X}$ and $\mathbf{Y}$, a variety of decompositions are possible. Although we primarily focus on two-term decompositions, it is fairly straightforward to extend the decompositions into more than two terms by successively applying some of the two-term decompositions.

(i) Decompositions of $\mathbf{P}_A$ (Rao and Yanai 1979):

$$\begin{aligned}
&\text{(a)}\ \ \mathbf{P}_A = \mathbf{P}_X + \mathbf{P}_Y \quad (\text{if and only if } \mathbf{X}'\mathbf{Y} = \mathbf{O}),\\
&\text{(b)}\ \ \mathbf{P}_A = \mathbf{P}_X + \mathbf{P}_Y - \mathbf{P}_X\mathbf{P}_Y \quad (\text{if and only if } \mathbf{P}_X\mathbf{P}_Y = \mathbf{P}_Y\mathbf{P}_X),\\
&\text{(c)}\ \ \mathbf{P}_A = \mathbf{P}_X + \mathbf{P}_{Q_XY} = \mathbf{P}_Y + \mathbf{P}_{Q_YX},\\
&\text{(d)}\ \ \mathbf{P}_A = \mathbf{P}_{X/Q_Y} + \mathbf{P}_{Y/Q_X} \quad (\text{if and only if } \mathrm{Sp}(\mathbf{X}) \cap \mathrm{Sp}(\mathbf{Y}) = \{\mathbf{0}\}),\\
&\text{(e)}\ \ \mathbf{P}_A = \mathbf{P}_{AB} + \mathbf{P}_{A(A'A)^-C} \quad (\text{where } \mathrm{Sp}(\mathbf{B}) \dot{\oplus} \mathrm{Sp}(\mathbf{C}) = \mathrm{Sp}(\mathbf{A}')).
\end{aligned} \tag{2.197}$$

Recall that "$\dot{\oplus}$" in (e) indicates an orthogonal direct sum (Section 2.1.3). The condition in (e) reduces to $\mathrm{Sp}(\mathbf{C}) = \mathrm{Ker}(\mathbf{B}')$ if $\mathbf{A}'\mathbf{A}$ is nonsingular.

Decomposition (a) holds if and only if $\mathbf{X}$ and $\mathbf{Y}$ are mutually orthogonal, so that $\mathbf{P}_X\mathbf{P}_Y = \mathbf{P}_Y\mathbf{P}_X = \mathbf{O}$.

Decomposition (b) holds if and only if $\mathbf{P}_X$ and $\mathbf{P}_Y$ commute. If they commute, then $(\mathbf{P}_X - \mathbf{P}_X\mathbf{P}_Y)(\mathbf{P}_Y - \mathbf{P}_X\mathbf{P}_Y) = (\mathbf{P}_Y - \mathbf{P}_X\mathbf{P}_Y)(\mathbf{P}_X - \mathbf{P}_X\mathbf{P}_Y) = \mathbf{O}$. That is, $\text{Sp}(\mathbf{X})$ and $\text{Sp}(\mathbf{Y})$ are orthogonal except for their intersection space $\text{Sp}(\mathbf{X}) \cap \text{Sp}(\mathbf{Y})$. (Decomposition (a) is a special case of (b) when the intersection space is null, i.e., $\text{Sp}(\mathbf{X}) \cap \text{Sp}(\mathbf{Y}) = \{\mathbf{0}\}$.) This decomposition plays an important role in two-way ANOVA without interactions. Suppose $\mathbf{X}$ and $\mathbf{Y}$ denote matrices of dummy variables representing two factors. Then $\mathbf{P}_X\mathbf{P}_Y = \mathbf{P}_Y\mathbf{P}_X = \mathbf{P}_{1_n}$, where $\mathbf{1}_n$ is the n-component vector of ones, and n is the total number of observations. In this case, we have $\mathbf{P}_A - \mathbf{P}_{1_n} = (\mathbf{P}_X - \mathbf{P}_{1_n}) + (\mathbf{P}_Y - \mathbf{P}_{1_n})$, where the first term on the right-hand side represents the main effect of $\mathbf{X}$ and the second term that of $\mathbf{Y}$.

Decomposition (c) is useful when either $\mathbf{X}$ or $\mathbf{Y}$ is fitted first, and then the remaining one is fitted to the residuals. The matrix $\mathbf{P}_{Q_XY}$ represents the orthogonal projector onto $\text{Sp}(\mathbf{Q}_X\mathbf{Y})$, the subspace spanned by $\mathbf{Q}_X\mathbf{Y}$ representing the portions of $\mathbf{Y}$ left unaccounted for by $\mathbf{X}$. The matrix $\mathbf{P}_{Q_YX}$ is similar. It can be easily verified that the two terms in the decompositions are mutually orthogonal, i.e., $\mathbf{P}_X\mathbf{P}_{Q_XY} = \mathbf{P}_{Q_XY}\mathbf{P}_X = \mathbf{O}$ and $\mathbf{P}_Y\mathbf{P}_{Q_YX} = \mathbf{P}_{Q_YX}\mathbf{P}_Y = \mathbf{O}$. It follows from this decomposition that

$$\mathbf{Q}_A = \mathbf{Q}_X\mathbf{Q}_{Q_XY} = \mathbf{Q}_Y\mathbf{Q}_{Q_YX}. \tag{2.198}$$

Decomposition (d) is identical to (2.196) and is useful when $\mathbf{X}$ and $\mathbf{Y}$ are fitted simultaneously, but it holds only when $\text{Sp}(\mathbf{X})$ and $\text{Sp}(\mathbf{Y})$ are disjoint. This decomposition can be directly derived by expanding

$$\mathbf{P}_A = [\mathbf{X}, \mathbf{Y}] \begin{bmatrix} \mathbf{X}'\mathbf{X} & \mathbf{X}'\mathbf{Y} \\ \mathbf{Y}'\mathbf{X} & \mathbf{Y}'\mathbf{Y} \end{bmatrix}^{-} \begin{bmatrix} \mathbf{X}' \\ \mathbf{Y}' \end{bmatrix}$$

using a generalized form of the inverse of a partitioned matrix (Section 2.1.5), or by Seber's trick as shown in the previous section. The matrix $\mathbf{P}_{X/Q_Y}$ is the projector onto $\text{Sp}(\mathbf{X})$ along $\text{Ker}(\mathbf{X}'\mathbf{Q}_Y) \oplus \text{Ker}(\mathbf{A}')$, and $\mathbf{P}_{Y/Q_X}$ is similar. If $\text{Sp}(\mathbf{X})$ and $\text{Sp}(\mathbf{Y})$ are disjoint, then

$$\mathbf{Q}_A = \mathbf{Q}_X\mathbf{Q}_{Y/Q_X} = \mathbf{Q}_Y\mathbf{Q}_{X/Q_Y}, \tag{2.199}$$

which is similar but subtly different from (2.198).

Note that the two terms in Decomposition (d) are not mutually orthogonal, and that all three projectors involved in this decomposition are associated with different metric matrices. Define

$$\mathbf{K}^* = \mathbf{Q}_X + \mathbf{Q}_Y + \tilde{\mathbf{A}}\Delta\tilde{\mathbf{A}}', \tag{2.200}$$

where $\tilde{\mathbf{A}}$ is such that $\text{Sp}(\tilde{\mathbf{A}}) = \text{Ker}(\mathbf{A}')$, and Δ is an arbitrary *nnd* matrix. Then

$$\begin{aligned} \mathbf{P}_A &= \mathbf{P}_{A/K^*}, \\ \mathbf{P}_{X/Q_Y} &= \mathbf{P}_{X/K^*}, \\ \mathbf{P}_{Y/Q_X} &= \mathbf{P}_{Y/K^*}, \end{aligned} \tag{2.201}$$

and

$$(\mathbf{P}_{X/K^*})'\mathbf{K}^*\mathbf{P}_{Y/K^*} = \mathbf{O}. \tag{2.202}$$

The $\mathbf{K}^*$ is a kind of orthogonalizing metric.

Decomposition (e) is identical to (2.186). It can be easily verified that the two terms in this decomposition are mutually orthogonal, since $\mathbf{B}'\mathbf{A}'\mathbf{A}(\mathbf{A}'\mathbf{A})^-\mathbf{C} = \mathbf{B}'\mathbf{A}'\mathbf{A}(\mathbf{A}'\mathbf{A})^-\mathbf{A}'\mathbf{C}^* = \mathbf{B}'\mathbf{A}'\mathbf{C}^* = \mathbf{B}'\mathbf{C} = \mathbf{O}$. This indicates that (e) is a special case of (a). Decomposition (c) may in turn be regarded as a special case of (e). Let $\mathbf{B} = \begin{bmatrix} \mathbf{P}_{X'} \\ \mathbf{O} \end{bmatrix}$ and $\mathbf{C} = \begin{bmatrix} \mathbf{O} \\ \mathbf{P}_{Y'} \end{bmatrix}$ be selection matrices such that $\mathbf{AB} = \mathbf{X}$ and $\mathbf{AC} = \mathbf{Y}$. Then, $\mathbf{A}(\mathbf{A}'\mathbf{A})^-\mathbf{C} = \mathbf{Q}_X\mathbf{Y}$, and $\mathbf{A}(\mathbf{A}'\mathbf{A})^-\mathbf{B} = \mathbf{Q}_Y\mathbf{X}$. Note that $\mathbf{B}'\mathbf{P}_{A'}\mathbf{C} = \mathbf{B}'\mathbf{C} = \mathbf{O}$. That is, $\mathbf{P}_{AB} = \mathbf{P}_X$ and $\mathbf{P}_{A(A'A)^-C} = \mathbf{P}_{Q_XY}$, and $\mathbf{P}_{AC} = \mathbf{P}_Y$ and $\mathbf{P}_{A(A'A)^-B} = \mathbf{P}_{Q_YX}$.

The decompositions above can be generalized as follows when a nonidentity metric matrix $\mathbf{K}$ is introduced. We assume, as before, that the rank conditions $\text{rank}(\mathbf{KX}) = \text{rank}(\mathbf{X})$ and $\text{rank}(\mathbf{KY}) = \text{rank}(\mathbf{Y})$, and consequently $\text{rank}(\mathbf{KA}) = \text{rank}(\mathbf{A})$ hold.

(ii) Decompositions of $\mathbf{P}_{A/K}$ (Takane and Yanai 1999):

$$\begin{aligned}
&(\text{a}')\ \mathbf{P}_{A/K} = \mathbf{P}_{X/K} + \mathbf{P}_{Y/K} \quad (\text{if and only if } \mathbf{X}'\mathbf{KY} = \mathbf{O}),\\
&(\text{b}')\ \mathbf{P}_{A/K} = \mathbf{P}_{X/K} + \mathbf{P}_{Y/K} - \mathbf{P}_{X/K}\mathbf{P}_{Y/K}\\
&\qquad\qquad (\text{if and only if } \mathbf{P}_{X/K}\mathbf{P}_{Y/K} = \mathbf{P}_{Y/K}\mathbf{P}_{X/K}),\\
&(\text{c}')\ \mathbf{P}_{A/K} = \mathbf{P}_{X/K} + \mathbf{P}_{Q_{X/K}Y/K} = \mathbf{P}_{Y/K} + \mathbf{P}_{Q_{Y/K}X/K},\\
&(\text{d}')\ \mathbf{P}_{A/K} = \mathbf{P}_{X/KQ_{Y/K}} + \mathbf{P}_{Y/KQ_{X/K}} \quad (\text{if and only if } \text{Sp}(\mathbf{X}) \cap \text{Sp}(\mathbf{Y}) = \{\mathbf{0}\}),\\
&(\text{e}')\ \mathbf{P}_{A/K} = \mathbf{P}_{AB/K} + \mathbf{P}_{A(A'KA)^-C/K} \quad (\text{where } \text{Sp}(\mathbf{B}) \dot{\oplus} \text{Sp}(\mathbf{C}) = \text{Sp}(\mathbf{A}'\mathbf{K})).
\end{aligned} \tag{2.203}$$

Among these decompositions the two terms on the right-hand side of (a′), (c′), and (e′) are K-orthogonal. In (b′), $\mathbf{P}_{X/K} - \mathbf{P}_{X/K}\mathbf{P}_{Y/K}$ and $\mathbf{P}_{Y/K} - \mathbf{P}_{X/K}\mathbf{P}_{Y,K}$ are K-orthogonal. The two terms in (d′) are K^*-orthogonal, where $\mathbf{K}^*$ is, for example, $\mathbf{K}^* = \mathbf{KQ}_{X/K} + \mathbf{KQ}_{Y/K} + \tilde{\mathbf{A}}\Delta\tilde{\mathbf{A}}'$, and we have $\mathbf{P}_{A/K} = \mathbf{P}_{A/K^*}$, $\mathbf{P}_{X/KQ_{Y/K}} = \mathbf{P}_{X/K^*}$, and $\mathbf{P}_{Y/KQ_{X/K}} = \mathbf{P}_{Y/K^*}$.

Analogous decompositions of $\mathbf{P}_{A:M^\perp}$ are possible. We assume the following rank conditions: $\text{rank}(\mathbf{X}'\mathbf{U}) = \text{rank}(\mathbf{X}) = \text{rank}(\mathbf{U})$, $\text{rank}(\mathbf{Y}'\mathbf{V}) = \text{rank}(\mathbf{Y}) = \text{rank}(\mathbf{V})$, and $\text{rank}(\mathbf{A}'\mathbf{M}) = \text{rank}(\mathbf{A}) = \text{rank}(\mathbf{M})$.

(iii) Decompositions of $\mathbf{P}_{A:M^\perp}$ (Takane and Yanai 1999):

$$\begin{aligned}
&(\text{a}'')\ \mathbf{P}_{A:M^\perp} = \mathbf{P}_{X:U^\perp} + \mathbf{P}_{Y:V^\perp} \quad (\text{if and only if } \mathbf{U}'\mathbf{Y} = \mathbf{O} \text{ and } \mathbf{V}'\mathbf{X} = \mathbf{O}),\\
&(\text{b}'')\ \mathbf{P}_{A:M^\perp} = \mathbf{P}_{X:U^\perp} + \mathbf{P}_{Y:V^\perp} - \mathbf{P}_{X:U^\perp}\mathbf{P}_{Y:V^\perp}\\
&\qquad\qquad (\text{if and only if } \mathbf{P}_{X:U^\perp}\mathbf{P}_{Y:V^\perp} = \mathbf{P}_{Y:V^\perp}\mathbf{P}_{X:U^\perp}),\\
&(\text{c}'')\ \mathbf{P}_{A:M^\perp} = \mathbf{P}_{X:U^\perp} + \mathbf{P}_{Q_{X:U^\perp}Y:(Q_{U:X^\perp}V)^\perp}\\
&\qquad\qquad = \mathbf{P}_{Y:V^\perp} + \mathbf{P}_{Q_{Y:V^\perp}X:(Q_{V:Y^\perp}U)^\perp},\\
&(\text{d}'')\ \mathbf{P}_{A:M^\perp} = \mathbf{P}_{X:(Q_{V:Y^\perp}U)^\perp} + \mathbf{P}_{Y:(Q_{U:X^\perp}V)^\perp}, \quad (\text{if } |\mathbf{I} - \mathbf{P}_{X:U^\perp}\mathbf{P}_{Y:V^\perp}| \neq 0),\\
&(\text{e}'')\ \mathbf{P}_{A:M^\perp} = \mathbf{P}_{AB:(MT)^\perp} + \mathbf{P}_{A(M'A)^-C:(M(A'M)^{-1}D)^\perp}\\
&\qquad (\text{where } \text{Sp}(\mathbf{B}) \dot{\oplus} \text{Sp}(\mathbf{D}) = \text{Sp}(\mathbf{A}'), \text{ and } \text{Sp}(\mathbf{T}) \dot{\oplus} \text{Sp}(\mathbf{C}) = \text{Sp}(\mathbf{M}')).
\end{aligned} \tag{2.204}$$

Note that none of the decompositions in (iii) are orthogonal unless special orthogonalizing metrics are used. Note also that in (d″), $|\mathbf{I}-\mathbf{P}_{X:U^\perp}\mathbf{P}_{Y,V^\perp}| \neq 0$ is a sufficient, but not a necessary condition, as opposed to the conditions in (d) and (d′). Proofs of the decompositions have been given in Takane and Yanai (1999).

2.2.12 *Ridge operators*

The estimators of regression coefficients discussed in Section 2.2.9 are all linear unbiased estimators. The BLUE is the best among them. If we lift the unbiasedness condition from this class of estimators, we may find better estimators in the sense that they are on average closer to true population parameters. Ridge LS (RLS) estimators are such estimators.

Consider minimizing

$$\mathbf{f}_\delta(\mathbf{b}) = \mathrm{SS}(\mathbf{z}-\mathbf{A}\mathbf{b}) + \delta \mathrm{SS}(\mathbf{b}), \tag{2.205}$$

with respect to $\mathbf{b}$, where δ is called the ridge parameter. The ridge parameter δ typically assumes a small positive value. Differentiating f_δ with respect to $\mathbf{b}$, we find

$$\frac{\partial f_\delta(\mathbf{b})}{\partial \mathbf{b}} = -2\mathbf{A}'(\mathbf{z}-\mathbf{A}\mathbf{b}) + 2\delta\mathbf{b}. \tag{2.206}$$

Setting the derivatives equal to zero, we find the RLS estimator of $\mathbf{b}$ by

$$\hat{\mathbf{b}}_{RLS} = (\mathbf{A}'\mathbf{A}+\delta\mathbf{I})^{-1}\mathbf{A}'\mathbf{z}. \tag{2.207}$$

This estimator is biased, but it is known (Hoerl and Kennard, 1970) that for a certain range of values of δ it has a smaller mean square error defined by

$$\mathrm{MSE}[\hat{\mathbf{b}}] = \mathrm{Ex}[(\hat{\mathbf{b}}-\mathbf{b})(\hat{\mathbf{b}}-\mathbf{b})'] \tag{2.208}$$

than the corresponding OLSE. The MSE (mean square error) can be decomposed into two parts:

$$\mathrm{MSE}[\hat{\mathbf{b}}] = \mathrm{Ex}[(\mathrm{Ex}[\hat{\mathbf{b}}]-\mathbf{b})(\mathrm{Ex}[\hat{\mathbf{b}}]-\mathbf{b})'] + \mathrm{Ex}[(\hat{\mathbf{b}}-\mathrm{Ex}[\hat{\mathbf{b}}])(\hat{\mathbf{b}}-\mathrm{Ex}[\hat{\mathbf{b}}])']. \tag{2.209}$$

The first term in this decomposition is called the squared bias (the mean departure of $\mathrm{Ex}[\hat{\mathbf{b}}]$ from the population parameters), and the second term as the variance of the estimators. The ridge estimators have small bias (for unbiased estimators, the first term in MSE (mean square error) is always zero), but may have much smaller variance than the OLSE, and so smaller overall MSE (mean square error). It is known (Hoerl and Kennard, 1970) that they are particularly attractive when predictor variables are collinear (i.e., have high correlations).

The ridge estimator (2.207) may be rewritten as:

$$\hat{\mathbf{b}}_{RLSE} = (\mathbf{A}'\mathbf{A}+\delta\mathbf{P}_{A'})^{-}\mathbf{A}'\mathbf{z}, \tag{2.210}$$

where $\mathbf{P}_{A'}$ is the orthogonal projector onto $\mathrm{Sp}(\mathbf{A}')$. Replacing $\mathbf{I}$ in (2.207) by $\mathbf{P}_{A'}$ can be done without loss of generality because it can always be assumed that $\mathbf{b} \in \text{Sp}(\mathbf{A}')$. This may be seen as follows: If $\mathbf{b} \notin \text{Sp}(\mathbf{A}')$, let $\mathbf{b} = \mathbf{b}_1 + \mathbf{b}_2$, where $\mathbf{b}_1 \in \text{Sp}(\mathbf{A}')$ and $\mathbf{b}_2 \in \text{Ker}(\mathbf{A})$. Then, $\mathbf{Ab} = \mathbf{Ab}_1 + \mathbf{Ab}_2 = \mathbf{Ab}_1$, so we may set $\mathbf{b}_1$ as $\mathbf{b}$ without affecting the prediction vector. It is also interesting to note that $(\mathbf{A}'\mathbf{A} + \delta\mathbf{I})^{-1}$ is a g-inverse of $\mathbf{A}'\mathbf{A} + \delta\mathbf{P}_{A'}$. This allows us to rewrite (2.210) as

$$\hat{\mathbf{b}}_{RLS} = (\mathbf{A}'\mathbf{M}_A(\delta)\mathbf{A})^{-}\mathbf{A}'\mathbf{z}, \tag{2.211}$$

where

$$\mathbf{M}_A(\delta) = \mathbf{P}_A + \delta(\mathbf{AA}')^{+} \tag{2.212}$$

is called the ridge metric matrix. Here, $\mathbf{P}_A$ is the orthogonal projector onto $\text{Sp}(\mathbf{A})$ and $(\mathbf{AA}')^{+}$ is the Moore-Penrose g-inverse of $\mathbf{AA}'$. See (2.123) for an explicit expression of $(\mathbf{AA}')^{+}$. The RLS estimate of the prediction vector is now obtained by

$$\mathbf{A}\hat{\mathbf{b}}_{RLSE} = \mathbf{A}(\mathbf{A}'\mathbf{M}_A(\delta)\mathbf{A})^{-}\mathbf{A}'\mathbf{z} = \mathbf{R}_A(\delta)\mathbf{z}, \tag{2.213}$$

where $\mathbf{R}_A(\delta) = \mathbf{A}(\mathbf{A}'\mathbf{M}_A(\delta)\mathbf{A})^{-}\mathbf{A}'$ is called a ridge operator. Ridge operators have many properties analogous to those of projectors (Takane and Yanai 2008).

The ridge operator defined above may be generalized to (Takane 2007b):

$$\begin{aligned}\mathbf{R}^{(L)}_{A/K}(\delta) &= \mathbf{A}(\mathbf{A}'\mathbf{KA} + \delta\mathbf{L})^{-}\mathbf{A}'\mathbf{K} \\ &= \mathbf{A}(\mathbf{A}'\mathbf{KM}^{(L)}_{A/K}(\delta)\mathbf{A})^{-}\mathbf{A}'\mathbf{K},,\end{aligned} \tag{2.214}$$

where $\mathbf{K}$ and $\mathbf{L}$ are *nnd* matrices such that $\text{rank}(\mathbf{KA}) = \text{rank}(\mathbf{A})$ and $\text{Sp}(\mathbf{L}) \supset \text{Sp}(\mathbf{A}')$, and

$$\mathbf{M}^{(L)}_{A/K}(\delta) = \mathbf{P}_{A/K} + \delta\mathbf{A}(\mathbf{A}'\mathbf{KA})^{-}\mathbf{L}(\mathbf{A}'\mathbf{KA})^{-}\mathbf{A}'\mathbf{K}. \tag{2.215}$$

The matrix $\mathbf{M}^{(L)}_{A/K}(\delta)$ is called a generalized ridge metric matrix (Takane 2007b). It is possible to derive decompositions of ridge operators similar to those of projection matrices. Some of these decompositions will be discussed where they are used in Section 3.3.4.

2.3 Singular Value Decomposition (SVD)

In this section we discuss singular value decomposition (SVD). The SVD underlies principal component analysis, which seeks to find the subspace inside the data space that captures the largest variation in the original data space. We begin this section by introducing the notion of suborthogonal matrices, which play a crucial role in ten Berge's (1983, 1993) theorem used to prove important optimality properties of SVD.

2.3.1 Suborthogonal matrices

As noted earlier, a square matrix $\mathbf{T}$ of order m that satisfies

$$\mathbf{T}'\mathbf{T} = \mathbf{TT}' = \mathbf{I}_m \tag{2.216}$$

is called a full orthogonal matrix. A tall matrix (of order $n \times m$ where $n > m$) that satisfies

$$\mathbf{T}'\mathbf{T} = \mathbf{I}_m \tag{2.217}$$

is called a columnwise orthogonal matrix. A flat matrix ($n < m$) that satisfies

$$\mathbf{T}\mathbf{T}' = \mathbf{I}_n \tag{2.218}$$

is called a rowwise orthogonal matrix. A columnwise or rowwise orthogonal matrix is called a semiorthogonal matrix.

A matrix that can be made an orthogonal matrix by appending appropriate rows and/or columns is called a suborthogonal matrix. Semiorthogonal matrices are special cases of suborthogonal matrices. No elements of a suborthogonal matrix exceeds one in absolute values. A product of two suborthogonal matrices is also suborthogonal. The notion of suborthogonal matrices is extremely useful in showing certain optimality properties of eigendecomposition and singular value decomposition (SVD).

2.3.2 Ten Berge's theorem

Kristof (1970) found an upper bound of

$$f(\mathbf{B}_1, \cdots, \mathbf{B}_p) \equiv \text{tr}(\prod_{j=1}^{p} \mathbf{B}_j \mathbf{C}_j), \tag{2.219}$$

where $\mathbf{B}_j$ $(j = 1, \cdots, p)$ is an orthogonal matrix of order m, and $\mathbf{C}_j$ $(j = 1, \cdots, p)$ is a diagonal matrix of order m. When $p = 1$ or $p = 2$, this problem had been solved by von Neumann (1937; see Rao (1980, Theorem 2.5)). This book only needs the case in which $p = 1$. Suppose, for simplicity, that $\mathbf{C} \equiv \mathbf{C}_1$ is *nnd*. We then have

$$f(\mathbf{B}) \equiv \text{tr}(\mathbf{BC}) \leq \text{tr}(\mathbf{C}), \tag{2.220}$$

where $\mathbf{B} \equiv \mathbf{B}_1$. This should be clear from $\text{tr}(\mathbf{BC}) = \sum_{i=1}^{m} b_{ii}c_i \leq \sum_{i=1}^{m} c_i = \text{tr}(\mathbf{C})$ (the matrix $\mathbf{B}$ being orthogonal implies $|b_{ii}| \leq 1$ for all i). Here, b_{ii} and c_i are the ith diagonal elements of $\mathbf{B}$ and $\mathbf{C}$, respectively.

Ten Berge (1983) generalized the theorem above in an important way. Let the diagonal elements of $\mathbf{C}$ be ordered as $c_1 \geq \cdots \geq c_m$, and let $\mathbf{B}$ be a suborthogonal matrix (the previous subsection) of rank r. Then we have

$$f(\mathbf{B}) \equiv \text{tr}(\mathbf{BC}) \leq \sum_{i=1}^{r} c_i. \tag{2.221}$$

This function takes its maximum when

$$\mathbf{B} = \begin{bmatrix} \mathbf{I}_r & \mathbf{O} \\ \mathbf{O} & \mathbf{O} \end{bmatrix}. \tag{2.222}$$

We call this ten Berge's theorem. See ten Berge (1983) for a proof. This theorem is extremely useful in showing the best low rank approximation property of the eigendecomposition of a symmetric matrix (the next section) and the SVD of a rectangular matrix (Section 2.3.4).

2.3.3 *Eigendecomposition of symmetric matrices*

In Section 2.1.9, we briefly encountered an eigenequation derived from a certain constrained optimization problem. Let $\mathbf{S}$ be an *nnd* matrix of order m and rank r. Recall that maximizing $\mathbf{w}'\mathbf{S}\mathbf{w}$ under the constraint that $\mathbf{w}'\mathbf{w} = 1$ lead to the eigenequation

$$\mathbf{S}\mathbf{w} - \lambda\mathbf{w} = (\mathbf{S} - \lambda\mathbf{I}_m)\mathbf{w} = \mathbf{0}. \tag{2.223}$$

The λ is called an eigenvalue, and $\mathbf{w}$ an eigenvector of $\mathbf{S}$. This equation has a nontrivial solution for $\mathbf{w}$ ($\mathbf{w} \neq \mathbf{0}$) only when $\mathbf{S} - \lambda\mathbf{I}_m$ is singular, i.e., when

$$\det(\mathbf{S} - \lambda\mathbf{I}_m) = 0, \tag{2.224}$$

which is a polynomial equation of order m in λ, and has m solutions (roots) for λ. When $\mathbf{S}$ is *nnd* of rank r, there will be r positive roots, while the remaining ones are all zero. In general, the positive roots may not all be distinct, although we assume, for simplicity, their distinctness throughout this book. Then, associated with each distinct eigenvalue, there is a unique normalized eigenvector (up to reflection).

Let Λ denote the diagonal matrix of r positive (and distinct) eigenvalues of $\mathbf{S}$ in its diagonal arranged in the descending order, and let $\mathbf{W}$ denote the matrix of the corresponding eigenvectors, arranged in the same order as the eigenvalues. For a symmetric matrix $\mathbf{S}$, eigenvectors corresponding to distinct eigenvalues are mutually orthogonal, i.e., $\mathbf{W}'\mathbf{W} = \mathbf{I}_r$. Using Λ and $\mathbf{W}$, we can express the matrix $\mathbf{S}$ as

$$\mathbf{S} = \mathbf{W}\Lambda\mathbf{W}'. \tag{2.225}$$

This is called the eigendecomposition (or spectral decomposition) of $\mathbf{S}$. Postmultiplying both sides of (2.225) by $\mathbf{W}$, we obtain

$$\mathbf{S}\mathbf{W} = \mathbf{W}\Lambda. \tag{2.226}$$

When $r < m$, let $\mathbf{W}_0$ represent a matrix of orthonormal basis vectors spanning $\text{Ker}(\mathbf{S})$. Define $\mathbf{W}_1 = [\mathbf{W}, \mathbf{W}_0]$, and $\Lambda_1 = \text{bdiag}(\Lambda, \mathbf{O})$, where $\text{bdiag}(\Lambda, \mathbf{O})$ indicates the block diagonal matrix with Λ and $\mathbf{O}$ as diagonal blocks. (The operator bdiag forms a block diagonal matrix with matrices in its argument as diagonal blocks.) We then have

$$\mathbf{S} = \mathbf{W}_1\Lambda_1\mathbf{W}_1' = [\mathbf{W}, \mathbf{W}_0]\begin{bmatrix} \Lambda & \mathbf{O} \\ \mathbf{O} & \mathbf{O} \end{bmatrix}\begin{bmatrix} \mathbf{W}' \\ \mathbf{W}_0' \end{bmatrix}. \tag{2.227}$$

The decomposition of $\mathbf{S}$ in (2.227) is called the complete eigendecomposition (or the spectral decomposition) of $\mathbf{S}$, while the decomposition in (2.225) is called the incomplete (or compact) eigendecomposition of $\mathbf{S}$.

If $\mathbf{S}$ is nonsingular ($r = m$), there are no zero eigenvalues of $\mathbf{S}$, and $\mathbf{S}^{-1}$ can be expressed as

$$\mathbf{S}^{-1} = \mathbf{W}\Lambda^{-1}\mathbf{W}', \tag{2.228}$$

where Λ^{-1} is the diagonal matrix with reciprocals of the diagonal elements of Λ in its diagonal. When $r < m$, the same expression gives the Moore-Penrose g-inverse $\mathbf{S}^+$. If $\mathbf{S}$ is *nnd*, its eigenvalues are all nonnegative, and we have

$$\mathbf{S}^{1/2} = \mathbf{W}\Lambda^{1/2}\mathbf{W}', \tag{2.229}$$

where $\mathbf{S}^{1/2}$ is a symmetric square root factor of $\mathbf{S}$ (i.e., $\mathbf{S}^{1/2}\mathbf{S}^{1/2} = \mathbf{S}$ and $(\mathbf{S}^{1/2})' = \mathbf{S}^{1/2}$), and $\Lambda^{1/2}$ is the unique square root factor of Λ (a diagonal matrix whose diagonal elements are the square roots of the diagonal elements of Λ). The symmetric square root factor of $\mathbf{S}$ is uniquely determined except for possible reflections (sign reversals) of its column vectors. Let $\mathbf{S}^{-1/2}$ denote $(\mathbf{S}^{-1})^{1/2}$ or $(\mathbf{S}^+)^{1/2}$. Then,

$$\mathbf{S}^{-1/2} = \mathbf{W}\Lambda^{-1/2}\mathbf{W}', \tag{2.230}$$

where $\Lambda^{-1/2}$ is the (regular) inverse of $\Lambda^{1/2}$.

Let $\mathbf{S}$ be *nnd* of rank r ($\leq m$), and consider maximizing

$$f(\mathbf{X}) = \text{tr}(\mathbf{X}'\mathbf{S}\mathbf{X}) \tag{2.231}$$

under the constraint that $\mathbf{X}'\mathbf{X} = \mathbf{I}_s$, where $s \leq r$. By the (compact) eigendecomposition of $\mathbf{S}$ and ten Berge's theorem, we have

$$f(\mathbf{X}) = \text{tr}(\mathbf{X}'\mathbf{W}\Lambda\mathbf{W}'\mathbf{X}) = \text{tr}(\mathbf{W}'\mathbf{X}\mathbf{X}'\mathbf{W}\Lambda) \leq \text{tr}(\Lambda_s), \tag{2.232}$$

where Λ_s is the portion of Λ corresponding to the s largest eigenvalues. (Recall that we assume that there are no identical eigenvalues.) Clearly, $\mathbf{W}'\mathbf{X}\mathbf{X}'\mathbf{W}$ is a suborthogonal matrix of rank s. In (2.232) the equality holds when

$$\mathbf{X} = \mathbf{W}_s\mathbf{T}, \tag{2.233}$$

where $\mathbf{W}_s$ is the portion of $\mathbf{W}$ corresponding to the s largest eigenvalues of $\mathbf{S}$, and $\mathbf{T}$ is an arbitrary orthogonal matrix of order s, that is, when

$$\mathbf{W}'\mathbf{X}\mathbf{X}'\mathbf{W} = \mathbf{W}'\mathbf{W}_s\mathbf{T}\mathbf{T}'\mathbf{W}_s'\mathbf{W} = \begin{bmatrix} \mathbf{I}_s & \mathbf{O} \\ \mathbf{O} & \mathbf{O} \end{bmatrix}. \tag{2.234}$$

The eigendecomposition of a symmetric matrix is not directly used in this book except in multidimensional scaling in Section 4.8. However, it will play an important role in showing the existence of SVD, and in investigating various properties of SVD, to which we now turn.

2.3.4 Singular value decomposition (SVD)

Let $\mathbf{A}$ be an $n \times m$ matrix of rank r. It is often assumed that $n \geq m$. If this is not the case, we may take $\mathbf{A}'$ as $\mathbf{A}$ in what follows. The matrix $\mathbf{A}$ can be decomposed into

$$\mathbf{A} = \mathbf{U}\mathbf{D}\mathbf{V}', \tag{2.235}$$

where $\mathbf{U}$ is the $n \times r$ matrix of left singular vectors corresponding to the r positive singular values of $\mathbf{A}$, $\mathbf{V}$ is the analogous $m \times r$ matrix of right singular vectors, and $\mathbf{D}$

the *pd* diagonal matrix of order r with positive singular values of $\mathbf{A}$ in its diagonal. This is called the (compact) singular value decomposition (SVD) of $\mathbf{A}$. The matrices $\mathbf{U}$ and $\mathbf{V}$ are columnwise orthogonal, i.e., $\mathbf{U}'\mathbf{U} = \mathbf{V}'\mathbf{V} = \mathbf{I}_r$. We assume without loss of generality that singular values are arranged in descending order of magnitude in the diagonal of $\mathbf{D}$, that is, $d_1 > \cdots > d_r$, where d_i is the ith diagonal element of $\mathbf{D}$, assuming that there are no identical singular values. This decomposition is unique up to simultaneous reflections of the corresponding left and right singular vectors.

We have

$$\mathbf{A}'\mathbf{A} = \mathbf{V}\mathbf{D}^2\mathbf{V}', \tag{2.236}$$

and

$$\mathbf{A}\mathbf{A}' = \mathbf{U}\mathbf{D}^2\mathbf{U}', \tag{2.237}$$

indicating that $\mathbf{V}$ is the matrix of eigenvectors of $\mathbf{A}'\mathbf{A}$ corresponding to its positive eigenvalues, $\mathbf{U}$ is the matrix of eigenvectors of $\mathbf{A}\mathbf{A}'$ corresponding to its positive eigenvalues, and $\mathbf{D}^2$ the diagonal matrix of positive eigenvalues of $\mathbf{A}'\mathbf{A}$ and $\mathbf{A}\mathbf{A}'$, which are identical. The positive singular values of $\mathbf{A}$ in the diagonal of $\mathbf{D}$ are positive square roots of the (positive) eigenvalues of $\mathbf{A}'\mathbf{A}$ and $\mathbf{A}\mathbf{A}'$ in the diagonal of $\mathbf{D}^2$. (The matrix $\mathbf{D}$ is a symmetric square root factor of $\mathbf{D}^2$.)

The existence of SVD can be shown as follows (ten Berge 1993). Since $\mathbf{A}'\mathbf{A}$ is *nnd*, it can be decomposed into

$$\mathbf{A}'\mathbf{A} = \mathbf{W}\Lambda\mathbf{W}' \tag{2.238}$$

using the eigendecomposition, where $\Lambda > \mathbf{O}$ (the diagonal elements of Λ are all positive). Define $\mathbf{V} = \mathbf{W}$, $\mathbf{D} = \Delta^{1/2}$, and $\mathbf{U} = \mathbf{A}\mathbf{V}\mathbf{D}^{-1}$. Then we have

$$\mathbf{U}\mathbf{D}\mathbf{V}' = \mathbf{A}\mathbf{V}\mathbf{D}^{-1}\mathbf{D}\mathbf{V}' = \mathbf{A}\mathbf{V}\mathbf{V}' = \mathbf{A}\mathbf{W}\mathbf{W}'. \tag{2.239}$$

We show that $\mathbf{A}\mathbf{W}\mathbf{W}' = \mathbf{A}$. We have $\mathbf{A}'\mathbf{A}\mathbf{W}\mathbf{W}' = \mathbf{A}'\mathbf{A}$ from (2.238), and hence $\mathbf{A}'\mathbf{A}(\mathbf{I} - \mathbf{W}\mathbf{W}') = \mathbf{O}$. By premultiplying both sides of this equation by $\mathbf{I} - \mathbf{W}\mathbf{W}'$, we find $(\mathbf{I} - \mathbf{W}\mathbf{W}')\mathbf{A}'\mathbf{A}(\mathbf{I} - \mathbf{W}\mathbf{W}') = \mathbf{O}$, leading to $\mathbf{A}(\mathbf{I} - \mathbf{W}\mathbf{W}') = \mathbf{O}$, or $\mathbf{A} = \mathbf{A}\mathbf{W}\mathbf{W}'$.

Now consider maximizing

$$f(\mathbf{X}) = \mathrm{tr}(\mathbf{A}'\mathbf{X}) \tag{2.240}$$

under the constraint that $\mathbf{X}'\mathbf{X} = \mathbf{I}_s$ where $s \le r$. Using the SVD of $\mathbf{A}$, (2.240) can be expanded as

$$f(\mathbf{X}) = \mathrm{tr}(\mathbf{V}\mathbf{D}\mathbf{U}'\mathbf{X}) = \mathrm{tr}(\mathbf{U}'\mathbf{X}\mathbf{V}\mathbf{D}). \tag{2.241}$$

By ten Berge's theorem, we have

$$f(\mathbf{X}) \le \mathrm{tr}(\mathbf{D}_s), \tag{2.242}$$

where $\mathbf{D}_s$ is the portion of $\mathbf{D}$ corresponding to the s largest singular values of $\mathbf{A}$. The upper bound of the left-hand side of (2.242) is attained by

$$\mathbf{X} = \mathbf{U}_s\mathbf{T}, \tag{2.243}$$

where $\mathbf{U}_s$ is the portion of $\mathbf{U}$ corresponding to the s largest singular values, and $\mathbf{T}$ is an arbitrary orthogonal matrix of order s.

Note 2.6 The SVD has many different names. It is called the Eckart–Young (1936) decomposition in psychometrics. According to Horn and Johnson (1990; see also Stewart (1993)), however, its origin can be traced as far back as Bertrami (1873) and Jordan (1874), followed by Schmidt (1907) who showed, in a somewhat broader context of integral equations, the existence and the best low rank approximation property of SVD, and by Weyl (1912) who gave an elegant proof for the latter as well as other important properties of SVD. The SVD is sometimes called the Eckart–Young–Mirsky decomposition or the Schmidt–Mirsky factorization (e.g., Golub et al. 1987; Stewart and Sun 1990) with the additional name Mirsky (1960), who showed that the best low rank approximation property of SVD held not only under the SS norm, but more generally under the orthogonally (unitarily) invariant norm. In pattern recognition literature in engineering, it is also known as the Karhunen–Loéve expansion (Fukunaga 1990).

The SVD plays fundamental roles in multivariate data analysis because it has the following two important optimality properties (numbered (i) and (ii) below):

(i) The best low rank approximation of a matrix: Consider approximating an $n \times m$ $(n \geq m)$ matrix $\mathbf{A}$ of rank r by another matrix $\mathbf{A}_0$ of the same size but of rank $s \leq r$. This problem can be rephrased as that of finding $\mathbf{A}_0$ that minimizes

$$f(\mathbf{A}_0) = \mathrm{SS}(\mathbf{A} - \mathbf{A}_0). \tag{2.244}$$

Since $\mathbf{A}_0$ has rank s, it can be expressed as $\mathbf{A}_0 = \mathbf{F}\mathbf{B}'$, where $\mathbf{F}$ is an $n \times s$ columnwise orthogonal matrix ($\mathbf{F}'\mathbf{F} = \mathbf{I}_s$), and $\mathbf{B}$ is an $m \times s$ matrix of full column rank. By minimizing $f(\mathbf{A}_0) = f(\mathbf{F}, \mathbf{B})$ with respect to $\mathbf{B}$ for fixed $\mathbf{F}$, we find

$$\hat{\mathbf{B}}' = (\mathbf{F}'\mathbf{F})^{-1}\mathbf{F}'\mathbf{A} = \mathbf{F}'\mathbf{A}. \tag{2.245}$$

If we put this estimate back in (2.244), we find

$$f^*(\mathbf{F}) = f(\mathbf{F}, \hat{\mathbf{B}}) = \min_{B|F} f(\mathbf{F}, \mathbf{B}) = \mathrm{SS}(\mathbf{A} - \mathbf{F}\mathbf{F}'\mathbf{A}). \tag{2.246}$$

The minimum of $f(\mathbf{A}_0)$ can be found by minimizing $f^*(\mathbf{F})$ further with respect to $\mathbf{F}$ under the restriction that $\mathbf{F}'\mathbf{F} = \mathbf{I}_s$. The right-hand side of (2.246) can be expanded as

$$f^*(\mathbf{F}) = \mathrm{tr}(\mathbf{A}'\mathbf{A}) - \mathrm{tr}(\mathbf{F}'\mathbf{A}\mathbf{A}'\mathbf{F}), \tag{2.247}$$

where $\mathrm{tr}(\mathbf{A}'\mathbf{A})$ is a constant, so minimizing $f^*(\mathbf{F})$ is equivalent to maximizing $\mathrm{tr}(\mathbf{F}'\mathbf{A}\mathbf{A}'\mathbf{F})$. Let $\mathbf{A} = \mathbf{U}\mathbf{D}\mathbf{V}'$ denote the (compact) SVD of $\mathbf{A}$. Recall that we assumed that all nonzero singular values were distinct. We then have $\mathbf{A}\mathbf{A}' = \mathbf{U}\mathbf{D}^2\mathbf{U}'$, and from (2.233) the maximum of $\mathrm{tr}(\mathbf{F}'\mathbf{A}\mathbf{A}'\mathbf{F})$ is found by

$$\hat{\mathbf{F}} = \mathbf{U}_s\mathbf{T}, \tag{2.248}$$

where $\mathbf{U}_s$ is the portion of $\mathbf{U}$ corresponding to the s largest singular values of $\mathbf{A}$, and $\mathbf{T}$ is an arbitrary orthogonal matrix of order s. Putting back the estimate of $\mathbf{F}$ above into (2.245), we find

$$\hat{\mathbf{B}}' = \mathbf{T}'\mathbf{U}_s\mathbf{A} = \mathbf{T}'\mathbf{D}_s\mathbf{V}_s', \tag{2.249}$$

where $\mathbf{D}_s$ and $\mathbf{V}_s$ are portions of $\mathbf{D}$ and $\mathbf{V}$ corresponding to the s largest singular values of $\mathbf{A}$. Setting $\mathbf{T} = \mathbf{I}_s$ leads to principal component analysis (PCA). With the choice of $\hat{\mathbf{F}}$ in (2.248) and $\hat{\mathbf{B}}$ in (2.249) we find

$$\hat{\mathbf{A}}_0 = \mathbf{U}_s\mathbf{D}_s\mathbf{V}_s', \tag{2.250}$$

and

$$f(\hat{\mathbf{A}}_0) = \sum_{i=s+1}^{r} d_i^2(\mathbf{A}), \tag{2.251}$$

where $d_i(\mathbf{A})$ is the ith largest singular value of $\mathbf{A}$.

The best low rank approximation property of SVD discussed above is essentially the same as Rao's (1980) Theorem 2.2 and what Stewart and Sun (1990) called the Schmidt–Mirsky theorem. Their theorem is slightly more general than the result presented above. In the former, the best low rank approximation property of SVD was shown under the orthogonally (unitarily) invariant norm, while in the latter, only under the SS norm. Rao's Theorem 2.2 is stated as

$$d_i(\mathbf{A} - \mathbf{A}_0) \geq d_{r+i}(\mathbf{A}), \tag{2.252}$$

where $d_i(\mathbf{M})$ indicates the ith largest singular value of a matrix $\mathbf{M}$. The equality in (2.252) holds when $\mathbf{A}_0 = \mathbf{U}_s\mathbf{D}_s\mathbf{V}_s'$. (Okamoto and Kanazawa (1968) proved optimality of PCA using a similar theorem for eigenvalues.) The inequality in (2.252) follows from another more general inequality (Weyl 1912). Let $\mathbf{A} = \mathbf{X} + \mathbf{Y}$. Then,

$$d_{i+j-1}(\mathbf{A}) \leq d_i(\mathbf{X}) + d_j(\mathbf{Y}). \tag{2.253}$$

The inequality (2.252) follows by setting $\mathbf{X} = \mathbf{A} - \mathbf{A}_0$ and $\mathbf{Y} = \mathbf{A}_0$, and $j = r + 1$ in (2.253).

There is another theorem closely related to (2.252), called the generalized Poincaré separation theorem (Rao 1980, Theorem 2.3), which states that

$$d_{n+m-k-r+i}(\mathbf{A}) \leq d_i(\mathbf{B}'\mathbf{A}\mathbf{C}) \leq d_i(\mathbf{A}), \tag{2.254}$$

where $\mathbf{A}$ is an $n \times m$ matrix as before, and $\mathbf{B}$ and $\mathbf{C}$ are $n \times r$ and $m \times k$ columnwise orthogonal matrices, respectively. The upper bound of $d_i(\mathbf{B}'\mathbf{A}\mathbf{C})$ is attained when $\mathbf{B} = \mathbf{U}_r$ and $\mathbf{C} = \mathbf{V}_k$. For a proof of (2.254), see Rao (1979, Theorem 2.2).

(ii) The best orthogonal approximation of a matrix: Consider approximating an $n \times m$ nonsingular matrix $\mathbf{A}$ by a columnwise orthogonal matrix $\mathbf{A}_1$ of the same size. That is, we wish to find $\mathbf{A}_1$ that minimizes

$$f(\mathbf{A}_1) = \mathrm{SS}(\mathbf{A} - \mathbf{A}_1) \tag{2.255}$$

under the restriction that $\mathbf{A}_1'\mathbf{A}_1 = \mathbf{I}_m$. Note first that (2.255) can be rewritten as

$$f(\mathbf{A}_1) = \text{tr}(\mathbf{A}'\mathbf{A}) + \text{tr}(\mathbf{A}_1'\mathbf{A}_1) - 2\text{tr}(\mathbf{A}'\mathbf{A}_1). \tag{2.256}$$

Since the first and second terms in (2.256) are constant, minimizing $f(\mathbf{A}_1)$ is equivalent to maximizing $\text{tr}(\mathbf{A}'\mathbf{A}_1)$. Let $\mathbf{A} = \mathbf{UDV}'$ denote the SVD of $\mathbf{A}$. (Again recall that we have assumed that all nonzero singular values are distinct.) Then we have

$$\text{tr}(\mathbf{A}'\mathbf{A}_1) = \text{tr}(\mathbf{VDU}'\mathbf{A}_1) = \text{tr}(\mathbf{U}'\mathbf{A}_1\mathbf{VD}) \le \text{tr}(\mathbf{D}). \tag{2.257}$$

From ten Berge's theorem, the equality in (2.257), and therefore the maximum of $\text{tr}(\mathbf{A}'\mathbf{A}_1)$, is attained when $\mathbf{U}'\mathbf{A}_1\mathbf{V} = \mathbf{I}_m$, or

$$\mathbf{A}_1 = \mathbf{UV}' = \mathbf{A}(\mathbf{A}'\mathbf{A})^{-1/2}. \tag{2.258}$$

The minimization problem above has been extended to the situation in which $\mathbf{A}_1$ is replaced by $\mathbf{GT}$, where $\mathbf{G}$ is a known matrix of the same size as $\mathbf{A}$, and $\mathbf{T}$ is fully orthogonal (square and orthogonal). This is called the orthogonal Procrustes rotation problem and will be discussed in details in Section 4.19.

We will not discuss standard ways of computing SVD in this book. There exists a good and reliable algorithm for SVD developed by Golub and Kahan (1965) and improved by Golub and Reinsch (1970). The latter is still the most heavily used algorithm for SVD. Cleve Moler, the founder of the MathWorks (the vendor of the MATLAB software) introduces some insightful applications of SVD in his article entitled "Professor SVD" featuring Gene Golub, to whom we still owe very much. This article can be accessed at: http://www.mathworks.com/company/newsletters/articles/professor-svd.html

2.3.5 *Complete SVD*

Let $\mathbf{U}_0$ represent a matrix of orthonormal basis vectors spanning $\text{Ker}(\mathbf{A}')$, and let $\mathbf{V}_0$ represent a matrix of orthonormal basis vectors spanning $\text{Ker}(\mathbf{A})$. By appending $\mathbf{U}_0$ to $\mathbf{U}$ and $\mathbf{V}_0$ to $\mathbf{V}$, we find

$$\mathbf{A} = [\mathbf{U}, \mathbf{U}_0] \begin{bmatrix} \mathbf{D} & \mathbf{O} \\ \mathbf{O} & \mathbf{O} \end{bmatrix} \begin{bmatrix} \mathbf{V}' \\ \mathbf{V}_0' \end{bmatrix} = \mathbf{U}_1\mathbf{D}_1\mathbf{V}_1', \tag{2.259}$$

where $\mathbf{U}_1 = [\mathbf{U}, \mathbf{U}_0]$, $\mathbf{V}_1 = [\mathbf{V}, \mathbf{V}_0]$, and $\mathbf{D}_1 = \text{bdiag}(\mathbf{D}, \mathbf{O})$. The decomposition above is called the complete SVD of $\mathbf{A}$. The matrices $\mathbf{U}_1$ and $\mathbf{V}_1$ are fully orthogonal.

Using (2.259), a g-inverse of $\mathbf{A}$ can be expressed as

$$\mathbf{A}^- = \mathbf{V}_1 \begin{bmatrix} \mathbf{D}^{-1} & \mathbf{B}_1 \\ \mathbf{B}_2 & \mathbf{B}_3 \end{bmatrix} \mathbf{U}_1', \tag{2.260}$$

where $\mathbf{B}_1$, $\mathbf{B}_2$, and $\mathbf{B}_3$ are arbitrary except for their sizes. By imposing various restrictions on $\mathbf{B}_j$'s, various g-inverses that were discussed in Section 2.2.4 can be generated: A reflexive g-inverse $\mathbf{A}_r^-$ is found by setting $\mathbf{B}_3 = \mathbf{B}_1\mathbf{D}\mathbf{B}_2$ in (2.260), a least squares g-inverse $\mathbf{A}_\ell^-$ is found by setting $\mathbf{B}_1 = \mathbf{O}$ (and consequently, $\mathbf{A}_{\ell r}^-$ by setting $\mathbf{B}_1 = \mathbf{O}$ and $\mathbf{B}_3 = \mathbf{O}$), a minimum norm g-inverse $\mathbf{A}_m^-$ by setting $\mathbf{B}_2 = \mathbf{O}$ (and $\mathbf{A}_{mr}^-$ by setting $\mathbf{B}_2 = \mathbf{O}$ and $\mathbf{B}_3 = \mathbf{O}$), and the Moore-Penrose g-inverse by setting all $\mathbf{B}_j$'s to zero matrices.

2.3.6 Generalized SVD (GSVD)

Singular value decomposition of an $n \times m$ matrix $\mathbf{A}$ of rank r under nonidentity metric matrices $\mathbf{K}$ and $\mathbf{L}$ is called generalized SVD (GSVD) and is denoted by $\text{GSVD}(\mathbf{A})_{K,L}$. The SVD of $\mathbf{A}$ under identity metrics, on the other hand, is denoted by $\text{GSVD}(\mathbf{A})_{I,I} = \text{SVD}(\mathbf{A})$. The $\text{GSVD}(\mathbf{A})_{K,L}$ is written as

$$\mathbf{A} = \mathbf{U}\mathbf{D}\mathbf{V}', \tag{2.261}$$

where $\mathbf{U}$ is an $n \times r$ matrix of left (generalized) singular vectors corresponding to the r positive (generalized) singular values of $\mathbf{A}$, $\mathbf{V}$ is an $m \times r$ matrix of right (generalized) singular vectors, and $\mathbf{D}$ is a *pd* diagonal matrix of order r with positive (generalized) singular values in its diagonal, arranged in descending order. The matrices $\mathbf{U}$ and $\mathbf{V}$ satisfy

$$\mathbf{U}'\mathbf{K}\mathbf{U} = \mathbf{V}'\mathbf{L}\mathbf{V} = \mathbf{I}_r, \tag{2.262}$$

that is, they are columnwise orthogonal with respect to the metric matrices $\mathbf{K}$ and $\mathbf{L}$, respectively. (We say that $\mathbf{U}$ is K-orthogonal, and $\mathbf{V}$ is L-orthogonal.) It is temporarily assumed that $\mathbf{K}$ and $\mathbf{L}$ are both *pd*.

The $\text{GSVD}(\mathbf{A})_{K,L}$ is calculated as follows. Let $\mathbf{K} = \mathbf{R}_K\mathbf{R}_K'$ and $\mathbf{L} = \mathbf{R}_L\mathbf{R}_L'$ be square root decompositions of $\mathbf{K}$ and $\mathbf{L}$, and define $\mathbf{A}^* = \mathbf{R}_K'\mathbf{A}\mathbf{R}_L$. Let

$$\mathbf{A}^* = \mathbf{U}^*\mathbf{D}^*\mathbf{V}^{*'} \tag{2.263}$$

denote the SVD of $\mathbf{A}^*$. Then

$$\begin{aligned} \mathbf{U} &= (\mathbf{R}_K')^{-1}\mathbf{U}^*, \\ \mathbf{V} &= (\mathbf{R}_L')^{-1}\mathbf{V}^*, \\ \mathbf{D} &= \mathbf{D}^*. \end{aligned} \tag{2.264}$$

Clearly, $\mathbf{U}$ and $\mathbf{V}$ above satisfy (2.262).

The situation gets a little more complicated when $\mathbf{K}$ and $\mathbf{L}$ are singular. In this case, $\mathbf{R}_K'$ and $\mathbf{R}_L'$ are not square and do not have regular inverses. By pre- and postmultiplying (2.263) by $(\mathbf{R}_K')^-$ and $\mathbf{R}_L^-$, we find

$$(\mathbf{R}_K')^-\mathbf{R}_K'\mathbf{A}\mathbf{R}_L\mathbf{R}_L^- = (\mathbf{R}_K')^-\mathbf{U}^*\mathbf{D}^*\mathbf{V}^{*'}\mathbf{R}_L^-, \tag{2.265}$$

so one idea is to set $\mathbf{U} = (\mathbf{R}_K')^-\mathbf{U}^*$, and $\mathbf{V} = (\mathbf{R}_L')^-\mathbf{V}^*$. However, then $\mathbf{U}$ and $\mathbf{V}$ are not unique. To find the unique $\mathbf{U}$ and $\mathbf{V}$, we may use the Moore-Penrose g-inverse $(\mathbf{R}_K')^+ = \mathbf{R}_K(\mathbf{R}_K'\mathbf{R}_K)^{-1}$ and $(\mathbf{R}_L')^+ = \mathbf{R}_L(\mathbf{R}_L'\mathbf{R}_L)^{-1}$ for $(\mathbf{R}_K')^-$ and $(\mathbf{R}_L')^-$. We then obtain

$$\mathbf{P}_{R_K}\mathbf{A}\mathbf{P}_{R_L} = (\mathbf{R}_K')^+\mathbf{U}^*\mathbf{D}^*\mathbf{V}^{*'}\mathbf{R}_L^+, \tag{2.266}$$

where $\mathbf{P}_{R_K} = \mathbf{R}_K(\mathbf{R}_K'\mathbf{R}_K)^{-1}\mathbf{R}_K'$ and $\mathbf{P}_{R_L} = \mathbf{R}_L(\mathbf{R}_L'\mathbf{R}_L)^{-1}\mathbf{R}_L'$ are the orthogonal projectors onto $\text{Sp}(\mathbf{R}_K)$ and $\text{Sp}(\mathbf{R}_L)$, respectively. If we define $\mathbf{U} = (\mathbf{R}_K')^+\mathbf{U}^*$ and $\mathbf{V} = (\mathbf{R}_L')^+\mathbf{V}^*$, $\mathbf{U}$ and $\mathbf{V}$ are uniquely determined. However, unless $\text{Sp}(\mathbf{K}) \supset \text{Sp}(\mathbf{A})$ and $\text{Sp}(\mathbf{L}) \supset \text{Sp}(\mathbf{A}')$, $\mathbf{P}_{R_K}\mathbf{A}\mathbf{P}_{R_L} \neq \mathbf{A}$, and only the projection of $\mathbf{A}$ onto $\text{Sp}(\mathbf{R}_K)$ and $\text{Sp}(\mathbf{R}_L)$ can be recovered by GSVD. The GSVD has optimality properties similar to those of SVD. Whereas SVD has the best low rank approximation and the best orthogonal approximation properties under the I-metrics, GSVD has similar properties under the KL-metrics.

Note that the decomposition we call GSVD in this book is different from the same terminology used by some numerical linear algebraists (Van Loan, 1976; Paige, 1985, 1986; Paige and Saunders, 1981). The latter is a technique to compute the generalized eigendecomposition of the form $(\mathbf{A}'\mathbf{A})\mathbf{w} = \lambda\mathbf{B}'\mathbf{B}\mathbf{w}$ (see the next section) without explicitly calculating $\mathbf{A}'\mathbf{A}$ and $\mathbf{B}'\mathbf{B}$, and has been renamed QSVD (quotient SVD) by De Moor and Golub (1991). Our use of the term GSVD follows the tradition of French data analysts (e.g., Cailliez and Pagés 1976).

Let $\mathbf{U}_0$ and $\mathbf{V}_0$ represent the matrices of K- and L-orthogonal basis vectors spanning K- and L-orthogonal spaces to $\text{Sp}(\mathbf{A})$ and $\text{Sp}(\mathbf{A}')$, respectively. By appending $\mathbf{U}_0$ and $\mathbf{V}_0$ to $\mathbf{U}$ and $\mathbf{V}$, and a zero diagonal block to $\mathbf{D}$, we find complete GSVD (as opposed to compact GSVD presented above) analogous to complete SVD described in the previous section. Complete GSVD is useful to represent generalized forms of g-inverses with nonidentity metrics such as $\mathbf{A}^-_{\ell(K)}$, $\mathbf{A}^-_{m(L)}$, and $\mathbf{A}^+_{(KL)}$. This is analogous to the use of complete SVD to represent various types of g-inverses defined by identity metrics.

Note 2.7 The decomposition that Bojanczyk et al. (1991; Ewerbring and Luk 1989) call KL-SVD refers to $\mathbf{U}$ and $\mathbf{V}$ that transform $\mathbf{A}$, $\mathbf{K}$, and $\mathbf{L}$ as follows:

$$\begin{aligned} \mathbf{U}'\mathbf{A}\mathbf{V} &= \mathbf{D}, \\ \mathbf{U}'\mathbf{K}\mathbf{U} &= \mathbf{I}, \\ \mathbf{V}'\mathbf{L}\mathbf{V} &= \mathbf{I}. \end{aligned} \tag{2.267}$$

Let

$$\mathbf{R}_K^+\mathbf{A}(\mathbf{R}_L')^+ = \mathbf{U}^*\mathbf{D}^*\mathbf{V}^{*'}$$

be the SVD of $\mathbf{R}_K^+\mathbf{A}(\mathbf{R}_L')^+$. Then the KL-SVD of $\mathbf{A}$ is obtained by $\mathbf{U} = \mathbf{R}_K^+\mathbf{U}^*$, $\mathbf{V} = \mathbf{R}_L^+\mathbf{V}^*$, and $\mathbf{D} = \mathbf{D}^*$. $\text{GSVD}(\mathbf{A})_{K,L}$ is equivalent to KL-SVD of $\mathbf{KAL}$.

The decomposition which Rao and Mitra (1971) call SVD with respect to $\mathbf{K}$ and $\mathbf{L}$ (they use $\mathbf{M}$ and $\mathbf{N}$ in their notation instead of $\mathbf{K}$ and $\mathbf{L}$) is written as

$$\mathbf{KAL} = \mathbf{U}^*\mathbf{D}^*\mathbf{V}^{*'}, \tag{2.268}$$

where $\mathbf{U}^{*'}\mathbf{K}\mathbf{U}^* = \mathbf{I}$, and $\mathbf{V}^{*'}\mathbf{L}\mathbf{V}^* = \mathbf{I}$. The $\mathbf{U}$, $\mathbf{V}$, and $\mathbf{D}$ in (2.261) and $\mathbf{U}^*$, $\mathbf{V}^*$, and $\mathbf{D}^*$ in (2.268) are related by

$$\begin{aligned} \mathbf{U}^* &= \mathbf{K}\mathbf{U}, \\ \mathbf{V}^* &= \mathbf{L}\mathbf{V}, \\ \mathbf{D}^* &= \mathbf{D}. \end{aligned} \tag{2.269}$$

2.3.7 *Generalized eigenequation*

In Section 2.1.9, we briefly touched on the problem of maximizing $\mathbf{w}'\mathbf{S}\mathbf{w}$ subject to the normalization restriction that $\mathbf{w}'\mathbf{T}\mathbf{w} = 1$, where $\mathbf{S}$ denotes a symmetric matrix of order m and $\mathbf{T}$ denotes an *nnd* matrix of the same size. This leads to the generalized eigenequation of the form

$$\mathbf{S}\mathbf{w} = \lambda\mathbf{T}\mathbf{w} \tag{2.270}$$

to be solved, where $\mathbf{w}$ is called a generalized eigenvector, and λ a generalized eigenvalue. Premultiplying both sides of (2.270) by $\mathbf{w}'$, we find $\mathbf{w}'\mathbf{S}\mathbf{w} = \lambda$ since $\mathbf{w}'\mathbf{T}\mathbf{w} = 1$, indicating that the maximum of $\mathbf{w}'\mathbf{S}\mathbf{w}$ coincides with the largest generalized eigenvalue of $\mathbf{S}$ with respect to $\mathbf{T}$.

When $\mathbf{T}$ is *pd*, the generalized eigenequation can easily be reduced to the simple eigenproblem. In this case, there is a nonsingular square root factor of $\mathbf{T}$ such that $\mathbf{T} = \mathbf{R}\mathbf{R}'$, and (2.270) can be rewritten as

$$(\mathbf{R}^{-1}\mathbf{S}(\mathbf{R}')^{-1})(\mathbf{R}'\mathbf{w}) = \lambda(\mathbf{R}'\mathbf{w}), \tag{2.271}$$

which is the usual eigenequation for $\mathbf{R}^{-1}\mathbf{S}(\mathbf{R}')^{-1}$ with the eigenvector $\mathbf{R}'\mathbf{w}$.

The situation is more complicated if $\mathbf{T}$ is singular. The following four cases should be distinguished:

(i) $\mathbf{w} \in \mathrm{Sp}(\mathbf{S}) \cap \mathrm{Sp}(\mathbf{T})$.
(ii) $\mathbf{w} \in \mathrm{Ker}(\mathbf{S}) \cap \mathrm{Sp}(\mathbf{T})$.
(iii) $\mathbf{w} \in \mathrm{Sp}(\mathbf{S}) \cap \mathrm{Ker}(\mathbf{T})$.
(iv) $\mathbf{w} \in \mathrm{Ker}(\mathbf{S}) \cap \mathrm{Ker}(\mathbf{T})$.

In Case (iii) we have $\mathbf{S}\mathbf{w} \neq \mathbf{0}$ and $\mathbf{T}\mathbf{w} = \mathbf{0}$, so there is no finite λ that satisfies (2.270). In Case (iv) we have $\mathbf{S}\mathbf{w} = \mathbf{T}\mathbf{w} = \mathbf{0}$, so any arbitrary scalar λ satisfies (2.270), and λ is indeterminate. Such an indeterminate λ is called an improper eigenvalue. In Case (ii) we have $\mathbf{S}\mathbf{w} = \mathbf{0}$ and $\mathbf{T}\mathbf{w} \neq \mathbf{0}$, so λ is always equal to zero. In Case (i) λ takes some finite value other than 0. In the first two cases, λ is called a proper eigenvalue, and the corresponding $\mathbf{w}$ a proper eigenvector. In this book we only deal with Case (i).

Let

$$r^* = \dim(\mathrm{Sp}(\mathbf{S}) \cap \mathrm{Sp}(\mathbf{T})). \tag{2.272}$$

Then there are r^* nonzero (generalized) eigenvalues λ_i $(i = 1, \cdots, r^*)$ of $\mathbf{S}$ with respect to $\mathbf{T}$, and the corresponding r^* (generalized) eigenvectors $\mathbf{w}_i$ $(i = 1, \cdots, r^*)$ of $\mathbf{S}$ with respect to $\mathbf{T}$. Let Λ denote a diagonal matrix of λ_i's in its diagonal arranged in descending order, and let $\mathbf{W}$ denote the matrix of $\mathbf{w}_i$ arranged conformably with the corresponding eigenvalues. Then, (2.270) can be collectively written as

$$\mathbf{S}\mathbf{W} = \mathbf{T}\mathbf{W}\Lambda \quad (\text{where } \mathbf{W}'\mathbf{T}\mathbf{W} = \mathbf{I}). \tag{2.273}$$

Note 2.8 From the viewpoint of matrix decomposition,

$$\mathbf{S} = \mathbf{T}\mathbf{W}\Lambda\mathbf{W}'\mathbf{T} \tag{2.274}$$

is called the generalized eigendecomposition of $\mathbf{S}$ with respect to $\mathbf{T}$.

Let $\mathbf{N}$ be a matrix such that $\mathrm{Sp}(\mathbf{N}) = \mathrm{Ker}(\mathbf{T})$, and assume that $\mathrm{rank}(\mathbf{N}'\mathbf{S}\mathbf{N}) = \mathrm{rank}(\mathbf{S}\mathbf{N})$ holds. (If $\mathbf{S}$ is *nnd*, the latter always holds.) Then, (2.273) can be rewritten as

$$\mathbf{R}^+\mathbf{Q}'_{N/S}\mathbf{S}\mathbf{Q}_{N/S}(\mathbf{R}')^+\mathbf{U} = \mathbf{U}\Lambda, \tag{2.275}$$

where $\mathbf{U}'\mathbf{U} = \mathbf{I}_{r^*}$, and $\mathbf{R}$ is a columnwise nonsingular square root factor of $\mathbf{T}$, and

$$\mathbf{Q}_{N/S} = \mathbf{I} - \mathbf{N}(\mathbf{N}'\mathbf{S}\mathbf{N})^-\mathbf{N}'\mathbf{S}. \tag{2.276}$$

It can be readily verified that $\mathbf{Q}_{N/S}$ has the following properties:

$$\mathbf{R}'\mathbf{Q}_{N/S} = \mathbf{R}', \tag{2.277}$$

$$\mathbf{Q}'_{N/S}\mathbf{S}\mathbf{Q}_{N/S} = \mathbf{Q}'_{N/S}\mathbf{S} = \mathbf{S}\mathbf{Q}_{N/S} \quad \text{(symmetric)}, \tag{2.278}$$

and

$$\mathbf{N}'\mathbf{S}\mathbf{Q}_{N/S} = \mathbf{O} \quad (\mathbf{S}\mathbf{Q}_{N/S}\mathbf{N} = \mathbf{O}). \tag{2.279}$$

From (2.279) $\mathbf{S}\mathbf{Q}_{N/S}$ can be expressed in the form of

$$\mathbf{S}\mathbf{Q}_{N/S} = \mathbf{R}\Delta\mathbf{R}' \tag{2.280}$$

for some Δ, and so

$$\mathbf{R}\mathbf{R}^+\mathbf{Q}'_{N/S}\mathbf{S} = \mathbf{S}\mathbf{Q}_{N/S}(\mathbf{R}')^+\mathbf{R}' = \mathbf{S}\mathbf{Q}_{N/S}. \tag{2.281}$$

The results above can be used to show the equivalence between (2.273) and (2.275). We first show that (2.273) implies (2.275). Premultiplying (2.273) by $\mathbf{Q}'_{N/S}$ and noting that $\mathbf{Q}'_{N/S}\mathbf{T} = \mathbf{T}$, we find, from (2.278),

$$\mathbf{Q}'_{N/S}\mathbf{S}\mathbf{Q}_{N/S}\mathbf{W} = \mathbf{R}\mathbf{R}'\mathbf{W}\Lambda. \tag{2.282}$$

Similarly, premultiplying (2.282) by $\mathbf{R}^+$ and considering (2.281), we find

$$\begin{aligned}\mathbf{R}^+&\mathbf{Q}'_{N/S}\mathbf{S}\mathbf{Q}_{N/S}(\mathbf{R}')^+\mathbf{R}'\mathbf{W} \\ &= \mathbf{R}^+\mathbf{R}\mathbf{R}'\mathbf{W}\Lambda = \mathbf{R}'(\mathbf{R}')^+\mathbf{R}'\mathbf{W}\Lambda = \mathbf{R}'\mathbf{W}\Lambda.\end{aligned} \tag{2.283}$$

By setting $\mathbf{U} = \mathbf{R}'\mathbf{W}$, we find (2.275). Conversely, by premultiplying both sides of (2.275) by $\mathbf{R}$, we find

$$\mathbf{R}\mathbf{R}^+\mathbf{Q}'_{N/S}\mathbf{S}\mathbf{Q}_{N/S}(\mathbf{R}')^+\mathbf{U} = \mathbf{R}\mathbf{U}\Lambda. \tag{2.284}$$

If we set $\mathbf{W} = \mathbf{Q}_{N/S}(\mathbf{R}')^+\mathbf{U}$, the left-hand side of (2.284) becomes $\mathbf{S}\mathbf{W}$ due to (2.281), and the right-hand side is equal to $\mathbf{R}'\mathbf{W} = \mathbf{R}'\mathbf{Q}_{N/S}(\mathbf{R}')^+\mathbf{U} = \mathbf{R}'(\mathbf{R}')^+\mathbf{U}$ due to (2.277). Hence, we find

$$\mathbf{R}\mathbf{U}\Lambda = \mathbf{R}\mathbf{R}^+\mathbf{R}\mathbf{U}\Lambda = \mathbf{R}\mathbf{R}'(\mathbf{R}')^+\mathbf{U}\Lambda = \mathbf{T}\mathbf{W}\Lambda, \tag{2.285}$$

showing that (2.275) implies (2.273).

Note 2.9 The generalized eigenproblem can also be formulated as the problem of simultaneous diagonalizations of two square matrices. In this case, eigenvalues do not appear on the scene, and consequently there is no problem of indeterminacy or incompatibility of eigenequations, and a more general formulation is possible. General solutions for simultaneous diagonalizations of a symmetric matrix **S** and an *nnd* matrix **T** have been obtained by de Leuuw (1982). Let

$$\mathbf{S}_1 \equiv \mathbf{R}^+\mathbf{Q}'_{N/S}\mathbf{S}\mathbf{Q}_{N/S}(\mathbf{R}')^+ = \mathbf{U}_1\Lambda_1\mathbf{U}'_1, \tag{2.286}$$

and

$$\mathbf{S}_2 \equiv \mathbf{N}'\mathbf{S}\mathbf{N} = \mathbf{V}_1\Lambda_2\mathbf{V}'_1 \tag{2.287}$$

denote the complete eigendecompositions of $\mathbf{S}_1$ and $\mathbf{S}_2$. (The matrix **U** in (2.275) is the portion of $\mathbf{U}_1$ corresponding to the nonzero eigenvalues of $\mathbf{S}_1$.) A general solution to the square nonsingular matrix **W** of order m is given by

$$\mathbf{W} = \left[\; \mathbf{Q}_{N/S}(\mathbf{R}')^+\mathbf{U}_1 + \mathbf{Q}_{N/S}\mathbf{N}\mathbf{Z} \;\middle|\; \mathbf{N}\mathbf{V}_1 \;\right], \tag{2.288}$$

where **Z** is arbitrary. Note that $\mathbf{I} - (\mathbf{N}'\mathbf{S}\mathbf{N})^-\mathbf{N}'\mathbf{S}\mathbf{N} = \mathbf{O}$ if $\mathbf{N}'\mathbf{S}\mathbf{N}$ is nonsingular (or $\mathrm{Sp}(\mathbf{S}) \cap \mathrm{Ker}(\mathbf{T}) = \{\mathbf{0}\}$). We have $\mathbf{W}'\mathbf{S}\mathbf{W} = \begin{bmatrix} \Lambda_1 & \mathbf{O} \\ \mathbf{O} & \Lambda_2 \end{bmatrix}$ and $\mathbf{W}'\mathbf{T}\mathbf{W} = \begin{bmatrix} \mathbf{I} & \mathbf{O} \\ \mathbf{O} & \mathbf{O} \end{bmatrix}$.

2.3.8 SVD of matrix products (PSVD)

Product SVD (PSVD) is a method of finding the SVD of matrix products without explicitly forming the product (Fernando and Hammarling 1988). Let **A** and **N** be $n \times m$ and $m \times p$ matrices. The pair of decompositions,

$$\mathbf{A} = \mathbf{U}\mathbf{D}\mathbf{X}', \quad \text{and} \quad \mathbf{N} = (\mathbf{X}')^{-1}\mathbf{J}\mathbf{V}' \tag{2.289}$$

is called PSVD of the matrix pair **A** and **N**, and is denoted by PSVD(**A**,**N**). Here, **U** and **V** are (full) orthogonal matrices of order n and m, respectively, **X** is a nonsingular matrix of order m, and

$$\mathbf{D} = \begin{bmatrix} {}_h\mathbf{D}^*_h & {}_h\mathbf{O}_s & {}_h\mathbf{O}_t & {}_h\mathbf{O} \\ {}_t\mathbf{O}_h & {}_t\mathbf{O}_s & {}_t\mathbf{I}_t & {}_t\mathbf{O} \\ \mathbf{O}_h & \mathbf{O}_s & \mathbf{O}_t & \mathbf{O} \end{bmatrix}, \quad \text{and} \quad \mathbf{J} = \begin{bmatrix} {}_h\mathbf{I}_h & {}_h\mathbf{O}_s & {}_h\mathbf{O} \\ {}_s\mathbf{O}_h & {}_s\mathbf{I}_s & {}_s\mathbf{O} \\ {}_t\mathbf{O}_h & {}_t\mathbf{O}_s & {}_t\mathbf{O} \\ \mathbf{O}_h & \mathbf{O}_s & \mathbf{O} \end{bmatrix}, \tag{2.290}$$

where $h = \mathrm{rank}(\mathbf{AN})$, $s = \mathrm{rank}(\mathbf{N}) - h$, $t = \mathrm{rank}(\mathbf{A}) - h$, and ${}_h\mathbf{D}^*_h$ is a *pd* diagonal matrix of order h. It follows that

$$\mathbf{AN} = \mathbf{U}(\mathbf{DJ})\mathbf{V}' \tag{2.291}$$

gives the SVD of **AN**, where **DJ** is the $n \times p$ diagonal matrix with nonzero singular values of **AN** in the leading diagonals. Note that portions of **X** corresponding to the last column block of **D** and the last row block of **J** are not unique.

The PSVD of a matrix pair can readily be extended to the product of matrix triplets, **A**, $\mathbf{M}'$ ($q \times n$), and **N**. Let $\mathbf{A} = \mathbf{X}_1\mathbf{D}\mathbf{X}'_2$, $\mathbf{M}' = \mathbf{U}\mathbf{E}\mathbf{X}_1^{-1}$, and $\mathbf{N} = (\mathbf{X}'_2)^{-1}\mathbf{J}\mathbf{V}'$.

Then, $\mathbf{M}'\mathbf{AN} = \mathbf{U}(\mathbf{EDJ})\mathbf{V}'$ gives the SVD of $\mathbf{M}'\mathbf{AN}$. The PSVD of the matrix triplets, $\mathbf{A}$, $\mathbf{M}$ and $\mathbf{N}$, is denoted by PSVD$(\mathbf{A},\mathbf{M},\mathbf{N})$. Here, $\mathbf{E}$, $\mathbf{D}$, and $\mathbf{J}$ are of the following forms:

$$\mathbf{E} = \begin{bmatrix} {}_h\mathbf{I}_h & {}_h\mathbf{O}_s & {}_h\mathbf{O}_k & {}_h\mathbf{O}_j & {}_h\mathbf{O}_i & {}_h\mathbf{O} \\ {}_j\mathbf{O}_h & {}_j\mathbf{O}_s & {}_j\mathbf{O}_k & {}_j\mathbf{I}_j & {}_j\mathbf{O}_i & {}_j\mathbf{O} \\ {}_i\mathbf{O}_h & {}_i\mathbf{O}_s & {}_i\mathbf{O}_k & {}_i\mathbf{O}_j & {}_i\mathbf{I}_i & {}_i\mathbf{O} \\ \mathbf{O}_h & \mathbf{O}_s & \mathbf{O}_k & \mathbf{O}_j & \mathbf{O}_i & \mathbf{O} \end{bmatrix},$$

$$\mathbf{D} = \begin{bmatrix} {}_h\mathbf{D}_h^* & {}_h\mathbf{O}_s & {}_h\mathbf{O}_t & {}_h\mathbf{O}_k & {}_h\mathbf{O}_j & {}_h\mathbf{O} \\ {}_s\mathbf{O}_h & {}_s\mathbf{I}_s & {}_s\mathbf{O}_t & {}_s\mathbf{O}_k & {}_s\mathbf{O}_j & {}_s\mathbf{O} \\ {}_k\mathbf{O}_h & {}_k\mathbf{O}_s & {}_k\mathbf{O}_t & {}_k\mathbf{I}_k & {}_k\mathbf{O}_j & {}_k\mathbf{O} \\ {}_j\mathbf{O}_j & {}_j\mathbf{O}_s & {}_j\mathbf{O}_t & {}_j\mathbf{O}_k & {}_j\mathbf{I}_j & {}_j\mathbf{O} \\ {}_i\mathbf{O}_h & {}_i\mathbf{O}_s & {}_i\mathbf{O}_t & {}_i\mathbf{O}_k & {}_i\mathbf{O}_j & {}_i\mathbf{O} \\ \mathbf{O}_h & \mathbf{O}_s & \mathbf{O}_t & \mathbf{O}_k & \mathbf{O}_j & \mathbf{O} \end{bmatrix},$$

$$\mathbf{J} = \begin{bmatrix} {}_h\mathbf{I}_h & {}_h\mathbf{O}_s & {}_h\mathbf{O}_t & {}_h\mathbf{O} \\ {}_s\mathbf{O}_h & {}_s\mathbf{I}_s & {}_s\mathbf{O}_t & {}_s\mathbf{O} \\ {}_t\mathbf{O}_h & {}_t\mathbf{O}_s & {}_t\mathbf{I}_t & {}_t\mathbf{O} \\ {}_k\mathbf{O}_h & {}_k\mathbf{O}_s & {}_k\mathbf{O}_t & {}_k\mathbf{O} \\ {}_j\mathbf{O}_h & {}_j\mathbf{O}_s & {}_j\mathbf{O}_t & {}_j\mathbf{O} \\ \mathbf{O}_h & \mathbf{O}_s & \mathbf{O}_t & \mathbf{O} \end{bmatrix},$$

where $h = \text{rank}(\mathbf{M}'\mathbf{AN})$, $s = \text{rank}(\mathbf{AN}) - h$, $t = \text{rank}(\mathbf{N}) - \text{rank}(\mathbf{AN})$, $k = \text{rank}(\mathbf{A}) - \text{rank}(\mathbf{M}'\mathbf{A}) - \text{rank}(\mathbf{AN}) + h$, $j = \text{rank}(\mathbf{M}'\mathbf{A}) - h$, and $i = \text{rank}(\mathbf{M}) - \text{rank}(\mathbf{M}'\mathbf{A})$.

The PSVD of a matrix triplet has been used to show what kind of g-inverses of $\mathbf{M}'\mathbf{AN}$ are necessary for $\text{rank}(\mathbf{A} - \mathbf{AN}(\mathbf{M}'\mathbf{AN})^-\mathbf{M}'\mathbf{A}) = \text{rank}(\mathbf{A}) - \text{rank}(\mathbf{AN}(\mathbf{M}'\mathbf{AN})^-\mathbf{M}'\mathbf{A})$ to hold (Takane and Yanai 2007). The result generalizes the Wedderburn–Guttman theorem (Guttman 1944, 1957; Wedderburn 1934). See Section 4.16.

Note 2.10 In contrast, quotient SVD (QSVD)) noted earlier (Section 2.3.6) obtains the SVD of a matrix quotient (i.e., $\mathbf{AN}^{-*}$ with a special kind of g-inverse $\mathbf{N}^{-*}$ of $\mathbf{N}$) without explicitly calculating $\mathbf{AN}^{-*}$ or $\mathbf{N}^{-*}$ (van Loan, 1976). The QSVD has been extended to restricted SVD (RSVD; Zha, 1991) that involves three matrices simultaneously. See Takane (2003) for comparative analysis of various SVDs.

Chapter 3

Constrained Principal Component Analysis (CPCA)

We now come to the main topic of this monograph. We discuss data requirements for analyses by CPCA, essential features, and various generalizations of CPCA, following Takane and Shibayama (1991), and Takane and Hunter (2001, 2011). Sections 3.1 and 3.2 are basic, while Section 3.3 is a little more advanced.

3.1 Data Requirements

In CPCA we first decompose a given data matrix into several additive components according to the external information about its rows and columns (External Analysis). We then apply PCA to the decomposed components, as interests dictate (Internal Analysis). We therefore need to have some external information as a prerequisite for CPCA. In the traditional PCA, an unweighted least squares (ULS or ordinary LS (OLS)) criterion is most often used for model fitting. In CPCA, we may use a weighted LS (WLS) criterion with specific weight matrices (called metric matrices) applied to the rows and columns of the data matrix. This provision plays a particularly important role in the analysis of frequency tables, as a special case of CPCA. We begin this chapter by introducing the input data, external information, and metric matrices.

3.1.1 Data matrices

We denote an $n \times m$ data matrix by $\mathbf{Z}$. We may suppose that n subjects are measured on m variables, so the rows of $\mathbf{Z}$ represent the n subjects, and the columns the m variables. In principle, any multivariate data matrix will do. Unless otherwise stated, we do not make any distributional assumptions on $\mathbf{Z}$. However, caution must be exercised to avoid having extreme outliers influence analysis results too much. We refer to Section 5.4 for possible treatments of missing data. The data may consist of numerical data measured on relatively continuous scales, discrete data coded in dummy variables, or mixtures of both kinds. They could also be contingency tables. (See examples in Chapter 1.)

The input data matrix may be raw data or may have been preprocessed in some way. By preprocessing, we mean such operations as centering, standardization, etc.

The centering and standardization of data are usually performed rowwise or columnwise rather than matrixwise. In any case, results of PCA are often heavily influenced by which preprocessing is applied to the data prior to the analysis. The investigator is responsible for deliberately choosing an appropriate preprocessing method in light of his/her interests. For example, the columnwise centering implies discarding column means from subsequent analyses under the assumption that the column means are not of interest. Similarly, the columnwise standardization implies that differences in dispersion among variables, which are often due to differences in measurement units, are regarded as meaningless, and are excluded from subsequent analyses. See Section 4.1 for other data preprocessing issues. The problem of data preprocessing is also closely related to the external information and metric matrices (to be discussed later in this section).

When the data matrix consists of both continuous and discrete variables, incompatibility of measurement units across the two types of variables may pose a nontrivial problem. In such cases, variables are often uniformly standardized. Kiers (1991), however, recommends orthonormalizing dummy variables corresponding to each discrete variable after columnwise centering. The resultant matrix may be set equal to $\mathbf{A}_1$ in (2.258). A bit of care must be exercised here. The columnwise centering of dummy variables reduces the rank of the matrix by one, and so $(\mathbf{A}'\mathbf{A})^{-1/2}$ in (2.258) must be replaced by $((\mathbf{A}'\mathbf{A})^{+})^{1/2}$, where $^{+}$ indicates the Moore-Penrose inverse. The matrices $\mathbf{U}$ and $\mathbf{V}$ are no longer square, but consist of left and right singular vectors corresponding to strictly positive (nonzero) singular values of $\mathbf{A}$, and so $\mathbf{A}_1'\mathbf{A}_1 = \mathbf{V}\mathbf{V}'$ is no longer an identity matrix, although it is an orthogonal projector.

3.1.2 External information

If no information other than the main data is available, we have no choice but to apply the usual (unconstrained) PCA to the main data. In many cases, however, some auxiliary information is available on the rows and columns of the data matrix. Even in the case in which such information does not assume a concrete form, it is often used informally to aid interpretation of PCA results. In CPCA, such information is regarded as an intrinsic part of the data, and is used in a more constructive way by directly incorporating it in the analysis. Such information is called external data in CPCA.

The incorporation of external data in the analysis may have several advantages (e.g., Takane et al. 1995), as stated below:

(i) Interpretation of analysis results may become easier. We may obtain more easily interpretable results, since relationships among variables and subjects are already structured by the external information.

(ii) More stable solutions may be obtained. Relationships among variables and subjects are represented by a smaller number of parameters, and consequently more reliable parameter estimates may be obtained. (Note that a more stable solution may not be a better solution if the incorporated external information is not consistent with the main data.)

(iii) Empirical validity of hypotheses (hypothesized structures) may be examined. external data may serve as hypotheses on the structure of the main data. By assessing the extent to which the external data capture the structure of the main data, the validity of hypothesized structures may be investigated.

(iv) Prediction may become possible. External data may be regarded as predictors (explanatory variables) to predict (explain) the structures in the data. We may predict unknown structures or future observations through predictor variables.

(v) Incidental parameters (those which increase as the sample size increases) may be eliminated. Parameters representing the rows of the data matrix are often incidental to subjects. In the presence of such parameters, consistent estimators may not be obtained even for structural (nonincidental) parameters in the model. (Estimators whose accuracy increases as the sample size increases are called consistent estimators.) This undesirable situation can be avoided if the incidental parameters can be reparameterized by a smaller number of parameters that do not increase in number with sample size.

Whatever the reasons for its use, external data play important roles in many practical data analysis situations. There are two kinds of external data used in CPCA, one for the rows and the other for the columns of the main data matrix. The former is called a row information matrix, and the latter a column information matrix. The row information matrix is denoted by an $n \times p$ matrix $\mathbf{G}$, and the column information matrix by an $m \times q$ matrix $\mathbf{H}$.

Note 3.1 In the literature, the row information matrix $\mathbf{G}$ is sometimes referred to as the column design matrix because it represents constraints in the column space of $\mathbf{Z}$. However, in this monograph, we call $\mathbf{G}$ the row information matrix because it specifies relationships among the rows of $\mathbf{Z}$. We similarly call the matrix $\mathbf{H}$ the column information matrix. There also are two kinds of metric matrices, one for rows and the other for columns. We use $\mathbf{K}$ to denote a row metric matrix (that goes with $\mathbf{G}$), and $\mathbf{L}$ to denote a column metric matrix (that goes with $\mathbf{H}$).

When there are no obvious row- and/or column-side external data, we may simply set $\mathbf{G} = \mathbf{I}_n$ and/or $\mathbf{H} = \mathbf{I}_m$.

As examples of the row information matrix $\mathbf{G}$, we may think of demographic information about the subjects such as gender, age, level of education, etc. In this case we may be interested in analyzing how these variables are related to variables in the main data. As another example, by taking $\mathbf{G} = \mathbf{1}_n$, where $\mathbf{1}_n$ is the n-component vector of ones, we can analyze the mean tendencies of the subjects. Generalizing this notion slightly, we may also consider a matrix of dummy variables that indicate groups to which the subjects belong, and analyze differences in mean tendencies among the groups. The groups indicated by $\mathbf{G}$ can be either intact groups such as males and females or artificially created groups such as treatment groups in designed experiments.

For **H**, we may consider something similar. As a concrete example, **H** may be a stimulus design or feature matrix, where the variables in **Z** represent some kind of stimuli produced according to a specific design. It could also be a contrast matrix, where the variables in **Z** represent different conditions in an experiment, or a matrix of coefficients of orthogonal polynomials, when repeated measurements are taken of the same attribute over time. In Section 4.11, we use a pair comparison design matrix for **H** indicating which variables in **Z** represent comparisons between which pair of stimuli.

Besse and Ramsay (1986) used a special **H** to isolate periodic and aperiodic changes in seasonal temperature in various cities in France. Ramsay and Silverman (2005) also present a comprehensive methodology for approximating data viewed as continuous functions in time by providing various basis functions in **H** used as building blocks for data approximations. See also Ramsay (1982) and Ramsay and Dalzell (1991). There are numerous other examples of **G** and **H** in the literature on redundancy analysis (see Section 4.2) and in the analysis of growth curves (see Section 4.13).

Incorporating a specific **G** and **H** as the external data in CPCA implies that we constrain the spaces of row and column representations of the data matrix **Z** to the spaces spanned by **G** and **H**. Specifying the subspaces in which the row and column representations lie also entails specifying the spaces orthogonal to them spanned by the residual matrix. This leads to the idea that we may also analyze the portion of the original data that remains after certain effects are partialled out. For example, analyzing a centered data matrix is equivalent to analyzing residuals from mean tendencies in the original data.

3.1.3 Metric matrices

A metric matrix may also exist on each side (row and column) of a data matrix. We use **K**, a symmetric *nnd* (nonnegative-definite) matrix of order n, to denote the row metric matrix, and **L**, a symmetric *nnd* matrix of order m, to denote the column metric matrix. If **K** and/or **L** are *psd* (positive-semidefinite) but not *pd* (positive-definite), we require the following conditions:

$$\text{rank}(\mathbf{KG}) = \text{rank}(\mathbf{G}), \tag{3.1}$$

and

$$\text{rank}(\mathbf{LH}) = \text{rank}(\mathbf{H}). \tag{3.2}$$

These conditions are essential for "projectors" in External Analysis (the next section) to have the necessary properties. As noted earlier (Section 2.2.5), they are automatically satisfied if **K** and **L** are nonsingular.

Metric matrices are closely related to the criterion used to fit a model to data. We need not use special metric matrices if the coordinate axes representing both rows and columns of the data matrix are orthogonal and have comparable units. In this case, we may simply set $\mathbf{K} = \mathbf{I}_n$ and $\mathbf{L} = \mathbf{I}_m$, and use the unweighted (or ordinary) LS criterion. However, we need a nonidentity metric matrix **L** if input variables are measured in uncomparable units such as height and weight. In the previous section we briefly discussed standardization of the data matrix as preprocessing of data. This operation has a similar effect to setting **L** to be a diagonal matrix whose diagonal elements are equal to the reciprocals of the variances of the variables.

We need a special nonidentity metric matrix **K** when the rows of the data matrix are correlated. In PCA, rows of the data matrices are usually assumed statistically independent. This assumption is not unreasonable when the rows of the data matrix represent a random sample of subjects from a target population. However, it is often violated in the case where the rows of the data matrix represent time points of measurements from a single subject. This means that the coordinate axes used to represent the rows of the data matrix are not mutually orthogonal. We need to orthogonalize them by estimating a matrix of correlations (or covariances) among them, and using its inverse as the metric matrix **K**. Escoufier (1987) estimated serial correlations among the rows of his data matrix using the first-order autoregressive model, and proposed to use its inverse matrix as the row metric matrix **K**.

If the rows are uncorrelated but have different reliabilities, a special nonconstant diagonal matrix is necessary to differentially weight the rows of **Z**. For example, suppose that the data are aggregated across subgroups of subjects. If the subgroups have different sample sizes, their means are not equally reliable, and a metric matrix that reflects the differences in reliability is in order. In correspondence analysis (CA: Section 4.6), it is conventional to use diagonal matrices of row and column totals of contingency tables as the row and column metric matrices after the contingency tables are scaled by the inverses of these matrices. This may be considered as another instance of adjustments for differential reliabilities.

Columns of a data matrix representing variables are usually correlated, but we usually do not orthogonalize them *a priori*, since PCA attempts to explain correlational structures among them by hypothesizing a few latent variables (components) underlying the observed variables. We may still use a nonidentity metric matrix **L** by estimating the variance-covariance structures of residuals, when they are correlated or have different variances across columns. This will have the effect of increasing the accuracy of parameter estimates by removing redundant information from the columns and equalizing the contributions of different variables. As a side benefit, scale invariance of parameter estimates may be gained by a nonidentity metric matrix **L** (Rao 1964, Section 9). Meredith and Millsap (1985) recommend estimating variances of measurement errors by reliability coefficients (e.g., test-retest reliability), or based on the image theory of Guttman (1953; see Section 4.20).

Besse and Ramsay (1986) used a specially defined metric matrix for **L** that has the effect of performing a PCA on the original continuous functional data, while actually analyzing discrete approximations to the original data. See also Winsberg (1988). As a somewhat far-fetched example, we also mention maximum likelihood common factor analysis. As discussed in the note below, it can be regarded as a special kind of PCA with the diagonal matrix of reciprocals of unique variances used as **L**.

Note 3.2 Maximum likelihood factor analysis (MLFA; e.g., Jöreskog, 1967) reduces to the eigenproblem of

$$\mathbf{U}^{-1}(\mathbf{C}-\mathbf{U}^2)\mathbf{U}^{-1} = \mathbf{U}^{-1}\mathbf{C}\mathbf{U}^{-1} - \mathbf{I}, \tag{3.3}$$

where $\mathbf{C}$ is the observed variance-covariance matrix, and $\mathbf{U}^2$ is the diagonal matrix of uniqueness assumed known temporarily. Let $\mathbf{Z}$ denote a columnwise centered data matrix. Then

$$\mathbf{C} = \mathbf{Z}'\mathbf{Z}. \tag{3.4}$$

(We assume here that the columns of $\mathbf{Z}$ are scaled in such a way that $\mathbf{Z}'\mathbf{Z}$ (rather than $\mathbf{Z}'\mathbf{Z}/n$) gives the observed variance-covariance matrix.) It is well known that $\mathbf{U}^{-1}\mathbf{C}\mathbf{U}^{-1} - \mathbf{I}$ has the same set of eigenvectors as $\mathbf{U}^{-1}\mathbf{C}\mathbf{U}^{-1}$, although the eigenvalues of the former are all one less than the corresponding eigenvalues of the latter. The eigendecomposition of the latter is related to the singular value decomposition (SVD) of $\mathbf{Z}\mathbf{U}^{-1}$ in a simple manner. In MLFA, $\mathbf{U}^2$ is typically estimated along with the factor loading matrix so as to maximize the likelihood criterion. If, however, $\mathbf{U}^2$ is estimated *a priori* by a noniterative method such as Ihara and Kano's (1986) method, MLFA may be viewed as a special case of CPCA.

It may be said that the importance of metric matrices in data analysis is not yet universally recognized among data analysis practitioners. Research into how they should be chosen has just started, and this is definitely an area in which further investigative efforts are essential.

3.2 CPCA: Method

The theory and methods of CPCA are closely tied to the mathematical foundations developed in Chapter 2, and we will frequently refer the reader back to the relevant sections during the exposition of this chapter. As has been alluded to earlier, there are two important ingredients in CPCA, External Analysis and Internal Analysis. We explain them in turn.

3.2.1 External analysis

As introduced in the previous section, let $\mathbf{Z}$ ($n \times m$) denote the main data matrix, and let $\mathbf{G}$ ($n \times p$) and $\mathbf{H}$ ($m \times q$) denote the matrices of external information. We may write the CPCA model for External Analysis as

$$\mathbf{Z} = \mathbf{G}\mathbf{M}\mathbf{H}' + \mathbf{B}\mathbf{H}' + \mathbf{G}\mathbf{C} + \mathbf{E}, \tag{3.5}$$

where $\mathbf{M}$ ($p \times q$), $\mathbf{B}$ ($n \times q$), and $\mathbf{C}$ ($p \times m$) are matrices of regression coefficients (weights) to be estimated, and $\mathbf{E}$ ($n \times m$) is the matrix of disturbance terms.

Note 3.3 There are four additive terms in the model above. However, one may omit any of them or combine some of them into one, as interests dictate. For example, we may isolate only the first term in (3.5), while putting all the other terms as part of a disturbance term. We then have a growth curve model (Potthoff and Roy 1964; Section 4.13) or two-way CANDELINC (Carroll et al. 1980; Section 4.12), which is written as

$$\mathbf{Z} = \mathbf{G}\mathbf{M}\mathbf{H}' + \mathbf{E}_1, \tag{3.6}$$

where $\mathbf{E}_1$ is the matrix of disturbance terms different from $\mathbf{E}$ in (3.5). We may also have only the second term in (3.5), or we may equivalently assume that $\mathbf{G} = \mathbf{I}_n$ and define $\mathbf{M} + \mathbf{B}$ as the new $\mathbf{B}$ in (3.5). We then obtain

$$\mathbf{Z} = \mathbf{B}\mathbf{H}' + \mathbf{E}_2. \tag{3.7}$$

Similarly, we may have only the third term in (3.5), or we may equivalently assume $\mathbf{H} = \mathbf{I}_m$ and set $\mathbf{M} + \mathbf{C}$ as the new $\mathbf{C}$. We then have

$$\mathbf{Z} = \mathbf{G}\mathbf{C} + \mathbf{E}_3. \tag{3.8}$$

This model is called the redundancy analysis model (Section 4.2). We will discuss these special cases further in later chapters. In this chapter, we primarily focus on the full CPCA model as presented in (3.5).

Note that the model given in (3.5) is unidentifiable. For example, adding $\mathbf{M}_0$ to $\mathbf{M}$ may be compensated by subtracting either $\mathbf{G}\mathbf{M}_0$ from $\mathbf{B}$ or $\mathbf{M}_0\mathbf{H}'$ from $\mathbf{C}$, and the model can be kept invariant. To identify the model, we impose the following orthogonality constraints:

$$\mathbf{G}'\mathbf{K}\mathbf{B} = \mathbf{O}, \tag{3.9}$$

and

$$\mathbf{C}\mathbf{L}\mathbf{H} = \mathbf{O}, \tag{3.10}$$

where $\mathbf{K}$ $(n \times n)$ and $\mathbf{L}$ $(m \times m)$ are the metric matrices, as introduced in the previous section. As has been noted earlier, $\mathbf{K}$ and $\mathbf{L}$ can be identity matrices of appropriate order if there is no special reason to use nonidentity metric matrices. The constraints (3.9) and (3.10) imply that $\mathbf{B}$ and $\mathbf{C}'$ are orthogonal to $\mathbf{G}$ and $\mathbf{H}$ with respect to $\mathbf{K}$ and $\mathbf{L}$, respectively. Due to these orthogonality constraints, $\mathbf{M}$, $\mathbf{B}$ and $\mathbf{C}$ are uniquely determined, and the terms in the model can be uniquely interpreted. The first term in the model represents the portions of $\mathbf{Z}$ that can be explained by both $\mathbf{G}$ and $\mathbf{H}$, the second term represents those that can be explained by $\mathbf{H}$ but not by $\mathbf{G}$, the third term represents those that can be explained by $\mathbf{G}$ but not by $\mathbf{H}$, and the fourth term represents those that can be explained by neither $\mathbf{G}$ nor $\mathbf{H}$.

We estimate parameters in the model ($\mathbf{M}$, $\mathbf{B}$, and $\mathbf{C}$) such that

$$f \equiv \text{SS}(\mathbf{E})_{K,L} = \text{tr}(\mathbf{E}'\mathbf{K}\mathbf{E}\mathbf{L}) = \text{SS}(\mathbf{R}_K'\mathbf{E}\mathbf{R}_L)_{I_n,I_p} \equiv \text{SS}(\mathbf{R}_K'\mathbf{E}\mathbf{R}_L), \tag{3.11}$$

is minimized with respect to the parameter estimates, where $\mathbf{R}_K$ and $\mathbf{R}_L$ are square root factors of $\mathbf{K}$ and $\mathbf{L}$, respectively, i.e., they are columnwise nonsingular matrices such that $\mathbf{K} = \mathbf{R}_K\mathbf{R}_K'$ and $\mathbf{L} = \mathbf{R}_L\mathbf{R}_L'$. Criterion (3.11) is a weighted least squares (WLS) criterion, which leads to the following estimates of $\mathbf{M}$, $\mathbf{B}$, $\mathbf{C}$, and $\mathbf{E}$: By differentiating f with respect to $\mathbf{M}$ and setting the result to zero (Sections 2.1.9 and 2.2.9), we obtain

$$-\frac{1}{2}\frac{\partial f}{\partial \mathbf{M}} = \mathbf{G}'\mathbf{K}(\mathbf{Z} - \mathbf{G}\hat{\mathbf{M}}\mathbf{H}' - \hat{\mathbf{B}}\mathbf{H}' - \mathbf{G}\hat{\mathbf{C}})\mathbf{L}\mathbf{H} \equiv \mathbf{O}. \tag{3.12}$$

Taking into account the orthogonality constraints (3.9) and (3.10), this leads to

$$\hat{\mathbf{M}} = (\mathbf{G}'\mathbf{K}\mathbf{G})^{-}\mathbf{G}'\mathbf{K}\mathbf{Z}\mathbf{L}\mathbf{H}(\mathbf{H}'\mathbf{L}\mathbf{H})^{-}, \tag{3.13}$$

where $^-$ indicates a g-inverse. This estimate of $\mathbf{M}$ is not unique, unless $\mathbf{G}'\mathbf{K}\mathbf{G}$ and $\mathbf{H}'\mathbf{L}\mathbf{H}$ are nonsingular. Similarly, we have

$$-\frac{1}{2}\frac{\partial f}{\partial \mathbf{B}} = \mathbf{K}(\mathbf{Z}-\mathbf{G}\hat{\mathbf{M}}\mathbf{H}'-\hat{\mathbf{B}}\mathbf{H}'-\mathbf{G}\hat{\mathbf{C}})\mathbf{L}\mathbf{H} \equiv \mathbf{O}, \tag{3.14}$$

which leads to

$$\begin{aligned}\hat{\mathbf{B}} &= \mathbf{K}^-\mathbf{K}\mathbf{Z}\mathbf{L}\mathbf{H}(\mathbf{H}'\mathbf{L}\mathbf{H})^- - \mathbf{K}^-\mathbf{K}\mathbf{G}\hat{\mathbf{M}} \\ &= \mathbf{K}^-\mathbf{K}\mathbf{Z}\mathbf{L}\mathbf{H}(\mathbf{H}'\mathbf{L}\mathbf{H})^- - \mathbf{K}^-\mathbf{K}\mathbf{G}(\mathbf{G}'\mathbf{K}\mathbf{G})^-\mathbf{G}'\mathbf{K}\mathbf{Z}\mathbf{L}\mathbf{H}(\mathbf{H}'\mathbf{L}\mathbf{H})^- \\ &= \mathbf{K}^-\mathbf{K}\mathbf{Q}_{G/K}\mathbf{Z}\mathbf{L}\mathbf{H}(\mathbf{H}'\mathbf{L}\mathbf{H})^-, \end{aligned} \tag{3.15}$$

where $\mathbf{Q}_{G/K} = \mathbf{I}_n - \mathbf{P}_{G/K}$, and $\mathbf{P}_{G/K} = \mathbf{G}(\mathbf{G}'\mathbf{K}\mathbf{G})^-\mathbf{G}'\mathbf{K}$. The estimate of $\mathbf{B}$ above is not unique unless $\mathbf{K}$ and $\mathbf{H}'\mathbf{L}\mathbf{H}$ are nonsingular. Similarly, we obtain

$$\hat{\mathbf{C}} = (\mathbf{G}'\mathbf{K}\mathbf{G})^-\mathbf{G}'\mathbf{K}\mathbf{Z}\mathbf{Q}'_{H/L}\mathbf{L}\mathbf{L}^-, \tag{3.16}$$

where $\mathbf{Q}_{H/L} = \mathbf{I}_m - \mathbf{P}_{H/L}$, and $\mathbf{P}_{H/L} = \mathbf{H}(\mathbf{H}'\mathbf{L}\mathbf{H})^-\mathbf{H}\mathbf{L}$. Here, $\mathbf{P}_{H/L}$ and $\mathbf{Q}_{H/L}$ are projectors onto Sp($\mathbf{H}$) along Ker($\mathbf{H}'\mathbf{L}$) and onto Ker($\mathbf{H}'\mathbf{L}$) along Sp($\mathbf{H}$), respectively. The estimate of $\mathbf{C}$ above is not unique, unless $\mathbf{L}$ and $\mathbf{G}'\mathbf{K}\mathbf{G}$ are nonsingular. Finally, the estimate of $\mathbf{E}$ is obtained by

$$\hat{\mathbf{E}} = \mathbf{P}_{G/K}\mathbf{Z}\mathbf{P}'_{H/L} - \mathbf{K}^-\mathbf{K}\mathbf{Q}_{G/K}\mathbf{Z}\mathbf{P}'_{H/L} - \mathbf{P}_{G/K}\mathbf{Z}\mathbf{Q}'_{H/L}\mathbf{L}\mathbf{L}^-. \tag{3.17}$$

This estimate of $\mathbf{E}$ is not unique unless $\mathbf{K}$ and $\mathbf{L}$ are nonsingular. Under the rank condition (3.1), $\mathbf{P}_{G/K}$ and $\mathbf{Q}_{G/K}$ are projectors with the following properties: $\mathbf{P}^2_{G/K} = \mathbf{P}_{G/K}$, $\mathbf{Q}^2_{G/K} = \mathbf{Q}_{G/K}$, $\mathbf{P}_{G/K}\mathbf{Q}_{G/K} = \mathbf{Q}_{G/K}\mathbf{P}_{G/K} = \mathbf{O}$, $\mathbf{P}'_{G/K}\mathbf{K}\mathbf{P}_{G/K} = \mathbf{P}'_{G/K}\mathbf{K} = \mathbf{K}\mathbf{P}_{G/K}$, and $\mathbf{Q}'_{G/K}\mathbf{K}\mathbf{Q}_{G/K} = \mathbf{Q}'_{G/K}\mathbf{K} = \mathbf{K}\mathbf{Q}_{G/K}$. The matrix $\mathbf{P}_{G/K}$ is the projector onto Sp($\mathbf{G}$) along Ker($\mathbf{G}'\mathbf{K}$). Note that $\mathbf{P}_{G/K}\mathbf{G} = \mathbf{G}$ and $\mathbf{G}'\mathbf{K}\mathbf{P}_{G/K} = \mathbf{G}'\mathbf{K}$. The matrix $\mathbf{Q}_{G/K}$, on the other hand, is the projector onto Ker($\mathbf{G}'\mathbf{K}$) along Sp($\mathbf{G}$), satisfying $\mathbf{G}'\mathbf{K}\mathbf{Q}_{G/K} = \mathbf{O}$ and $\mathbf{Q}_{G/K}\mathbf{G} = \mathbf{O}$. Similar properties hold for $\mathbf{P}_{H/L}$ and $\mathbf{Q}_{H/L}$ under (3.2). These projectors reduce to the usual I-orthogonal projectors when $\mathbf{K} = \mathbf{I}_n$ and $\mathbf{L} = \mathbf{I}_m$. It is interesting to note that $\tilde{\mathbf{Q}}_{G/K} = \mathbf{K}^-\mathbf{K}\mathbf{Q}_{G/K}$ is also a projector with the property that $\mathbf{K}\tilde{\mathbf{Q}}_{G/K} = \mathbf{K}\mathbf{Q}_{G/K}$. A similar property holds for $\tilde{\mathbf{Q}}_{H/L} = \mathbf{L}^-\mathbf{L}\mathbf{Q}_{H/L}$. See Section 2.2 for more general discussions and properties of projectors.

The effective numbers of parameters are pq in $\mathbf{M}$, $(n-p)q$ in $\mathbf{B}$, $p(m-q)$ in $\mathbf{C}$, and $(n-p)(m-q)$ in $\mathbf{E}$, assuming that $\mathbf{Z}$, $\mathbf{G}$, and $\mathbf{H}$ all have full column ranks, and $\mathbf{K}$ and $\mathbf{L}$ are nonsingular. These numbers add up to nm. The effective numbers of parameters in $\mathbf{M}$, $\mathbf{B}$, and $\mathbf{C}$ are less than the actual numbers of parameters in these matrices because the orthogonality constraints, (3.9) and (3.10), needed to be imposed to make them unique.

Putting the estimates of $\mathbf{M}$, $\mathbf{B}$, $\mathbf{C}$, and $\mathbf{E}$ given above in (3.5) yields the following decomposition of the data matrix $\mathbf{Z}$:

$$\begin{aligned}\mathbf{Z} = {} & \mathbf{P}_{G/K}\mathbf{Z}\mathbf{P}'_{H/L} + \mathbf{K}^-\mathbf{K}\mathbf{Q}_{G/K}\mathbf{Z}\mathbf{P}'_{H/L} + \mathbf{P}_{G/K}\mathbf{Z}\mathbf{Q}'_{H/L}\mathbf{L}\mathbf{L}^- \\ & + (\mathbf{Z} - \mathbf{P}_{G/K}\mathbf{Z}\mathbf{P}'_{H/L} + \mathbf{K}^-\mathbf{K}\mathbf{Q}_{G/K}\mathbf{Z}\mathbf{P}'_{H/L} + \mathbf{P}_{G/K}\mathbf{Z}\mathbf{Q}'_{H/L}\mathbf{L}\mathbf{L}^-). \end{aligned} \tag{3.18}$$

This decomposition is not unique, unless $\mathbf{K}$ and $\mathbf{L}$ are nonsingular. To make it unique, we may use the Moore-Penrose inverses, $\mathbf{K}^+$ and $\mathbf{L}^+$ for $\mathbf{K}^-$ and $\mathbf{L}^-$. The four terms in (3.18) are mutually orthogonal in the metric matrices $\mathbf{K}$ and $\mathbf{L}$, so we have

$$\mathrm{SS}(\mathbf{Z})_{K,L} = \mathrm{SS}(\mathbf{G}\hat{\mathbf{M}}\mathbf{H}')_{K,L} + \mathrm{SS}(\hat{\mathbf{B}}\mathbf{H}')_{K,L} + \mathrm{SS}(\mathbf{G}\hat{\mathbf{C}})_{K,L} + \mathrm{SS}(\hat{\mathbf{E}})_{K,L}. \quad (3.19)$$

That is, the sum of squares (SS) of elements in $\mathbf{Z}$ in the metrics of $\mathbf{K}$ and $\mathbf{L}$ is uniquely decomposed into the sum of the sums of squares of the four terms in (3.18).

Let $\mathbf{R}_K$ and $\mathbf{R}_L$ be square root factors of $\mathbf{K}$ and $\mathbf{L}$, as introduced earlier (for (3.11)), and let

$$\mathbf{Z}^* = \mathbf{R}_K'\mathbf{Z}\mathbf{R}_L, \quad (3.20)$$

$$\mathbf{G}^* = \mathbf{R}_K'\mathbf{G}, \quad (3.21)$$

and

$$\mathbf{H}^* = \mathbf{R}_L'\mathbf{H}. \quad (3.22)$$

We then have, corresponding to decomposition (3.18),

$$\mathbf{Z}^* = \mathbf{P}_{G^*}\mathbf{Z}^*\mathbf{P}_{H^*} + \mathbf{Q}_{G^*}\mathbf{Z}^*\mathbf{P}_{H^*} + \mathbf{P}_{G^*}\mathbf{Z}^*\mathbf{Q}_{H^*} + \mathbf{Q}_{G^*}\mathbf{Z}^*\mathbf{Q}_{H^*}, \quad (3.23)$$

where $\mathbf{P}_{G^*} = \mathbf{G}^*(\mathbf{G}^{*\prime}\mathbf{G}^*)^-\mathbf{G}^{*\prime}$, $\mathbf{Q}_{G^*} = \mathbf{I}_n - \mathbf{P}_{G^*}$, $\mathbf{P}_{H^*} = \mathbf{H}^*(\mathbf{H}^{*\prime}\mathbf{H}^*)^-\mathbf{H}^{*\prime}$, and $\mathbf{Q}_{H^*} = \mathbf{I}_m - \mathbf{P}_{H^*}$ are I-orthogonal projectors (Section 2.2.3). Note that $\mathbf{R}_K'\mathbf{K}^-\mathbf{K} = \mathbf{R}_K'$, and $\mathbf{R}_L'\mathbf{L}^-\mathbf{L} = \mathbf{R}_L'$. The four terms in (3.23) are I-orthogonal, so we obtain, corresponding to (3.19),

$$\begin{aligned}\mathrm{SS}(\mathbf{Z}^*)_{I_n,I_m} &= \mathrm{SS}(\mathbf{Z}^*) = \mathrm{SS}(\mathbf{P}_{G^*}\mathbf{Z}^*\mathbf{P}_{H^*}) \\ &+ \mathrm{SS}(\mathbf{Q}_{G^*}\mathbf{Z}^*\mathbf{P}_{H^*}) + \mathrm{SS}(\mathbf{P}_{G^*}\mathbf{Z}^*\mathbf{Q}_{H^*}) + \mathrm{SS}(\mathbf{Q}_{G^*}\mathbf{Z}^*\mathbf{Q}_{H^*}). \quad (3.24)\end{aligned}$$

These identities ((3.23) and (3.24)) show how to reduce nonidentity metrics, $\mathbf{K}$ and $\mathbf{L}$, to identity metrics in External Analysis.

When $\mathbf{K}$ and $\mathbf{L}$ are nonsingular, $\mathbf{K}^-\mathbf{K} = \mathbf{I}_n$ and $\mathbf{L}^-\mathbf{L} = \mathbf{I}_m$, so the decomposition (3.18) reduces to

$$\mathbf{Z} = \mathbf{P}_{G/K}\mathbf{Z}\mathbf{P}'_{H/L} + \mathbf{Q}_{G/K}\mathbf{Z}\mathbf{P}'_{H/L} + \mathbf{P}_{G/K}\mathbf{Z}\mathbf{Q}'_{H/L} + \mathbf{Q}_{G/K}\mathbf{Z}\mathbf{Q}'_{H/L}, \quad (3.25)$$

and (3.19) to

$$\begin{aligned}\mathrm{SS}(\mathbf{Z})_{K,L} &= \mathrm{SS}(\mathbf{P}_{G/K}\mathbf{Z}\mathbf{P}'_{H/L})_{K,L} + \mathrm{SS}(\mathbf{Q}_{G/K}\mathbf{Z}\mathbf{P}'_{H/L})_{K,L} \\ &+ \mathrm{SS}(\mathbf{P}_{G/K}\mathbf{Z}\mathbf{Q}'_{H/L})_{K,L} + \mathrm{SS}(\mathbf{Q}_{G/K}\mathbf{Z}\mathbf{Q}'_{H/L})_{K,L}. \quad (3.26)\end{aligned}$$

Decomposition (3.25) is unique.

3.2.2 *Internal analysis*

The decomposed terms obtained in External Analysis may be subjected to PCA either separately or with some of the terms combined. This phase of analysis is called Internal Analysis. There are four terms in the full model (3.5). There are five additional terms obtained by combining some or all of them, as argued in Note 3.3. All of these nine terms can be good candidates for Internal Analysis:

(1) $\mathbf{Z}$ (the original data matrix)
(2) $\mathbf{ZP}'_{H/L}$
(3) $\mathbf{ZQ}'_{H/L}$
(4) $\mathbf{P}_{G/K}\mathbf{Z}$
(5) $\mathbf{Q}_{G/K}\mathbf{Z}$
(6) $\mathbf{P}_{G/K}\mathbf{ZP}'_{H/L}$
(7) $\mathbf{Q}_{G/K}\mathbf{ZP}'_{H/L}$
(8) $\mathbf{P}_{G/K}\mathbf{ZQ}'_{H/L}$
(9) $\mathbf{Q}_{G/K}\mathbf{ZQ}'_{H/L}$

Which term or terms are subjected to PCA is up to the investigator's interests and discretionary considerations. For example, PCA of (6) above reveals the most dominant tendency in the data that can be explained by both $\mathbf{G}$ and $\mathbf{H}$, while that of (9) gives a kind of residual analysis (e.g., Gabriel 1978; Rao 1980; Yanai 1970) useful to explore structures in the data that are left unaccounted for by $\mathbf{G}$ and $\mathbf{H}$.

PCA with nonidentity metric matrices requires generalized singular value decomposition (GSVD; Section 2.3.6). Suppose that we would like to obtain the GSVD of $\mathbf{A}$ with the metric matrices $\mathbf{K}$ and $\mathbf{L}$, which is written as $\text{GSVD}(\mathbf{A})_{K,L}$. This GSVD can be easily obtained from $\text{SVD}(\mathbf{R}'_K\mathbf{AR}_L)$ as follows: Let

$$\mathbf{A} = \mathbf{UDV}' \tag{3.27}$$

denote the $\text{GSVD}(\mathbf{A})_{K,L}$, and let

$$\mathbf{R}'_K\mathbf{AR}_L = \mathbf{U}^*\mathbf{D}^*\mathbf{V}'^* \tag{3.28}$$

denote the $\text{SVD}(\mathbf{R}'_K\mathbf{AR}_L)$. Then $\mathbf{U}$, $\mathbf{V}$, and $\mathbf{D}$ in the GSVD of $\mathbf{A}$ can be obtained from the latter by

$$\mathbf{U} = (\mathbf{R}'_K)^-\mathbf{U}^*, \tag{3.29}$$

$$\mathbf{V} = (\mathbf{R}'_L)^-\mathbf{V}^*, \tag{3.30}$$

and

$$\mathbf{D} = \mathbf{D}^*. \tag{3.31}$$

The matrices $\mathbf{U}$ and $\mathbf{V}$ obtained above are not unique unless $\mathbf{K}$ and $\mathbf{L}$ are nonsingular. Unique $\mathbf{U}$ and $\mathbf{V}$ may be obtained by using the Moore-Penrose inverses for $(\mathbf{R}'_K)^-$ and $(\mathbf{R}'_L)^-$ in (3.29) and (3.30). Once $\mathbf{U}$, $\mathbf{V}$, and $\mathbf{D}$ are obtained, we retain only the portions of these matrices corresponding to the r most dominant singular values. Let these matrices be denoted by $\mathbf{U}_r$, $\mathbf{V}_r$, and $\mathbf{D}_r$. We then rescale $\mathbf{U}_r$ and $\mathbf{V}_r$ by $\sqrt{n}\mathbf{U}_r$ and $\mathbf{V}_r\mathbf{D}_r/\sqrt{n}$ to obtain matrices of component scores and component loadings, respectively. (In what follows, we may simply refer to $\mathbf{U}$ and $\mathbf{V}$ as component score and loading matrices.)

The method given above is a standard way of obtaining GSVD from SVD of a modified matrix. Substantial computational efficiency can be gained by taking into account special structures of the matrices whose GSVD are obtained in CPCA. Suppose that we would like to obtain the GSVD of $\mathbf{P}_{G/K}\mathbf{Z}\mathbf{P}'_{H/L} = \mathbf{G}\hat{\mathbf{M}}\mathbf{H}'$ with metric matrices $\mathbf{K}$ and $\mathbf{L}$. Let this GSVD be denoted as

$$\mathbf{P}_{G/K}\mathbf{Z}\mathbf{P}'_{H/L} = \mathbf{G}\hat{\mathbf{M}}\mathbf{H}' = \mathbf{U}\mathbf{D}\mathbf{V}'. \tag{3.32}$$

As shown above, this GSVD can be obtained by first obtaining the SVD of $\mathbf{R}'_K\mathbf{P}_{G/K}\mathbf{Z}\mathbf{P}'_{H/L}\mathbf{R}_L = \mathbf{P}_{G^*}\mathbf{Z}^*\mathbf{P}_{H^*}$, where $\mathbf{Z}^*$, $\mathbf{G}^*$, and $\mathbf{H}^*$ are as defined in (3.20), (3.21), and (3.22), respectively, and $\mathbf{P}_{G^*}$ and $\mathbf{P}_{H^*}$ are I-orthogonal projectors. Let this SVD be denoted as

$$\mathbf{R}'_K\mathbf{P}_{G/K}\mathbf{Z}\mathbf{P}'_{H/L}\mathbf{R}_L = \mathbf{P}_{G^*}\mathbf{Z}^*\mathbf{P}_{H^*} = \mathbf{U}^*\mathbf{D}^*\mathbf{V}^{*'}. \tag{3.33}$$

Then, $\mathbf{U} = (\mathbf{R}'_K)^-\mathbf{U}^*$, $\mathbf{V} = (\mathbf{R}'_L)^-\mathbf{V}^*$, and $\mathbf{D} = \mathbf{D}^*$. Let

$$\mathbf{G}^* = \mathbf{F}_{G^*}\mathbf{R}'_{G^*}, \tag{3.34}$$

and

$$\mathbf{H}^* = \mathbf{F}_{H^*}\mathbf{R}'_{H^*} \tag{3.35}$$

represent the compact QR decompositions (Section 2.1.12) of $\mathbf{G}^*$ and $\mathbf{H}^*$, where $\mathbf{F}_{G^*}$ and $\mathbf{F}_{H^*}$ are columnwise orthogonal, and $\mathbf{R}_{G^*}$ and $\mathbf{R}_{H^*}$ are upper trapezoidal (upper triangular if $\mathbf{G}^*$ and $\mathbf{H}^*$ are columnwise nonsingular). Then $\mathbf{P}_{G^*}$ and $\mathbf{P}_{H^*}$ can be expressed as

$$\mathbf{P}_{G^*} = \mathbf{F}_{G^*}\mathbf{F}'_{G^*}, \tag{3.36}$$

and

$$\mathbf{P}_{H^*} = \mathbf{F}_{H^*}\mathbf{F}'_{H^*}. \tag{3.37}$$

Let the SVD of $\mathbf{J} = \mathbf{F}'_{G^*}\mathbf{Z}^*\mathbf{F}_{H^*}$ be denoted as

$$\mathbf{J} = \mathbf{F}'_{G^*}\mathbf{Z}^*\mathbf{F}_{H^*} = \mathbf{U}_J\mathbf{D}_J\mathbf{V}'_J. \tag{3.38}$$

Then, the SVD of $\mathbf{P}_{G^*}\mathbf{Z}^*\mathbf{P}_{H^*}$ can be obtained by $\mathbf{U}^* = \mathbf{F}_{G^*}\mathbf{U}_J$, $\mathbf{V}^* = \mathbf{F}_{H^*}\mathbf{V}_J$, and $\mathbf{D}^* = \mathbf{D}_J$ (Takane and Hunter 2001, Theorem 1). Note that $\mathbf{J}$ is usually much smaller in size than $\mathbf{P}_{G^*}\mathbf{Z}^*\mathbf{P}_{H^*}$.

In some cases, we may be more directly interested in the PCA of $\hat{\mathbf{M}}$ rather than that of $\mathbf{G}\hat{\mathbf{M}}\mathbf{H}'$. This PCA amounts to $\text{GSVD}(\hat{\mathbf{M}})_{G^{*\prime}G^*,H^{*\prime}H^*}$. Let this GSVD be denoted as

$$\hat{\mathbf{M}} = \mathbf{U}_M\mathbf{D}_M\mathbf{V}'_M, \tag{3.39}$$

which can be obtained from the SVD of $\mathbf{J}$ by

$$\mathbf{U}_M = (\mathbf{R}'_{G^*})^-\mathbf{U}_J, \tag{3.40}$$

$$\mathbf{V}_M = (\mathbf{R}'_{H^*})^-\mathbf{V}_J, \tag{3.41}$$

and

$$\mathbf{D}_M = \mathbf{D}_J. \tag{3.42}$$

(The g-inverses in (3.40) and (3.41) may be replaced by the Moore-Penrose g-inverses to obtain unique $\mathbf{U}_M$ and $\mathbf{V}_M$.)

Tracing back the derivations in the preceding paragraph, we can establish the relationship between GSVD$(\mathbf{G}\hat{\mathbf{M}}\mathbf{H}')_{K,L}$ and GSVD$(\hat{\mathbf{M}})_{G'KG,H'L'H}$ (Takane and Hunter 2001). That is,

$$\mathbf{U} = \mathbf{G}\mathbf{U}_M, \text{ or } \mathbf{U}_M = (\mathbf{G}'\mathbf{G})^-\mathbf{G}'\mathbf{U}, \tag{3.43}$$

$$\mathbf{V} = \mathbf{H}\mathbf{V}_M, \text{ or } \mathbf{V}_M = (\mathbf{H}'\mathbf{H})^-\mathbf{H}'\mathbf{V}, \tag{3.44}$$

and

$$\mathbf{D} = \mathbf{D}_M. \tag{3.45}$$

This suggests that $\mathbf{U}_M$ and $\mathbf{V}_M$ are matrices of weights applied to $\mathbf{G}$ and $\mathbf{H}$ to derive component score and loading matrices $\mathbf{U}$ and $\mathbf{V}$. In some cases, the weights ($\mathbf{U}_M$ and $\mathbf{V}_M$) are easier to interpret, while in other cases, the scores and loadings ($\mathbf{U}$ and $\mathbf{V}$) are easier to interpret. Although the former case is rather rare, we will see a concrete example of this in vector preference models (Section 4.11). In general, investigators may choose whichever is more convenient for them to interpret. Furthermore, they are related in a simple manner ((3.43) through (3.45)), and one can be derived from the other quite easily. Table 3.1 summarizes the relationships among various GSVDs and SVDs mentioned above.

Table 3.1 *Relationships among various SVDs and GSVDs.*

	(1) GSVD$(\mathbf{P}_{G/K}\mathbf{Z}\mathbf{P}'_{H/L})_{K,L}$ $\mathbf{UDV}'$	(2) SVD$(\mathbf{P}_{G^*}\mathbf{Z}^*\mathbf{P}_{H^*})$ $\mathbf{U}^*\mathbf{D}^*\mathbf{V}^{*'}$	(3) SVD$(\mathbf{J})$ $\mathbf{U}_J\mathbf{D}_J\mathbf{V}'_J$	(4) GSVD$(\hat{M})_{\mathbf{G}^{*'}\mathbf{G}^*,\mathbf{H}^{*'}\mathbf{H}^*}$ $\mathbf{U}_M\mathbf{D}_J\mathbf{V}'_M$
(1)		$\mathbf{U}^* = \mathbf{R}'_K\mathbf{U}$ $\mathbf{V}^* = \mathbf{R}'_L\mathbf{V}$	$\mathbf{U}_J = \mathbf{F}'_{G^*}\mathbf{R}'_K\mathbf{U}$ $\mathbf{V}_J = \mathbf{F}'_{H^*}\mathbf{R}'_L\mathbf{V}$	$\mathbf{U}_M = (\mathbf{G}'\mathbf{G})^-\mathbf{G}'\mathbf{U}$ $\mathbf{V}_M = (\mathbf{H}'\mathbf{H})^-\mathbf{H}'\mathbf{V}$
(2)	$\mathbf{U} = (\mathbf{R}'_K)^-\mathbf{U}^*$ $\mathbf{V} = (\mathbf{R}'_L)^-\mathbf{V}^*$		$\mathbf{U}_J = \mathbf{F}'_{G^*}\mathbf{U}^*$ $\mathbf{V}_J = \mathbf{F}'_{H^*}\mathbf{V}^*$	$\mathbf{U}_M = (\mathbf{G}^{*'}\mathbf{G}^*)^-\mathbf{G}^{*'}\mathbf{U}^*$ $\mathbf{V}_M = (\mathbf{H}^{*'}\mathbf{H}^*)^-\mathbf{H}^{*'}\mathbf{V}^*$
(3)	$\mathbf{U} = (\mathbf{R}'_K)^-\mathbf{F}_{G^*}\mathbf{U}_J$ $\mathbf{V} = (\mathbf{R}'_L)^-\mathbf{F}_{H^*}\mathbf{V}_J$	$\mathbf{U}^* = \mathbf{F}_{G^*}\mathbf{U}_J$ $\mathbf{V}^* = \mathbf{F}_{H^*}\mathbf{V}_J$		$\mathbf{U}_M = (\mathbf{R}'_{G^*})^-\mathbf{U}_J$ $\mathbf{V}_M = (\mathbf{R}'_{H^*})^-\mathbf{V}_J$
(4)	$\mathbf{U} = \mathbf{G}\mathbf{U}_M$ $\mathbf{V} = \mathbf{H}\mathbf{V}_M$	$\mathbf{U}^* = \mathbf{G}^*\mathbf{U}_M$ $\mathbf{V}^* = \mathbf{H}^*\mathbf{V}_M$	$\mathbf{U}_J = \mathbf{R}'_{G^*}\mathbf{U}_M$ $\mathbf{V}_J = \mathbf{R}'_{H^*}\mathbf{V}_M$	

Notation: $\mathbf{K} = \mathbf{R}_K\mathbf{R}'_K$, $\mathbf{L} = \mathbf{R}_L\mathbf{R}'_L$, $\mathbf{G}^* = \mathbf{R}'_K\mathbf{G} = \mathbf{F}_{G^*}\mathbf{R}'_{G^*}$, $\mathbf{H}^* = \mathbf{R}'_L\mathbf{H} = \mathbf{F}_{H^*}\mathbf{R}'_{H^*}$, $\mathbf{Z}^* = \mathbf{R}'_K\mathbf{Z}\mathbf{R}_L$, and $\mathbf{J} = \mathbf{F}'_{G^*}\mathbf{Z}^*\mathbf{F}_{H^*}$

Note 3.4 The relationship between the weight matrix and score matrix in CPCA is analogous to that between the weight matrices (canonical weights) and the score matrices (canonical scores or variates) in canonical correlation analysis (CANO; Section 4.3). As will be discussed later, there are two alternative formulations of CANO, one in which the canonical weights are obtained directly from which the scores can be calculated, and the other in which the scores are derived directly from which the weights can be deduced. This parallels the distinction between (3.32) and (3.39). The loading matrix, on the other hand, represents the correlation matrix between the data matrix and score matrix.

In general, when we have a product of several matrices, say, $\mathbf{ABC}$, SVD$(\mathbf{ABC})$ can be related to a number of different GSVDs, notably

(1) GSVD$(\mathbf{I})_{C'B'A'ABC,I}$,
(2) GSVD$(\mathbf{A})_{I,BCC'B'}$,
(3) GSVD$(\mathbf{AB})_{I,CC'}$,
(4) GSVD$(\mathbf{B})_{A'A,CC'}$,
(5) GSVD$(\mathbf{BC})_{A'A,I}$,
(6) GSVD$(\mathbf{I})_{I,ABCC'B'A'}$.

This extends to products of four or more matrices.

Note 3.5 It is instructive to review the history of CPCA at this point. Special cases of the decomposition (3.25) have been proposed repeatedly over the past years. For example, Corsten (1976), Corsten and van Eijnsbergen (1972), Gabriel (1978), and Rao (1980) proposed a decomposition in which $\mathbf{K} = \mathbf{I}$ and $\mathbf{L} = \mathbf{I}$, and either the first and the second terms, or the first and third terms in (3.25) are combined. They reduce to one of the following decompositions:

$$\begin{aligned}\mathbf{Z} &= \mathbf{P}_G\mathbf{Z} + \mathbf{Q}_G\mathbf{Z}\mathbf{P}_H + \mathbf{Q}_G\mathbf{Z}\mathbf{Q}_H,\\ &= \mathbf{Z}\mathbf{P}_H + \mathbf{P}_G\mathbf{Z}\mathbf{Q}_H + \mathbf{Q}_G\mathbf{Z}\mathbf{Q}_H,\\ &= \mathbf{P}_G\mathbf{Z} + \mathbf{Z}\mathbf{P}_H - \mathbf{P}_G\mathbf{Z}\mathbf{P}_H + \mathbf{Q}_G\mathbf{Z}\mathbf{Q}_H.\end{aligned} \tag{3.46}$$

The FANOVA (factor analysis of variance) model proposed by Gollob (1968) corresponds to the special case of the decomposition in CPCA, where it is assumed that $\mathbf{G} = \mathbf{1}_n$ and $\mathbf{H} = \mathbf{1}_m$. Yanai (1970) proposed a PCA with external criteria in which the effect of gender differences (or more generally, the effects of group differences) are eliminated in PCA. Okamoto (1972) set $\mathbf{G} = \mathbf{1}_n$ and $\mathbf{H} = \mathbf{1}_m$ as in Gollob, and proposed PCA of the following four matrices: $\mathbf{Z}$, $\mathbf{Q}_G\mathbf{Z}$, $\mathbf{Z}\mathbf{Q}_H$, and $\mathbf{Q}_G\mathbf{Z}\mathbf{Q}_H$.

Most of the proposals above recommend PCA of residual terms. There also are several proposals that suggest PCA of structural parts, the portions of $\mathbf{Z}$ that can be explained by external information. For example, Rao (1964) gave a solution to a constrained eigendecomposition. He also proposed PCA of instrumental variables (Rao 1964), which is also known as redundancy analysis (van den Wollenberg 1977; see Section 4.2). Golub (1973) gave a solution to the problem of maximizing a bilinear form $\mathbf{x}'\mathbf{A}\mathbf{y}/||\mathbf{x}||||\mathbf{y}||$ under the constraints $\mathbf{R}'\mathbf{x} = \mathbf{0}$ and $\mathbf{C}'\mathbf{y} = \mathbf{0}$. Similar methods have also been proposed for discrete data (contingency tables) by ter Braak (1986; CCA, canonical correspondence analysis) and by Böckenholt and Böckenholt (1990). (See Section 4.7.) Nishisato and his collaborators (Nishisato 1980, 1988; Nishisato and Lawrence 1989) proposed a similar method called ANOVA of categorical data. The two-way CANDELINC (canonical decomposition with linear constraints; see Section 4.12) proposed by Carroll et al. (1980) analyzes only the first term in decomposition (3.25). The traditional growth curve model (GMANOVA; Khatri 1966; Potthoff and Roy 1964; Rao 1965, 1966, 1967) also fits only the first term in (3.5). (See Section 4.13.) In GMANOVA, however, the focus of interest usually lies in testing the bilinear hypothesis of the form $\mathbf{R}'\mathbf{MC} = \mathbf{O}$ rather than applying PCA to the fitted term. (See, however, more recent developments in this area by Reinsel and Velu (2003) and by Takane et al. (2011).)

3.3 Generalizations

Within the basic framework of CPCA presented in the previous section, various extensions are possible. We introduce some of them here.

3.3.1 *Finer decompositions*

Decomposition (3.18) or (3.25) is a very basic one. When more than one set of external constraints are available on either side of the data matrix, it is possible to decompose the data matrix into finer components. The problem of fitting multiple sets of constraints can be viewed as the decomposition of a projector defined on the joint space of all constraints into the sum of projectors defined on subspaces corresponding to the different subsets of constraints, as discussed in Section 2.2.11. The following decompositions can be readily obtained by setting $\mathbf{A} = \mathbf{G} = [\mathbf{X}, \mathbf{Y}]$ in (2.203). When $\mathbf{K} = \mathbf{I}_n$, they reduce to the decompositions in (2.197). As indicated in Section 2.2.11, a variety of decompositions are possible, depending on the relationship between $\mathbf{X}$ and $\mathbf{Y}$ (e.g., Rao and Yanai 1979).

When $\mathbf{X}$ and $\mathbf{Y}$ are mutually orthogonal (in the metric of $\mathbf{K}$), we have

$$\mathbf{P}_{G/K} = \mathbf{P}_{X/K} + \mathbf{P}_{Y/K}. \tag{3.47}$$

This simply partitions the joint effect of $\mathbf{X}$ and $\mathbf{Y}$ into the sum of the separate effects of $\mathbf{X}$ and $\mathbf{Y}$. Since $\mathbf{X}$ and $\mathbf{Y}$ are orthogonal (in the metric $\mathbf{K}$), the decomposition is simple and unique. When $\mathbf{X}$ and $\mathbf{Y}$ are not completely orthogonal, but are orthogonal outside their intersection space, $\mathbf{P}_{X/K}$ and $\mathbf{P}_{Y/K}$ commute multiplicatively (i.e., $\mathbf{P}_{X/K}\mathbf{P}_{Y/K} = \mathbf{P}_{Y/K}\mathbf{P}_{X/K}$, where the product of the two projectors is also a projector), and

$$\mathbf{P}_{G/K} = \mathbf{P}_{X/K} + \mathbf{P}_{Y/K} - \mathbf{P}_{X/K}\mathbf{P}_{Y/K} \tag{3.48}$$

holds. This decomposition, when $\mathbf{K} = \mathbf{I}_n$, plays an important role in ANOVA for factorial designs. When $\mathbf{X}$ and $\mathbf{Y}$ are not mutually orthogonal in any sense, two decompositions are possible:

$$\mathbf{P}_{G/K} = \mathbf{P}_{X/K} + \mathbf{P}_{Q_{Y/K}X/K} \tag{3.49}$$

$$= \mathbf{P}_{Y/K} + \mathbf{P}_{Q_{X/K}Y/K}, \tag{3.50}$$

where $\mathbf{P}_{Q_{Y/K}X/K}$ and $\mathbf{P}_{Q_{X/K}Y/K}$ are projectors onto $\mathrm{Sp}(\mathbf{Q}_{Y/K}\mathbf{X})$ (the space spanned by the portion of $\mathbf{X}$ left unaccounted for by $\mathbf{Y}$), and $\mathrm{Sp}(\mathbf{Q}_{X/K}\mathbf{Y})$ (the space spanned by the portion of $\mathbf{Y}$ left unaccounted for by $\mathbf{X}$), respectively. The above decompositions are useful when one of $\mathbf{X}$ and $\mathbf{Y}$ is fitted first, and the other is fitted to the residuals.

When $\mathrm{Sp}(\mathbf{X})$ and $\mathrm{Sp}(\mathbf{Y})$ are disjoint, but not orthogonal, we may use

$$\mathbf{P}_{G/K} = \mathbf{X}(\mathbf{X}'\mathbf{K}\mathbf{Q}_{Y/K}\mathbf{X})^{-}\mathbf{X}'\mathbf{K}\mathbf{Q}_{Y/K} + \mathbf{Y}(\mathbf{Y}'\mathbf{K}\mathbf{Q}_{X/K}\mathbf{Y})^{-}\mathbf{Y}'\mathbf{K}\mathbf{Q}_{X/K}. \tag{3.51}$$

Note that $\mathbf{K}\mathbf{Q}_{Y/K}$ and $\mathbf{K}\mathbf{Q}_{X/K}$ are both symmetric. This decomposition is useful when $\mathbf{X}$ and $\mathbf{Y}$ are fitted simultaneously. The first term on the right-hand side of (3.51) is the projector onto $\mathrm{Sp}(\mathbf{X})$ along $\mathrm{Sp}(\mathbf{Q}_{G/K}) \oplus \mathrm{Sp}(\mathbf{Y})$ where $\oplus$ indicates the direct sum of two disjoint spaces, and the second term the projector onto $\mathrm{Sp}(\mathbf{Y})$ along $\mathrm{Sp}(\mathbf{Q}_{G/K}) \oplus \mathrm{Sp}(\mathbf{X})$. Note that unlike all the previous decompositions discussed in this section, the two terms in this decomposition are not mutually orthogonal. Takane and Yanai (1999), however, discuss a special metric $\mathbf{K}^*$ under which the two terms in (3.51) are mutually orthogonal, and are such that $\mathbf{P}_{G/K} = \mathbf{P}_{G/K^*}$, $\mathbf{P}_{X/KQ_{Y/K}} = \mathbf{P}_{X/K^*}$ and $\mathbf{P}_{Y/KQ_{X/K}} = \mathbf{P}_{Y/K*}$. An example of such a metric is $\mathbf{K}^* = \mathbf{K}\mathbf{Q}_{X/K} + \mathbf{K}\mathbf{Q}_{Y/K} + \mathbf{T}\mathbf{S}\mathbf{T}'$, where $\mathbf{T}$ is such that $\mathrm{Sp}(\mathbf{T}) = \mathrm{Ker}(\mathbf{G}')$ and $\mathbf{S}$ is an arbitrary symmetric *nnd* matrix.

When additional information is given as constraints on the weight matrix $\mathbf{U}_G$ on $\mathbf{G}$, the following decomposition is useful. Suppose the constraints can be expressed as $\mathbf{U}_G = \mathbf{B}\mathbf{U}_A$ for a given matrix $\mathbf{B}$. Then

$$\mathbf{P}_{G/K} = \mathbf{P}_{GB/K} + \mathbf{P}_{G(G'KG)^{-}C/K}, \tag{3.52}$$

where $\mathbf{B}'\mathbf{C} = \mathbf{O}$, $\mathrm{Sp}(\mathbf{B}) \oplus \mathrm{Sp}(\mathbf{C}) = \mathrm{Sp}(\mathbf{G}'\mathbf{K})$, so that $\mathbf{C} = \mathbf{G}'\mathbf{K}\mathbf{W}$ for some $\mathbf{W}$ (Takane et al. 1991; Yanai and Takane 1992). The first term in this decomposition is the projector onto $\mathrm{Sp}(\mathbf{GB})$, which is a subspace of $\mathrm{Sp}(\mathbf{G})$, and the second term onto the subspace of $\mathrm{Sp}(\mathbf{G})$ orthogonal to $\mathrm{Sp}(\mathbf{GB})$. Since $\mathbf{C}'(\mathbf{G}'\mathbf{K}\mathbf{G})^{-}\mathbf{G}'\mathbf{K}\mathbf{G}\mathbf{U}_B = \mathbf{O}$ for $\mathbf{C}$ such that $\mathbf{C} = \mathbf{G}'\mathbf{K}\mathbf{W}$ for some $\mathbf{W}$, the constraint $\mathbf{U}_G = \mathbf{B}\mathbf{U}_B$ can also be expressed as $\mathbf{C}'\mathbf{U}_G = \mathbf{O}$. This decomposition is an example of higher-order structures to be discussed in the next section. It is often used when we have a specific hypothesis about $\mathbf{M}$, such as in model (3.5), and we would like to obtain an estimate of $\mathbf{M}$ under this hypothesis. Some applications of this decomposition will be given in Section 3.3.2.

When $\mathbf{B}$ and $\mathbf{C}$ are selection matrices such that $\mathbf{GB} = \mathbf{X}$ and $\mathbf{GC} = \mathbf{Y}$, the decomposition above reduces to (3.49). For example, if $\mathbf{G} = [\mathbf{X}, \mathbf{Y}]$, we choose $\mathbf{B} = \begin{bmatrix} \mathbf{I} \\ \mathbf{O} \end{bmatrix}$, and $\mathbf{C} = \begin{bmatrix} \mathbf{O} \\ \mathbf{I} \end{bmatrix}$. When the roles of $\mathbf{B}$ and $\mathbf{C}$ are interchanged, we obtain (3.50).

It is obvious that similar decompositions apply to $\mathbf{H}$ as well. (Set $\mathbf{A} = \mathbf{H}$ and $\mathbf{K} = \mathbf{L}$ in (2.20).) It is also straightforward to extend the decompositions to more than two sets of constraints on each side of a data matrix. The above decompositions can further be generalized to oblique projectors (Takane and Yanai 1999; see also (2.204)), which are useful for instrumental variable (IV) estimation often used in econometrics (e.g., Johnston 1984; Amemiya 1985) when predictor variables and disturbance terms are correlated. See also Section 4.16 on the Wedderburn–Guttman theorem and the accompanying decomposition (Takane and Yanai 2005, 2007, 2009). Takane and Jung (2009b; see also Section 4.8) and Takane and Zhou in press; see also

Section 4.7) extensively use these decompositions in decompositions of chi-square statistics in the analysis of contingency tables.

Decompositions into finer components may generally be expressed as (Nishisato and Lawrence, 1989):

$$\mathbf{Z} = (\sum_i \mathbf{P}_{G_i/\tilde{K}})\mathbf{Z}(\sum_j \mathbf{P}'_{H_j/\tilde{L}}), \tag{3.53}$$

where $\sum_i \mathbf{P}_{G_i/\tilde{K}} = \mathbf{I}_n$ and $\sum_j \mathbf{P}_{H_j/\tilde{L}} = \mathbf{I}_m$, and where $\mathbf{P}_{G_i/\tilde{K}}$ and $\mathbf{P}_{H_j/\tilde{L}}$ are projectors onto $\mathrm{Sp}(\mathbf{G}_i)$ and $\mathrm{Sp}(\mathbf{H}_j)$, respectively, in the metrics of $\tilde{\mathbf{K}}$ and $\tilde{\mathbf{L}}$. The $\tilde{\mathbf{K}}$ and $\tilde{\mathbf{L}}$ are orthogonalizing metrics, which are simply $\mathbf{K}$ and $\mathbf{L}$, except in (3.51) where $\tilde{\mathbf{K}} = \mathbf{K}^*$ and $\tilde{\mathbf{L}} = \mathbf{L}^*$. Because of the orthogonality of the terms in the decomposition (3.53), the SS in $\mathbf{Z}$ is uniquely partitioned into the sum of part SSs, each pertaining to the terms in (3.53).

3.3.2 Higher-order structures

When oblique factor solutions are obtained in common factor analysis, higher-order factors may be assumed to explain correlations among them. Something analogous may be done in CPCA. We have already seen one such example in decomposition (3.52). In this section we give other examples.

For simplicity, we consider the model in which we have only the first term in (3.5), that is, (3.9). Suppose the main data represent pair comparison judgments between stimuli constructed by systematically varying the values of attributes defining the stimuli. Let $\mathbf{H}$ denote the design matrix for pair comparisons (Section 4.11), and let $\mathbf{T}$ denote the design matrix for the stimuli. Then, we may consider

$$\mathbf{Z} = \mathbf{GMH}' + \mathbf{E}, \tag{3.54}$$

where

$$\mathbf{M} = \mathbf{WT}' + \mathbf{E}^*. \tag{3.55}$$

Here, $\mathbf{W}$ is the matrix of regression weights applied to $\mathbf{T}$, and $\mathbf{E}$ and $\mathbf{E}^*$ are matrices of disturbance terms at two different levels. By substituting (3.55) for $\mathbf{M}$ in (3.54), we obtain the model

$$\mathbf{Z} = \mathbf{GWT}'\mathbf{H}' + \mathbf{GE}^*\mathbf{H}' + \mathbf{E}, \tag{3.56}$$

where $\mathbf{Z}$ is decomposed into the sum of three terms. The first term in (3.56) represents the portions of $\mathbf{Z}$ that can be explained by both $\mathbf{G}$ and $\mathbf{HT}$, the second term by $\mathbf{G}$ and the portions of $\mathbf{H}$ that cannot be explained by $\mathbf{HT}$, and the third term the residuals from the first two. The third term corresponds to the sum of the second, third and fourth terms in model (3.5). See Section 4.13 for the estimation of parameters for this model. The terms in (3.56) can, of course, be subjected to PCA for Internal Analysis. Other similar examples can be found in the analysis of growth curves detailed in Section 4.13.

It is also possible to apply PCA to $\hat{\mathbf{M}}$ and obtain $\mathbf{U}_M$ or $\mathbf{U} = \mathbf{GU}_M$, to which external constraints may be imposed. In the former case the model can be written as

$$\begin{aligned}\mathbf{Z} &= \mathbf{G}(\mathbf{U}_M^*\mathbf{D}_M^*\mathbf{V}_M^{*'} + \mathbf{E}^*)\mathbf{H}' + \mathbf{E} \\ &= \mathbf{G}((\mathbf{TW} + \tilde{\mathbf{E}})\mathbf{D}_M^*\mathbf{V}_M^{*'} + \mathbf{E}^*)\mathbf{H}' + \mathbf{E},\end{aligned} \tag{3.57}$$

where $\mathbf{U}_M^*\mathbf{D}_M^*\mathbf{V}_M^{*'}$ is the best fixed-rank approximation of $\hat{\mathbf{M}}$. In this model, a submodel is postulated for $\mathbf{U}_M^*$, that is, $\mathbf{U}_M^* = \mathbf{TW} + \tilde{\mathbf{E}}$, but no submodel is postulated for $\mathbf{D}_M^*$ or $\mathbf{V}_M^*$. When external constraints are imposed on $\mathbf{U}$, the model may be written as

$$\begin{aligned}\mathbf{Z} &= \mathbf{U}^*\mathbf{D}^*\mathbf{V}^{*'} + \mathbf{GE}^*\mathbf{H}' + \mathbf{E} \\ &= (\mathbf{TW} + \tilde{\mathbf{E}})\mathbf{D}^*\mathbf{V}^{*'} + \mathbf{GE}^*\mathbf{H}' + \mathbf{E},\end{aligned} \tag{3.58}$$

where $\mathbf{U}^*\mathbf{D}^*\mathbf{V}^{*'}$ is the best fixed-rank approximation of $\mathbf{G}\hat{\mathbf{M}}\mathbf{H}'$. In (3.57) and (3.58), estimates of $\mathbf{W}$ are obtained by

$$\hat{\mathbf{W}} = (\mathbf{T}'\mathbf{G}'\mathbf{KGT})^-\mathbf{T}'\mathbf{G}'\mathbf{KGU}_M^*, \tag{3.59}$$

and

$$\hat{\mathbf{W}} = (\mathbf{T}'\mathbf{KT})^-\mathbf{T}'\mathbf{KU}^*, \tag{3.60}$$

respectively. In multidimensional scaling (MDS), correspondence analysis (CA), and PCA, external information is often embedded into a configuration of variables, subjects, and/or stimuli, which has been obtained separately (e.g., Carroll 1972; Lebart et al. 1984). This kind of *a posteriori* mapping is a special case of (3.58).

Higher-order structures may generally be represented as

$$\mathbf{Z} = (\prod_i \mathbf{A}_i)\mathbf{M}^*(\prod_j \mathbf{B}_j)' \tag{3.61}$$

with properly defined block matrices, $\mathbf{A}_i$, $\mathbf{B}_j$, and $\mathbf{M}^*$ (Faddeev and Faddeeva 1963, p. 486). This expression looks similar to COSAN, a structural equation model by McDonald (1978). However, whereas COSAN analyzes covariance matrices, CPCA analyzes data matrices. If we regard $\mathbf{Z}$ as a covariance matrix in COSAN, it must be symmetric and *nnd*, and $\mathbf{A}_i$ and $\mathbf{B}_j$ for $i = j$ should be identical. There are no such restrictions in CPCA. (Even the numbers of $\mathbf{A}_i$ and $\mathbf{B}_j$ could be different.)

Note 3.6 We show, as an example, that (3.56) can be written in the form of (3.61). We let

$$\mathbf{M}^* = \begin{bmatrix} \mathbf{W} & \mathbf{O} & \mathbf{O} \\ \mathbf{O} & \mathbf{E}^* & \mathbf{O} \\ \mathbf{O} & \mathbf{O} & \mathbf{E} \end{bmatrix},$$

$$\mathbf{A}_1 = [\mathbf{G}, \mathbf{I}], \quad \mathbf{A}_2 = \begin{bmatrix} \mathbf{I} & \mathbf{I} & \mathbf{O} \\ \mathbf{O} & \mathbf{O} & \mathbf{I} \end{bmatrix},$$

$$\mathbf{B}_1 = [\mathbf{H}, \mathbf{I}], \text{ and } \mathbf{B}_2 = \begin{bmatrix} \mathbf{T} & \mathbf{I} & \mathbf{O} \\ \mathbf{O} & \mathbf{O} & \mathbf{I} \end{bmatrix}.$$

3.3.3 Alternative decompositions

Takane and Hunter (2011) proposed alternative decompositions of the data matrix $\mathbf{Z}$ for External Analysis of CPCA in view of the fact that the terms in decompositions (3.18) and (3.25) may not be columnwise orthogonal, and that they may not be in the column space of $\mathbf{Z}$. Instead of projecting rows of $\mathbf{Z}$ directly onto $\text{Sp}(\mathbf{H})$ for a given column information matrix $\mathbf{H}$, they proposed to set $\mathbf{G} = \mathbf{ZH}$ and project columns of $\mathbf{Z}$ onto $\text{Sp}(\mathbf{G}) = \text{Sp}(\mathbf{ZH})$. They also proposed, instead of projecting columns of $\mathbf{Z}$ directly onto $\text{Sp}(\mathbf{G})$ for a given row information matrix $\mathbf{G}$, to set $\mathbf{H} = \mathbf{Z}'\mathbf{G}$ and project columns of $\mathbf{Z}$ onto $\text{Sp}(\mathbf{ZH}) = \text{Sp}(\mathbf{ZZ}'\mathbf{G})$. These operations ensure that the decomposed data matrices are all columnwise orthogonal, and that they all stay inside $\text{Sp}(\mathbf{Z})$. Takane and Hunter also considered applying similar operations to the matrix of dual basis of $\mathbf{Z}$, denoted by $\mathbf{Z}^* = \mathbf{Z}(\mathbf{Z}'\mathbf{KZ})^-$, rather than to $\mathbf{Z}$ itself. This lead to additional interesting decompositions of the data matrix. In particular, $\mathbf{H}$ used with $\mathbf{Z}$ regulates the weights applied to $\mathbf{Z}$ to derive component scores, while that used with $\mathbf{Z}^*$ regulates the covariances between $\mathbf{Z}$ and the component scores. For example, when $\mathbf{H} = \mathbf{1}_m$ is used with $\mathbf{Z}$, the weights used to derive component scores from $\mathbf{Z}$ are enforced to be constant, while when it is used with $\mathbf{Z}^*$, a component which covaries equally with all the observed variables is obtained.

Takane and Hunter first derived decompositions of the projector $\mathbf{P}_{Z/K}$ defined by the data matrix $\mathbf{Z}$. Once the decompositions of $\mathbf{P}_{Z/K}$ are obtained, corresponding decompositions of the data matrix can be derived mechanically by premultiplying $\mathbf{Z}$ by the decomposed projectors. Along the lines suggested above, they specifically derived four two-term decompositions of $\mathbf{P}_{Z/K}$:

$$\mathbf{P}_{Z/K} = \mathbf{P}_{ZH/K} + \mathbf{P}_{Z^*\tilde{H}/K}, \tag{3.62}$$

where $\tilde{\mathbf{H}}$ is such that $\mathbf{H}'\tilde{\mathbf{H}} = \mathbf{O}$ and $\text{Sp}(\mathbf{H}) \oplus \text{Sp}(\tilde{\mathbf{H}}) = \text{Sp}(\mathbf{Z}'\mathbf{K})$ (this condition reduces to $\text{Sp}(\tilde{\mathbf{H}}) = \text{Ker}(\mathbf{H}')$ if $\mathbf{KZ}$ is columnwise nonsingular),

$$\mathbf{P}_{Z/K} = \mathbf{P}_{Z^*H/K} + \mathbf{P}_{Z\tilde{H}/K}, \tag{3.63}$$

$$\mathbf{P}_{Z/K} = \mathbf{P}_{ZZ'KG/K} + \mathbf{P}_{Z^*T/K}, \tag{3.64}$$

where $\mathbf{T}$ is such that $\mathbf{G}'\mathbf{KZT} = \mathbf{O}$ and $\text{Sp}(\mathbf{Z}'\mathbf{KG}) \oplus \text{Sp}(\mathbf{T}) = \text{Sp}(\mathbf{Z}'\mathbf{K})$ (this condition reduces to $\text{Sp}(\mathbf{T}) = \text{Ker}(\mathbf{G}'\mathbf{KZ})$ if $\mathbf{KZ}$ is columnwise nonsingular), and

$$\mathbf{P}_{Z/K} = \mathbf{P}_{P_{Z/K}G/K} + \mathbf{P}_{Z^*T/K}. \tag{3.65}$$

The first decomposition (3.62) follows from (3.52) by setting $\mathbf{G} = \mathbf{Z}$, $\mathbf{B} = \mathbf{H}$, and $\mathbf{C} = \tilde{\mathbf{H}}$, the second decomposition (3.63) by interchanging the roles of $\mathbf{Z}$ and $\mathbf{Z}^*$ (or the roles of $\mathbf{H}$ and $\tilde{\mathbf{H}}$) in (3.62), the third decomposition (3.64) by setting $\mathbf{H} = \mathbf{Z}'\mathbf{G}$ in (3.62), and the last decomposition (3.65) by interchanging the roles of $\mathbf{Z}$ and $\mathbf{Z}^*$ (or the roles of $\mathbf{Z}'\mathbf{G}$ and $\mathbf{T}$) in (3.62). Note that $\mathbf{Z}^*\mathbf{Z}'\mathbf{K} = \mathbf{Z}\mathbf{Z}^{*'}\mathbf{K} = \mathbf{P}_{Z/K}$.

Since $\mathbf{P}_{Z/K}\mathbf{Z} = \mathbf{Z}$, we obtain the corresponding decompositions of the data matrix $\mathbf{Z}$ by premultiplying $\mathbf{Z}$ by the decompositions above. For example, if we apply (3.62) to $\mathbf{Z}$, we obtain $\mathbf{Z} = \mathbf{P}_{ZH/K}\mathbf{Z} + \mathbf{P}_{Z^*\tilde{H}/K}\mathbf{Z}$. The two terms in this decomposition may then be subjected to PCA for Internal Analysis. This amounts to GSVD$(\mathbf{P}_{ZH/K}\mathbf{Z})_{K,I}$ and GSVD$(\mathbf{P}_{Z^*\tilde{H}/K}\mathbf{Z})_{K,I}$. The columnwise orthogonality of the two terms in this decomposition in the metric matrix $\mathbf{K}$ is assured by the orthogonality of the applied decomposed projectors. That they both stay inside Sp$(\mathbf{Z})$ is assured by the fact that they can all be expressed as $\mathbf{ZW}$ for some $\mathbf{W}$. Similar properties hold for the other decompositions.

By combining some of the above decompositions, we obtain four-term decompositions of $\mathbf{P}_{Z/K}$, which, when premultiplied to the data matrix $\mathbf{Z}$, yield the corresponding four-term decompositions of $\mathbf{Z}$. Let $\mathbf{B}$ and $\mathbf{C}$ be such that $\mathbf{G}'\mathbf{KZHB} = \mathbf{O}$ and Sp$(\mathbf{H}'\mathbf{Z}'\mathbf{KG}) \oplus$ Sp$(\mathbf{B}) =$ Sp$(\mathbf{H}'\mathbf{Z}'\mathbf{K})$, and $\mathbf{G}'\mathbf{KZ}^*\tilde{\mathbf{H}}\mathbf{C} = \mathbf{O}$ and Sp$(\tilde{\mathbf{H}}'\mathbf{Z}^{*\prime}\mathbf{KG}) \oplus$ Sp$(\tilde{\mathbf{H}}'\mathbf{Z}^{*\prime}\mathbf{K})$. Then, by combining (3.62) and (3.64), we obtain

$$\begin{aligned}\mathbf{P}_{Z/K} &= \mathbf{P}_{ZHH'Z'KG/K} + \mathbf{P}_{ZH(H'Z'KZH)^-B/K} \\ &\quad + \mathbf{P}_{Z^*\tilde{H}\tilde{H}'Z^{*\prime}G/K} + \mathbf{P}_{Z^*\tilde{H}(\tilde{H}'(Z'KZ)^-\tilde{H})^-C/K},\end{aligned} \tag{3.66}$$

and by combining (3.62) and (3.65), we obtain

$$\mathbf{P}_{Z/K} = \mathbf{P}_{P_{ZH/K}G/K} + \mathbf{P}_{ZHB/K} + \mathbf{P}_{P_{Z^*\tilde{H}/K}G/K} + \mathbf{P}_{Z^*\tilde{H}C/K}. \tag{3.67}$$

Let $\mathbf{B}^*$ and $\mathbf{C}^*$ be such that $\mathbf{G}'\mathbf{KZ}^*\mathbf{HB}^* = \mathbf{O}$ and Sp$(\mathbf{H}'\mathbf{Z}^{*\prime}\mathbf{KG}) \oplus$ Sp$(\mathbf{B}^*) =$ Sp$(\mathbf{H}'\mathbf{Z}^{*\prime}\mathbf{K})$, and $\mathbf{G}'\mathbf{KZ}\tilde{H}\mathbf{C}^* = \mathbf{O}$ and Sp$(\tilde{\mathbf{H}}'\mathbf{Z}'\mathbf{KG}) \oplus$ Sp$(\mathbf{C}^*) =$ Sp$(\tilde{\mathbf{H}}'\mathbf{Z}'\mathbf{K})$. Then, by combining (3.63) and (3.64), we obtain

$$\begin{aligned}\mathbf{P}_{Z/K} &= \mathbf{P}_{Z^*HH'Z^{*\prime}G/K} + \mathbf{P}_{Z^*H(H'(Z'KZ)^-H)^-B^*/K} \\ &\quad + \mathbf{P}_{Z\tilde{H}\tilde{H}'Z'KG/K} + \mathbf{P}_{Z\tilde{H}(\tilde{H}'Z'KZ\tilde{H})^-C^*/K},\end{aligned} \tag{3.68}$$

and by combining (3.63) and (3.65), we obtain

$$\mathbf{P}_{Z/K} = \mathbf{P}_{P_{Z^*H/K}G/K} + \mathbf{P}_{Z^*HB^*/K} + \mathbf{P}_{P_{Z\tilde{H}/K}G/K} + \mathbf{P}_{Z\tilde{H}C^*/K}. \tag{3.69}$$

Again, these decompositions of $\mathbf{P}_{Z/K}$ may be used to derive corresponding decompositions of the data matrix $\mathbf{Z}$, and then may be subjected to PCA for Internal Analysis.

Note 3.7 The derivation of the combined decompositions above may be a bit difficult to grasp. To obtain the decomposition (3.66), for example, we start with the decomposition (3.64). First, the data matrix $\mathbf{Z}$ in the first term of (3.64) is split into what can and cannot be explained by $\mathbf{ZH}$ using the decomposition (3.62) (or (3.52)), and then $\mathbf{Z}^*$ in the second term of (3.64) is split into what can and cannot be explained by $\mathbf{Z}^*\tilde{\mathbf{H}}$ using the decomposition (3.62) (or (3.52)).

3.3.4 Regularized CPCA

If we use the ridge LS (RLS) estimation in External Analysis instead of the usual LS, we obtain the generalized weighted ridge operator defined by

$$\begin{aligned}\mathbf{R}^{(S)}_{Z/K}(\delta) &= \mathbf{Z}(\mathbf{Z}'\mathbf{KZ} + \delta\mathbf{S})^-\mathbf{Z}'\mathbf{K} \\ &= \mathbf{Z}(\mathbf{Z}'\mathbf{KM}^{(S)}_{Z/K}(\delta)\mathbf{Z})^-\mathbf{Z}'\mathbf{K},\end{aligned} \tag{3.70}$$

where δ is the ridge parameter assuming a small positive value, $\mathbf{S}$ is a symmetric *nnd* matrix, and

$$\mathbf{M}^{(S)}_{Z/K}(\delta) = \mathbf{P}_{Z/K} + \delta\mathbf{Z}(\mathbf{Z}'\mathbf{K}\mathbf{Z})^{-}\mathbf{S}(\mathbf{Z}'\mathbf{K}\mathbf{Z})^{-}\mathbf{Z}'\mathbf{K} \tag{3.71}$$

is called the generalized weighted ridge metric matrix. See Section 2.2.12 for a more general account of the RLS estimation. See also Takane (2007b) and Takane and Yanai (2008). Since $\mathbf{S}$ and δ are common in all ridge operators and ridge metric matrices, we omit them hereafter to avoid clutter in notation. We thus use $\mathbf{R}_{Z/K}$ for $\mathbf{R}^{(S)}_{Z/K}(\delta)$, and $\mathbf{M}_{Z/K}$ for $\mathbf{M}^{(S)}_{Z/K}(\delta)$.

The generalized weighted ridge operator may be decomposed, just as the projection operator $\mathbf{P}_{Z/K}$ was decomposed in the previous subsection. Let $\tilde{\mathbf{H}}$ and $\mathbf{T}$ be as defined in the previous subsection. Let $\mathbf{Z}^* = \mathbf{Z}(\mathbf{Z}'\mathbf{K}\mathbf{M}_{Z/K}\mathbf{Z})^{-}$ be the matrix of dual basis of $\mathbf{Z}$ with respect to the metric matrix $\mathbf{K}\mathbf{M}_{Z/K}$. Then, the following four decompositions of $\mathbf{R}_{Z/K}$ are possible, analogously to (3.62) through (3.65):

$$\mathbf{R}_{Z/K} = \mathbf{R}_{ZH/K} + \mathbf{R}_{Z^*\tilde{H}/K}, \tag{3.72}$$

$$\mathbf{R}_{Z/K} = \mathbf{R}_{Z\tilde{H}/K} + \mathbf{R}_{Z^*H/K}, \tag{3.73}$$

$$\mathbf{R}_{Z/K} = \mathbf{R}_{ZZ'KG/K} + \mathbf{R}_{Z^*T/K}, \tag{3.74}$$

and

$$\mathbf{R}_{Z/K} = \mathbf{R}_{R_{Z/K}G/K} + \mathbf{R}_{ZT/K}. \tag{3.75}$$

As in the previous subsection, we may combine some of the above decompositions of $\mathbf{R}_{Z/K}$ to obtain more elaborate decompositions. (A trick used in combining the decompositions is similar to that described in the previous section.) Let $\mathbf{B}$ and $\mathbf{C}$ be as defined in the previous subsection. Then, by combining (3.72) and (3.74), we obtain

$$\begin{aligned}\mathbf{R}_{Z/K} = {} & \mathbf{R}_{ZHH'Z'KG/K} + \mathbf{R}_{ZH(H'Z'KM_{Z/K}ZH)^{-}B} \\ & + \mathbf{R}_{Z^*\tilde{H}\tilde{H}'Z^{*\prime}KG/K} + \mathbf{R}_{Z^*\tilde{H}(\tilde{H}'Z^{*\prime}KM_{Z/K}Z^*\tilde{H})^{-}C},\end{aligned} \tag{3.76}$$

and by combining (3.72) and (3.74), we obtain

$$\mathbf{R}_{Z/K} = \mathbf{R}_{R_{ZH/K}G/K} + \mathbf{R}_{ZHB/K} + \mathbf{R}_{R_{Z^*\tilde{H}/K}G/K} + \mathbf{R}_{Z^*\tilde{H}C^*/K}. \tag{3.77}$$

Let $\mathbf{B}^*$ and $\mathbf{C}^*$ be as defined in the previous subsection. Then, by combining (3.73) and (3.74), we obtain

$$\begin{aligned}\mathbf{R}_{Z/K} = {} & \mathbf{R}_{Z^*HH'Z^{*\prime}KG/K} + \mathbf{R}_{Z^*H(H'Z^{*\prime}M_{Z/K}Z^*H)^{-}B^*} \\ & + \mathbf{R}_{Z\tilde{H}\tilde{H}'Z'KG/K} + \mathbf{R}_{Z\tilde{H}(\tilde{H}'Z'KM_{Z/K}Z\tilde{H})^{-}C^*/K},\end{aligned} \tag{3.78}$$

and by combining (3.73) and (3.74), we obtain

$$\mathbf{R}_{Z/K} = \mathbf{R}_{R_{Z^*H/K}G/K} + \mathbf{R}_{Z^*HB^*/K} + \mathbf{R}_{R_{ZH/K}G/K} + \mathbf{R}_{ZHC^*/K}. \tag{3.79}$$

By premultiplying the above decompositions of $\mathbf{R}_{Z/K}$ to the data matrix $\mathbf{Z}$, we obtain the corresponding decompositions of $\mathbf{Z}$. The decomposed parts may then be subjected to PCA for Internal Analysis. For example, if we apply decomposition (3.72), we obtain $\mathbf{R}_{Z/K}\mathbf{Z} = \mathbf{R}_{ZH/K}\mathbf{Z} + \mathbf{R}_{Z^*\tilde{H}/K}\mathbf{Z}$. The two terms on the right may then be subjected to GSVD with the metric matrices $\mathbf{K}$ and $\mathbf{S}^-$. The other decompositions can be similarly applied. Premultiplying ridge operators to $\mathbf{Z}$ has the effect of shrinking the (generalized) singular values. We show this result for $\mathbf{R}_{Z/K}\mathbf{Z}$: Let $\mathbf{Z} = \mathbf{UDV}'$ represent the $\text{GSVD}(\mathbf{Z})_{K,S^-}$. Then, $\text{GSVD}(\mathbf{R}_{Z/K}\mathbf{Z})_{K,S^-}$ is given by

$$\mathbf{R}_{Z/K}\mathbf{Z} = \mathbf{UD}^3(\mathbf{D}^2 + \delta\mathbf{I})^{-1}\mathbf{V}', \tag{3.80}$$

where $\mathbf{D}^3(\mathbf{D}^2 + \delta\mathbf{I})^{-1}$ is the diagonal matrix of generalized singular values of $\mathbf{R}_{Z/K}\mathbf{Z}$, which are smaller than those of the original data matrix $\mathbf{Z}$ for a positive value of δ by the factor of $\mathbf{D}^2(\mathbf{D}^2 + \delta\mathbf{I})^{-1}$. That is, singular values are shrunk toward 0, while singular vectors remain intact.

Assume that $\mathbf{K} = \mathbf{I}_n$ and $\mathbf{S} = \mathbf{P}_{Z'}$ (the orthogonal projector onto the row space of $\mathbf{Z}$), as in the case of ordinary ridge LS. Let $\mathbf{Z} = \mathbf{UDV}'$ represent the SVD of $\mathbf{Z}$. Then, the SVD of $\mathbf{R}_{Z/I}\mathbf{Z}$ is given by

$$\mathbf{R}_{Z/I}\mathbf{Z} = \mathbf{UD}^3(\mathbf{D}^2 + \delta\mathbf{I})^{-1}\mathbf{V}'. \tag{3.81}$$

We thus obtain essentially the same equation as above. Note, however, that the meaning of $\mathbf{U}$, $\mathbf{D}$, and $\mathbf{V}$ is different in the two contexts. In the former, they represent matrices in the GSVD of $\mathbf{Z}$ in the metrics of $\mathbf{K}$ and $\mathbf{S}^-$, while in the latter they are those in the simple SVD of $\mathbf{Z}$. The important thing to observe is that premultiplying ridge operators to the data matrix $\mathbf{Z}$ in general has the effect of shrinking estimates of model parameters through shrinking singular values, thus the name regularized (C)PCA. The use of RLS estimation in special cases of CPCA will be described in several sections of Chapter 4.

Chapter 4

Special Cases and Related Methods

In this chapter we discuss special cases of CPCA and related techniques. These topics will be particularly interesting if the reader already has some knowledge about these techniques. Others may be interested in learning more about them after learning how they relate to CPCA. Takeuchi et al. (1982) and/or Chapter 6 of Yanai et al. (2011) are recommended for further readings. Additional references will be provided where applicable.

4.1 Pre- and Postprocessings

As is well known, unconstrained PCA was originated by Pearson (1901), who discovered many other important tools in statistics being used today. These include, among others, the product moment correlation coefficient and Pearson's chi-square statistic for contingency tables (Pearson 1900). According to de Leeuw (1983), Pearson (1904, 1906) also had a basic idea leading to correspondence analysis (CA; Section 4.6), a popular technique for analysis of contingency tables.

CPCA reduces to unconstrained PCA when there is no additional row or column information to be incorporated in the analysis. In this case we set $\mathbf{G} = \mathbf{I}_n$ and $\mathbf{H} = \mathbf{I}_m$. We also usually assume that $\mathbf{K} = \mathbf{I}_n$ and $\mathbf{L} = \mathbf{I}_m$. Examples of unconstrained PCA have been given in Chapter 1 (Examples 1.1 and 1.2), and technical underpinnings have been provided in Section 2.3.4. Let $\mathbf{Z} = \mathbf{UDV}'$ represent the (compact) SVD of a data matrix $\mathbf{Z}$. The key idea is that if we retain r components from the SVD of $\mathbf{Z}$, $\mathbf{U}_r\mathbf{D}_r\mathbf{V}_r'$ gives the best rank r approximation to $\mathbf{Z}$, where $\mathbf{U}_r$, $\mathbf{D}_r$ and $\mathbf{V}_r$ are portions of $\mathbf{U}$, $\mathbf{D}$, and $\mathbf{V}$ pertaining to the r dominant singular values.

In this section, we discuss two topics of interest relevant to PCA in general. One concerns preprocessing of data before PCA is applied, and the other concerns postprocessing after PCA is applied, namely rotations of components. We also briefly discuss criteria for extracting components other than those used in PCA, i.e., accounting for the maximum variance in the data. There are a host of other interesting ramifications of PCA that can be further explored, but will not be covered in this book. We refer to Jolliffe (2002) for some other topics that may be of interest (e.g., variable selection, multimode PCA, relations to factor analysis, etc.).

As has been emphasized earlier, results of PCA typically depend on how the data are preprocessed. It is important therefore to discuss some options for data preprocessing and their implications. The best preprocessing method also depends on the purpose of the analysis. The data analyst must have a clear vision of the goals of the data analysis when deciding on which method of preprocessing to apply. A few options are given below. In what follows, $\mathbf{Z}^*$ indicates the raw data matrix in standard PCA situations, while $\mathbf{F}$ indicates the raw contingency table in correspondence analysis (CA; Section 4.6) situations. Symbols $\mathbf{I}$ and $\mathbf{1}$ indicate the identity matrix and the vector of ones, respectively, of size indicated by their subscripts.

(i) Columnwise centering: $\mathbf{Z}_c = (\mathbf{I}_n - \mathbf{1}_n\mathbf{1}_n'/n)\mathbf{Z}^*$. Information regarding column means is discarded by this operation. It has the effect of emphasizing differences between rows (e.g., subjects) of $\mathbf{Z}^*$.

(ii) Rowwise centering: $\mathbf{Z}_r = \mathbf{Z}^*(\mathbf{I}_m - \mathbf{1}_m\mathbf{1}_m'/m)$. Row means are eliminated by this operation, which has the effect of enhancing differences between columns (e.g., variables) of the data matrix.

(iii) Double centering: $\mathbf{Z}_{rc} = (\mathbf{I}_n - \mathbf{1}_n\mathbf{1}_n'/n)\mathbf{Z}^*(\mathbf{I}_m - \mathbf{1}_m\mathbf{1}_m'/m)$.

(iv) Centering with respect to nonidentity metric matrices: In CA, for example, it is customary to transform an original two-way contingency table $\mathbf{F}$ by

$$\mathbf{F}^* = (\mathbf{I}_R - \mathbf{D}_R\mathbf{1}_R\mathbf{1}_R'/n)\mathbf{F}(\mathbf{I}_C - \mathbf{1}_C\mathbf{1}_C'\mathbf{D}_C/n), \tag{4.1}$$

where $\mathbf{D}_R$ and $\mathbf{D}_C$ are diagonal matrices of row and column totals of $\mathbf{F}$, R and C are the numbers of rows and columns in $\mathbf{F}$, and n is the total number of observations. This eliminates the so-called trivial solution in CA.

(v) Columnwise standardization: Let $\mathbf{S} = \mathrm{diag}(\mathbf{Z}_c'\mathbf{Z}_c/n)$, where $\mathbf{Z}_c$ is as defined in (i). Then define $\mathbf{Z} = \mathbf{Z}_c\mathbf{S}^{-1/2}$. This eliminates differences in variance across the columns. It is the standard preprocessing option in PCA when the columns of the data matrix have uncomparable measurement units (scales). Note that $\mathbf{S}$ is often defined as $\mathbf{S} = \mathrm{diag}(\mathbf{Z}_c'\mathbf{Z}_c)$, so that the resultant $\mathbf{Z}$ gives the correlation matrix $\mathbf{R} = \mathbf{Z}'\mathbf{Z}$ without further dividing it by n.

(vi) Rowwise standardization: Similar to (v). See Eqs. (1.5) and (1.6).

(vii) Double standardization: In CA, it is customary to transform $\mathbf{F}^*$ in (iv) further by $\mathbf{Z} = \mathbf{D}_R^{-1}\mathbf{F}^*\mathbf{D}_C^{-1}$. This is equal to (1.15).

(viii) Some prescribed nonlinear transformations of variables such as the log, power, logit, and arcsine transformations. These transformations are often applied to linearize the relationships among the variables. Others that involve optimizations of parameters, such as the LS monotonic transformation, will be discussed separately (Section 5.6).

We now turn to the rotation problem. It was noted in Section 2.3.4 that $\hat{\mathbf{F}}$ in (2.248) and $\hat{\mathbf{B}}'$ in (2.249) that minimized (2.244) were not uniquely determined due to arbitrariness in the transformation matrix $\mathbf{T}$. The argument there assumed the columnwise orthogonality of $\hat{\mathbf{F}}$. If this condition is abandoned, the range of indeterminacy is even broader. Let $\mathbf{T}$ be any arbitrary nonsingular matrix of order s. Then,

$$\mathbf{F}\mathbf{B}' = \mathbf{F}\mathbf{T}\mathbf{T}^{-1}\mathbf{B}' = \mathbf{F}^*\mathbf{B}^{*'}, \tag{4.2}$$

where $\mathbf{F}^* = \mathbf{F}\mathbf{T}$ and $\mathbf{B}^* = \mathbf{B}(\mathbf{T}^{-1})'$, implying that $\mathbf{F}^*$ and $\mathbf{B}^*$ are as good as the original **F** and **B**, as minimizers of (2.245). (This indeterminacy will again be relevant in biplots to be discussed in Section 5.7.)

The nonuniqueness of **F** and **B** may appear troublesome. However, it may also provide the opportunity to transform PCA solutions to more readily interpretable ones. If the orthogonality of components should be maintained, i.e., $\mathbf{F}^{*\prime}\mathbf{F}^{*} = \mathbf{I}$, then **T** must be (fully) orthogonal. This is called an orthogonal transformation. If, on the other hand, the transformed components can be correlated but still with unit variances, **T** must satisfy

$$\text{diag}(\mathbf{T}'\mathbf{T}) = \mathbf{I}. \tag{4.3}$$

This is called an oblique transformation. A variety of rotation methods have been proposed in the field of factor analysis, a technique closely related to PCA. These methods can also be used in component analysis with no essential modifications. Interested readers are referred to books on factor analysis (e.g., Mulaik 1972) or the review article on rotation methods by Browne (2001). In the programs referred to in Chapter 7, three representative rotation methods, namely varimax, promax, and orthogonal Procrustes rotation, are provided. The first two are so-called "simple structure" rotation methods that facilitate the interpretability of components. The simple structure here roughly means that each component is well defined by a nonoverlapping subset of variables in the data matrix. The Procrustes rotation method rotates the original loading matrix to the best agreement with a prescribed target matrix.

Traditionally, the predominant approach has been to find PCA solutions first, which are then rotated to have certain desirable properties (such as the simple structure). However, there is little reason why the desirable components cannot be extracted directly from the data. SCoTLASS (Jolliffe and Uddin 2000; Jolliffe et al. 2003; Trendafilov and Jolliffe 2006), for example, directly extracts components satisfying a simple structure using the "lasso" penalty function (the least absolute shrinkage and selection operator; Tibshirani 1996). The lasso uses the sum of absolute values (the L_1 norm) of parameters for penalty, which tends to create "sparsity" (many zero elements) in the estimate of parameter vectors that resemble simple structures. There have been a surge of techniques proposed recently in similar veins. See Zou et al. (2006), Shen and Huang (2008), Huang et al. (2009), Witten et al. (2009), Lee et al. (2010), Johnstone and Lu (2009), d'Aspremont et al. (2007), and Fowler 2009). These techniques derive SVD-like decompositions of data matrices with sparsity in parameter vectors and matrices. Also, see Vines (2000), and Chipman and Gu (2005).

There are other criteria for extracting components. Independent component analysis (ICA; Hyvärinen and Oja 2000; Hyvärinen et al. 2001; Shimizu et al. 2006) extracts components which are as independent from each other as possible. In ICA, components are not only uncorrelated, but also all higher-order central moments among the components are as zero as possible. This sometimes reveals substantively interesting components. Hyvärinen and Oja (2000) show interesting connections between independent components and nonnormality. The maximization of the latter was proposed a long time ago in the name of projection pursuit (Friedman 1987; Friedman and Tukey 1974) as a method of exploratory data analysis. In ICA the process called whitening is typically applied before desired components are extracted, and PCA is often used for this purpose. Thus, ICA may also be conceived as a rotation method (Jennrich and Trendafilov 2005), in which a PCA solution is extracted first, and then rotated to satisfy the requirements of independent components.

It may also be pointed out that Boik and his collaborators (Boik et al. 2010; Boik in press) recently proposed a procedure for eigen-like decompositions of correlation or covariance matrices, in which elements of "eigenvectors" were constrained to satisfy certain prescribed relations. This method can also be regarded as a component extraction method based on a nontraditional criterion.

4.2 Redundancy Analysis (RA)

Redundancy analysis (RA) is a useful technique for multivariate predictions. It extracts a series of orthogonal components from predictor variables that successively account for the maximum variability in criterion variables. It maximizes the proportion of the total sum of squares in the criterion variables that can be accounted for by each successive component. The set of components thus obtained defines, in the space of the predictor variables, a subspace best predictive of the criterion variables. This is in contrast with canonical correlation analysis (CANO; the next section) between two sets of variables, in which components are extracted from each set that are maximally correlated with each other. A large canonical correlation, however, does not imply that the two sets of variables are highly correlated as a whole (Lambert et al. 1988).

Let $\mathbf{Z}$ denote an $n \times m$ ($n \geq m$) matrix of criterion variables, and let $\mathbf{G}$ denote an $n \times p$ (typically, $n \geq p$) matrix of predictor variables. Consider the following multivariate linear regression model:

$$\mathbf{Z} = \mathbf{GB} + \mathbf{E}, \tag{4.4}$$

where $\mathbf{B}$ ($p \times m$) is the matrix of regression coefficients, and $\mathbf{E}$ ($n \times m$) is the matrix of disturbance terms. The criterion variables in $\mathbf{Z}$ are often highly correlated. To capture redundancies among the criterion variables in a parsimonious way, we impose a rank restriction on $\mathbf{B}$ such that

$$\text{rank}(\mathbf{B}) = s \leq \min(n, p). \tag{4.5}$$

Model (4.4) along with (4.5) constitutes a RA model. We find a LS estimate of $\mathbf{B}$ that minimizes $\text{SS}(\mathbf{Z} - \mathbf{GB})$ subject to the rank restriction. Such a $\mathbf{B}$ can be found as follows: Note first that the LS criterion above can be decomposed as

$$\begin{aligned} \text{SS}(\mathbf{Z} - \mathbf{GB}) &= \text{SS}(\mathbf{GB} - \mathbf{G\hat{B}}) + \text{SS}(\mathbf{Z} - \mathbf{G\hat{B}}) \\ &= \text{SS}(\mathbf{B} - \mathbf{\hat{B}})_{G'G,I} + \text{SS}(\mathbf{Z})_{Q_G,I}, \end{aligned} \tag{4.6}$$

where $\mathbf{\hat{B}} = (\mathbf{G}'\mathbf{G})^{-}\mathbf{G}'\mathbf{Z}$ is a rank-free OLSE of $\mathbf{B}$, and $\mathbf{Q}_G = \mathbf{I} - \mathbf{G}(\mathbf{G}'\mathbf{G})^{-}\mathbf{G}'$. Since the second term on the right-hand side of (4.6) is unrelated to $\mathbf{B}$, the estimate of $\mathbf{B}$ under the rank restriction is found by minimizing the first term, which is obtained via $\text{SVD}(\mathbf{G\hat{B}})$ or via $\text{GSVD}(\mathbf{\hat{B}})_{G'G,I}$. By now it should be clear that RA follows from

CPCA by setting $\mathbf{H} = \mathbf{I}$, $\mathbf{K} = \mathbf{I}$ and $\mathbf{L} = \mathbf{I}$. Examples of application of RA have been given in Sections 1.1 and 1.2.

The description of RA above shows that it is a kind of reduced rank regression (RRR; Anderson 1951), which will be more fully described in the next section. Rao (1964) called essentially the same method as above "principal components of instrumental variables." Stewart and Love (1968) called $\text{tr}(\mathbf{Z}'\mathbf{P}_G\mathbf{Z})$ (the portion of SS in $\mathbf{Z}$ that can be explained by $\mathbf{G}$) an index of redundancy of $\mathbf{Z}$ with respect to $\mathbf{G}$. Note that this is different from the redundancy of $\mathbf{G}$ with respect to $\mathbf{Z}$. Van den Wollenberg (1977) proposed a method for representing $\mathbf{P}_G\mathbf{Z}$ in a low dimensional space and called it redundancy analysis (RA), which we follow in this book. Ten Berge (1985) showed that RA was essentially equivalent to Fortier's (1966) simultaneous prediction method.

There are newer developments as well in RA. Van der Leeden (1990) proposed to incorporate structured error covariance matrices in RA. More recently, Takane and Jung (2008) extended the ordinary RA to partial and constrained RA (PRA and CRA), which may be motivated as follows. In Example 1.2, we tried to predict cancer mortality rates from people's eating habits. In the example, however, only food variables were considered as the predictor variables. This could be problematic if we want to assess the unique contributions of the food variables. There may be many other factors than food variables that may influence cancer mortality rates, e.g., average wealth, accessibility to health care systems, etc. The effects of these factors must be eliminated for separating the "pure" impact of the food variables. This calls for partial RA (PRA). The food variables may also be grouped into several distinct categories. Such group structures may be incorporated as additional constraints in the analysis, calling for constrained RA (CRA). In PRA and CRA, the matrix of criterion variables $\mathbf{Z}$ is decomposed further in a manner similar to $\mathbf{z}$ in Section 2.2.10, which leads to decompositions of the total SS similar to (4.6). These decompositions all have the following structure in common: They all have a part that is a function of the regression coefficients, and a part unrelated to the regression coefficients once their rank-free estimates are given. The reduced rank estimates of regression coefficients can be found by minimizing only the first part, which is obtained via GSVD of the rank-free estimates of regression coefficients.

As noted above, the rank reduction feature of RA helps capture dependencies among the criterion variables in a concise manner. However, it does not help alleviate the problem of multicollinearity among the predictor variables. To address this issue, Takane and Hwang (2007) introduced a ridge type of regularized LS estimation (RLS; Section 2.2.12) into RA. It has been shown that similar strategies to the above (nonregularized cases) can be used for estimating parameters in the regularized cases. That is, the reduced rank RLS estimates of regression coefficients can be found by first obtaining rank-free RLS estimates of regression coefficients, which are then subjected to rank reduction by GSVD. Takane and Jung (2008) further extended the RLS estimation to PRA and CRA.

Takane and Hwang (2007) also proposed kernel RA (KRA), in which the space of predictor variables is expanded by nonlinear transformations of the original variables. The key question in this context is which nonlinear transformations to apply. Do we have to know the exact forms of the transformations? Kernel methods (Schölkopf et al. 1997) circumvent this problem by a clever trick, called the kernel trick, which allows us to carry out nonlinear RA, while using computational tools developed for linear models.

Let

$$\mathbf{Z} = \mathbf{G}^*\mathbf{B} + \mathbf{E} \tag{4.7}$$

represent a nonlinear RA model, where $\mathbf{G}^*$ is presumed to have been derived by some nonlinear transformations of the original predictor variables. Notice that this is a linear model once $\mathbf{G}^*$ is derived. It may be assumed without loss of generality that the matrix of regression coefficients $\mathbf{B}$ is in the row space of $\mathbf{G}^*$. (This may be seen as follows: Suppose that $\mathbf{B}$ is not in $\text{Sp}(\mathbf{G}^{*'})$. Then, $\mathbf{B} = \mathbf{B}_1 + \mathbf{B}_2$, where $\text{Sp}(\mathbf{B}_1) \subset \text{Sp}(\mathbf{G}^{*'})$, and $\text{Sp}(\mathbf{B}_2) \subset \text{Ker}(\mathbf{G}^*)$. It follows that $\mathbf{G}^*\mathbf{B} = \mathbf{G}^*\mathbf{B}_1 + \mathbf{G}^*\mathbf{B}_2 = \mathbf{G}^*\mathbf{B}_1$. We may hence reset $\mathbf{B} = \mathbf{B}_1$ without affecting the prediction matrix $\mathbf{G}^*\mathbf{B}$.) This implies that $\mathbf{B}$ can be written as $\mathbf{B} = \mathbf{G}^{*'}\mathbf{A}$ for some $\mathbf{A}$. If we substitute this expression of $\mathbf{B}$ into (4.7), we obtain

$$\mathbf{Z} = \mathbf{G}^*\mathbf{G}^{*'}\mathbf{A} + \mathbf{E} = \mathbf{K}\mathbf{A} + \mathbf{E}, \tag{4.8}$$

where $\mathbf{K} = \mathbf{G}^*\mathbf{G}^{*'}$ is called a kernel matrix. In kernel methods, we directly define $\mathbf{K}$ (instead of $\mathbf{G}^*$), which is presumed to have arisen from a certain desirable (but unknown) $\mathbf{G}^*$. Assume for the moment that such a $\mathbf{K}$ has already been derived. We then estimate $\mathbf{A}$ in (4.7) based on the given $\mathbf{K}$. Note, however, that the ordinary LS estimation cannot be used in this situation, since for any nonsingular matrix $\mathbf{K}$, this leads to $\hat{\mathbf{A}} = \mathbf{K}^{-1}\mathbf{Z}$, but $\hat{\mathbf{Z}} = \mathbf{K}\mathbf{K}^{-1}\mathbf{Z} = \mathbf{Z}$. That is, $\mathbf{Z}$ (the data at hand) is always perfectly predicted. It is unlikely that this provides satisfactory predictions for future observations. We may hence use an RLS estimation instead and obtain

$$\hat{\mathbf{A}} = (\mathbf{K} + \delta\mathbf{I})^{-1}\mathbf{Z}, \tag{4.9}$$

where δ is the ridge parameter, an optimal value of which may be determined by some cross validation method (see Section 5.3). The kernel matrix $\mathbf{K} = [k_{ij}]$ is a similarity matrix between cases and may be defined, for example, by $k_{ij} = \exp(-d_{ij}^2/\sigma)$, where d_{ij} is the Euclidean distance between the ith and jth row vectors in $\mathbf{G}$, and σ is the scale parameter regulating how quickly k_{ij} decreases as a function of the squared distance between them. The kernel matrix defined above is called a Gaussian kernel. The matrix of predictions $\mathbf{K}(\mathbf{K} + \lambda\mathbf{I})^{-1}\mathbf{Z}$ may further be subjected to rank reduction by SVD (Takane and Hwang 2007).

4.3 Canonical Correlation Analysis (CANO)

Canonical correlation analysis (CANO), originated by Hotelling (1936), analyzes relationships between two sets of variables. (We use a somewhat unconventional acronym CANO for canonical correlation analysis, while reserving the more conventional acronym CCA for canonical correspondence analysis (Section 4.7).) As noted in the previous section, CANO extracts a pair of linear combinations of variables called canonical variates, one from each set, which are maximally correlated with each other. If one pair is not sufficient to capture the majority of the total association between the two sets of variables, another pair of canonical variates, which are orthogonal to the first but are maximally correlated with each other, are extracted. This process is continued until no significant association is left unaccounted for in the two sets of variables.

Let $\mathbf{G}$ and $\mathbf{H}$ denote two sets of columnwise centered multivariate data. Computationally, CANO amounts to SVD($\mathbf{P}_G\mathbf{P}_H$) or to GSVD($(\mathbf{G}'\mathbf{G})^-\mathbf{G}'\mathbf{H}(\mathbf{H}'\mathbf{H})^-)_{G'G,H'H}$, where $\mathbf{P}_G$ and $\mathbf{P}_H$ are the orthogonal projectors onto Sp($\mathbf{G}$) and Sp($\mathbf{H}$), respectively. Let $\mathbf{UDV}'$ denote the SVD of the former, and let $\mathbf{U}^*\mathbf{D}^*\mathbf{V}^{*'}$ denote the GSVD of the latter. Then, $\mathbf{U} = \mathbf{GU}^*$, $\mathbf{V} = \mathbf{HV}^*$, and $\mathbf{D} = \mathbf{D}^*$ (or $\mathbf{U}^* = (\mathbf{G}'\mathbf{G})^-\mathbf{G}'\mathbf{U}$, $\mathbf{V}^* = (\mathbf{H}'\mathbf{H})^-\mathbf{H}'\mathbf{V}$, and $\mathbf{D}^* = \mathbf{D}$). The former finds canonical variates (scores) directly, while the latter finds weights applied to $\mathbf{G}$ and $\mathbf{H}$ to obtain canonical variates. (Generalized) singular values in the diagonal of $\mathbf{D} = \mathbf{D}^*$ are called canonical correlation coefficients, indicating the correlations between the corresponding pairs of canonical variates for the two sets of variables. The sum of the the squared canonical correlations is given by $\text{tr}(\mathbf{P}_X\mathbf{P}_Y) = \text{tr}(\mathbf{D}^2)$, indicating the total association between the canonical variates. CANO can be derived from CPCA in two different ways. One is by setting $\mathbf{Z} = \mathbf{I}$, $\mathbf{K} = \mathbf{I}$, and $\mathbf{L} = \mathbf{I}$. The other is by setting $\mathbf{Z} = (\mathbf{G}'\mathbf{G})^-\mathbf{G}'\mathbf{H}(\mathbf{H}'\mathbf{H})^-$, $\mathbf{K} = \mathbf{G}'\mathbf{G}$, $\mathbf{L} = \mathbf{H}'\mathbf{H}$, $\mathbf{G} = \mathbf{I}$, and $\mathbf{H} = \mathbf{I}$.

A more traditional way of deriving solutions in CANO is through the eigendecomposition of $\mathbf{P}_G\mathbf{P}_H$ (or $\mathbf{P}_H\mathbf{P}_G$). Let the SVD of $\mathbf{P}_G\mathbf{P}_H$ be as given above. We then have

$$\mathbf{P}_G\mathbf{P}_H\mathbf{P}_G = \mathbf{UD}^2\mathbf{U}'. \tag{4.10}$$

By postmultiplying both sides of (4.10) by $\mathbf{U}$, we obtain the following eigenequation:

$$\mathbf{P}_G\mathbf{P}_H\mathbf{P}_G\mathbf{U} = \mathbf{P}_G\mathbf{P}_H\mathbf{U} = \mathbf{UD}^2, \tag{4.11}$$

where $\mathbf{D}^2$ is the diagonal matrix of squared canonical correlation coefficients. Note that $\mathbf{P}_G\mathbf{U} = \mathbf{U}$ (i.e., Sp($\mathbf{U}$) $\subset$ Sp($\mathbf{G}$)). The case of $\mathbf{P}_H\mathbf{P}_G$ is similar. (Note, however, that $\mathbf{P}_G\mathbf{P}_H$ and $\mathbf{P}_H\mathbf{P}_G$ are usually not identical, i.e., $\mathbf{P}_G$ and $\mathbf{P}_H$ do not commute. Nor are they generally symmetric.)

RA discussed in the previous section obtains SVD($\mathbf{P}_G\mathbf{H}$) (using the notation of the present section), while CANO obtains SVD($\mathbf{P}_G\mathbf{P}_H$). As noted earlier, this difference stems from the fact that in RA the main objective is to predict one set of variables ($\mathbf{H}$) from the other ($\mathbf{G}$), while CANO analyzes mutual relationships between the two. The matrix $\mathbf{P}_G\mathbf{H}$ in RA is obtained by regressing each criterion variable separately onto Sp($\mathbf{G}$) regardless of correlations among the criterion variables, while $\mathbf{P}_G\mathbf{P}_H$ in CANO takes into account covariances between the criterion variables. What effect does this difference have in analysis results? In Example 1.2, we applied RA to predict cancer mortality rates from food variables. What would happen if we apply CANO instead? It turns out that the results are strikingly similar to those obtained by RA. The largest canonical correlation is found to be .90, indicating a strong relationship between the first canonical variates. Correlations between the observed variables ($\mathbf{G}$ and $\mathbf{H}$) and the corresponding canonical variates are all high, showing similar patterns of correlations to predictor loadings (correlations between $\mathbf{G}$ and redundancy components) and criterion loadings (correlations between $\mathbf{H}$ and redundancy components) in RA. These results, however, may be peculiar to this example, and should not be generalized too far.

Yanai and Takane (1992) proposed constrained CANO, called CANOLC (CANO with linear constraints). In CANOLC, the (canonical) weights are constrained by $\mathbf{U}^* = \mathbf{X}\tilde{\mathbf{U}}$ and $\mathbf{V}^* = \mathbf{Y}\tilde{\mathbf{V}}$, where $\mathbf{X}$ and $\mathbf{Y}$ represent matrices of additional constraints on $\mathbf{U}^*$ and $\mathbf{V}*$, respectively. As has been noted in Section 2.1.10, these constraints can equivalently be stated as $\mathbf{R}'\mathbf{U}^* = \mathbf{O}$ and $\mathbf{C}'\mathbf{V}^* = \mathbf{O}$, where $\mathbf{R}$ and $\mathbf{C}$ are such that $\text{Sp}(\mathbf{R}) \oplus \text{Sp}(\mathbf{X}) = \text{Sp}(\mathbf{G}')$, and $\text{Sp}(\mathbf{C}) \oplus \text{Sp}(\mathbf{Y}) = \text{Sp}(\mathbf{H}')$. CANOLC may be motivated in a similar manner to constrained RA (CRA), as described in the previous section. One obvious difference is that there are two sets of weights in CANOLC, and both of them may be constrained, while there is only one set of weights in CRA. In Example 1.2, for example, separate constraints may be imposed on both food variables and cancer variables. Computationally, CANOLC reduces to the usual (unconstrained) CANO between $\mathbf{GX}$ and $\mathbf{HY}$ or between $\mathbf{GQ}_R$ and $\mathbf{HQ}_C$.

Takane and Hwang (2002), and Takane et al. (2006) further extended CANO/CANOLC to various kinds of partial CANO/CANOLC. Suppose that $\mathbf{G}$ and $\mathbf{H}$ each consist of two subsets of variables, say, $\mathbf{G} = [\mathbf{G}_1, \mathbf{G}_2]$ and $\mathbf{H} = [\mathbf{H}_1, \mathbf{H}_2]$, where the relationships between $\mathbf{G}_1$ and $\mathbf{H}_1$ are of primary interest, while $\mathbf{G}_2$ and $\mathbf{H}_2$ are nuisance variables (covariates) whose effects are to be partialled out. Then we may eliminate the effects of $\mathbf{G}_2$ from $\mathbf{G}_1$ and the effects of $\mathbf{H}_2$ from $\mathbf{H}_1$, and apply CANO between the residual matrices $\mathbf{Q}_{G_2}\mathbf{G}_1$ and $\mathbf{Q}_{H_2}\mathbf{H}_1$. This is called bipartial CANO (Timm 1975; Timm and Carlson 1976), if $\mathbf{G}_2$ and $\mathbf{H}_2$ are distinct. It is called partial CANO, when $\mathbf{G}_2$ and $\mathbf{H}_2$ coincide. CANO in which the effects of nuisance variables are eliminated from only one set of variables is called semipartial CANO. For example, CANO between $\mathbf{Q}_{G_2}\mathbf{G}_1$ and $\mathbf{H}_1$, and that between $\mathbf{G}_1$ and $\mathbf{Q}_{H_2}\mathbf{H}_1$ are semipartial CANO. The various kinds of partial CANO/CANOLC can be motivated in a manner similar to partial RA in the previous section. We may, for example, analyze the relationships between food variables and cancer mortality rates after eliminating the effects of average wealth ($\mathbf{G}_2$) from the food variables, and the effects of accessibility to health care systems ($\mathbf{H}_2$) from the cancer mortality rates. We may also combine the average wealth and accessibility into one, and use it in partial CANO or semipartial CANO. Takane et al. (2006) developed a comprehensive framework for decompositions of the total association ($\text{tr}(\mathbf{P}_G\mathbf{P}_H)$) between $\mathbf{G}$ and $\mathbf{H}$ that include all of the analyses mentioned above as special cases. In their approach, $\mathbf{P}_G$ and $\mathbf{P}_H$ are decomposed separately into the sum of projectors orthogonal to each other, CANO is applied between pairs of projectors, one from each decomposition of $\mathbf{P}_G$ and $\mathbf{P}_H$.

As noted in the previous section, RA is a kind of reduced rank regression (RRR) analysis. In deriving RA, we used the ordinary (unweighted) LS criterion, which resulted in the "asymmetric" treatment of $\mathbf{H}$ (the criterion variables) and $\mathbf{G}$ (the predictor variables). Strictly speaking, however, this treatment can only be justified when all the elements in the matrix of disturbance terms $\mathbf{E}$ are independently and identically distributed (*iid*). Suppose that the disturbance terms are correlated across columns of $\mathbf{E}$. Let $\text{Var}[\mathbf{e}_i] = \Sigma$ for all i, where $\mathbf{e}_i$ is the ith row of $\mathbf{E}$, and assume for the moment that Σ is known. Then we may use a weighted LS criterion

$$f(\mathbf{B}) = \mathrm{SS}(\mathbf{H} - \mathbf{GB})_{I,\Sigma^{-1}} \tag{4.12}$$

to be minimized. (Note, however, that in this case, the nonidentity weight matrix is applied to the column side, while in the conventional unweighted LS, it is applied to the row side.) To minimize (4.12), we first rewrite it as

$$f(\mathbf{B}) = \mathrm{SS}(\mathbf{B} - \hat{\mathbf{B}})_{G'G,\Sigma^{-1}} + \mathrm{SS}(\mathbf{H} - \mathbf{G}\hat{\mathbf{B}})_{I,\Sigma^{-1}}, \tag{4.13}$$

where $\hat{\mathbf{B}} = (\mathbf{G}'\mathbf{G})^{-}\mathbf{G}'\mathbf{H}$ is a rank-free OLSE of $\mathbf{B}$. Since the second term on the right-hand side of (4.13) is unrelated to $\mathbf{B}$, the reduced rank estimate of $\mathbf{B}$ can be obtained by minimizing the first term, which is achieved by $\mathrm{GSVD}(\hat{\mathbf{B}})_{G'G,\Sigma^{-1}}$. Note that this GSVD problem is different from the one in RA due to Σ despite the fact that the rank-free estimate of $\mathbf{B}$ remains the same. The error covariance matrix is usually unknown and must be estimated from the data. One "natural" choice of $n\hat{\Sigma}$ is $\mathbf{S} = \mathbf{H}'\mathbf{Q}_G\mathbf{H}$, which may be substituted into the GSVD problem above. This GSVD, however, is essentially the same as the one solved in CANO between $\mathbf{G}$ and $\mathbf{H}$, since $\mathbf{H}'\mathbf{H} = \mathbf{H}'\mathbf{P}_G\mathbf{H} + \mathbf{H}'\mathbf{Q}_G\mathbf{H}$. This approach represents another school of RRR originated by Anderson (1951) and followed by Izenman (1975), Tso (1981), and Davies and Tso (1982). It may be pointed out that the performance of this approach relative to that of RA, which assumes $\Sigma = \mathbf{I}$, depends on how far Σ deviates from $\mathbf{I}$, and how accurately Σ can be estimated.

The curds and whey (CW) method (Breiman and Friedman 1997) represents another interesting approach to multivariate prediction. As in CANO, this approach takes into account covariances among the criterion variables. Let $\mathbf{UDV}'$ denote the SVD of $\mathbf{P}_G\mathbf{P}_H$ in CANO. Then we have $\mathbf{P}_G\mathbf{H} = \mathbf{P}_G\mathbf{P}_H\mathbf{H} = \mathbf{UDV}'\mathbf{H}$. That is, $\mathbf{UDV}'$ works as an operator that turns $\mathbf{H}$ into its prediction $\hat{\mathbf{H}}$. In the CW method, canonical correlations in $\mathbf{D}$ are separately shrunken by a *nnd* diagonal matrix $\mathbf{C} \le \mathbf{I}$. Optimal values of the shrinkage factors are determined by some cross validation procedures (e.g., full and generalized cross validation methods; see Section 5.3). Breiman and Friedman (1997) compared a number of approaches to multivariate prediction and demonstrated that the CW method with the full cross validation method (the leaving-one-out method) worked best.

The shrinkage estimation similar to the above may also be realized by a ridge type of regularization method. Let $\mathbf{A} = [\mathbf{G}, \mathbf{H}]$, where $\mathbf{G}$ and $\mathbf{H}$ are assumed disjoint. We define a ridge metric matrix by $\mathbf{M}_A(\delta) = \mathbf{P}_A + \delta(\mathbf{AA}')^{+}$. Then regularized CANO (RCANO) amounts to $\mathrm{SVD}(\mathbf{R}_G(\delta)\mathbf{R}_H(\delta))$, where

$$\mathbf{R}_G(\delta) = \mathbf{G}(\mathbf{G}'\mathbf{M}_A(\delta)\mathbf{G})^{-}\mathbf{G}' = \mathbf{G}(\mathbf{G}'\mathbf{G} + \delta\mathbf{P}_{G'})^{-}\mathbf{G}', \tag{4.14}$$

and

$$\mathbf{R}_H(\delta) = \mathbf{H}(\mathbf{H}'\mathbf{M}_A(\delta)\mathbf{H})^{-}\mathbf{H}' = \mathbf{H}(\mathbf{H}'\mathbf{H} + \delta\mathbf{P}_{H'})^{-}\mathbf{H}' \tag{4.15}$$

are ridge operators (Section 2.2.12) associated with $\mathrm{Sp}(\mathbf{G})$ and $\mathrm{Sp}(\mathbf{H})$. Note that $\mathbf{G}'\mathbf{M}_A(\lambda)\mathbf{H} = \mathbf{G}'\mathbf{H}$. RCANO is a special case of the regularized multiple-set CANO proposed by Takane et al. (2008).

4.4 Canonical Discriminant Analysis (CDA)

CANO subsumes a number of interesting techniques for multivariate data analysis as special cases. Indeed, some argue that the interest in CANO lies mainly in its special cases. We discuss one such instance in this section. When one of two sets of variables in CANO consists of dummy-coded categorical variables, CANO reduces to canonical discriminant analysis (CDA; Fisher 1936).

Let $\mathbf{G}$ denote a matrix of dummy variables indicating groups to which cases (subjects) belong, and let $\mathbf{H}$ denote a matrix of predictor variables consisting of measures of attributes of the cases deemed useful to distinguish the groups. (Note that the roles of $\mathbf{G}$ and $\mathbf{H}$ are temporarily reversed. This is just for convenience and makes no essential differences in the analysis.) We assume that $\mathbf{H}$ is columnwise centered and normalized so that $\mathbf{C}_T = \mathbf{H}'\mathbf{H}$ gives the matrix of total covariances among the predictor variables. Let $\mathbf{C}_B = \mathbf{H}'\mathbf{P}_G\mathbf{H}$ denote the between-group covariance matrix, and $\mathbf{C}_W = \mathbf{H}'\mathbf{Q}_G\mathbf{H}$ the within-group covariance matrix. Then, $\mathbf{C}_T$ can be decomposed into the sum of $\mathbf{C}_B$ and $\mathbf{C}_W$, i.e., $\mathbf{C}_T = \mathbf{C}_B + \mathbf{C}_W$. We wish to find linear combinations of $\mathbf{H}$ (called discriminant functions) which maximally discriminate the groups. Let $\mathbf{V}^*$ denote the matrix of weights in the linear combinations. Then, we seek to find $\mathbf{V}^*$ that maximizes $\text{tr}(\mathbf{V}^{*'}\mathbf{C}_B\mathbf{V}^*)$ subject to the orthonormalization restriction that $\mathbf{V}^{*'}\mathbf{C}_T\mathbf{V}^* = \mathbf{I}$. This leads to the following generalized eigenequation (Section 2.3.7):

$$\mathbf{C}_B\mathbf{V}^* = \mathbf{C}_T\mathbf{V}^*\Lambda, \tag{4.16}$$

from which the matrix of discriminant functions $\mathbf{V}$ can be derived by $\mathbf{V} = \mathbf{H}\mathbf{V}^*$. The centroids of the groups are calculated by $\mathbf{M} = (\mathbf{G}'\mathbf{G})^{-1}\mathbf{G}'\mathbf{V}$. (The weight matrix $\mathbf{U}^*$ for $\mathbf{G}$ is simply $\mathbf{U}^* = \mathbf{M}\Lambda^{-1/2}$. That is, $\mathbf{M}$ must be scaled by $\Lambda^{-1/2}$ to satisfy $\mathbf{U}'\mathbf{U} = \mathbf{I}$, where $\mathbf{U} = \mathbf{G}\mathbf{U}^*$.) Cases are classified into groups whose centroids are closest in the space spanned by the discriminant functions. Future observations (not used in the estimation of $\mathbf{V}^*$) are first mapped into the discriminant space by $\mathbf{V}^*$, and then classified into groups with nearest centroids.

Essentially the same solution can also be obtained by applying CANO between $\mathbf{G}$ and $\mathbf{H}$, that is, by $\text{SVD}(\mathbf{P}_G\mathbf{P}_H)$ or by $\text{GSVD}((\mathbf{G}'\mathbf{G})^-\mathbf{G}'\mathbf{H}(\mathbf{H}'\mathbf{H})^-)_{G'G,H'H}$. Note that in this particular case, $\mathbf{G}$ does not need to be columnwise centered once $\mathbf{H}$ has been columnwise centered. See the note below for why.

Note 4.1 The $n \times q$ matrix of dummy variables $\mathbf{G}$ has the following property when there are no missing data:

$$\mathbf{G}\mathbf{1}_q = \mathbf{1}_n. \tag{4.17}$$

It follows that

$$\tilde{\mathbf{G}} \equiv \mathbf{Q}_{1_n}\mathbf{G} = \mathbf{G}\mathbf{Q}_{1_q/G'G}, \tag{4.18}$$

where $\mathbf{Q}_{1_n} = \mathbf{I}_n - \mathbf{1}_n\mathbf{1}_n'/n$ and $\mathbf{Q}_{1_q/G'G} = \mathbf{I}_q - \mathbf{1}_q\mathbf{1}_q'\mathbf{G}'\mathbf{G}/n$. Hence,

$$\tilde{\mathbf{G}}'\tilde{\mathbf{G}} = \mathbf{Q}_{1_q/G'G}'\mathbf{G}'\mathbf{G} = \mathbf{G}'\mathbf{G}\mathbf{Q}_{1_q/G'G}, \tag{4.19}$$

and so

$$(\mathbf{G}'\mathbf{G})^{-1}\mathbf{Q}_{1_q/G'G}' = \mathbf{Q}_{1_q/G'G}(\mathbf{G}'\mathbf{G})^{-1} \in \{(\tilde{\mathbf{G}}'\tilde{\mathbf{G}})_r^-\}. \tag{4.20}$$

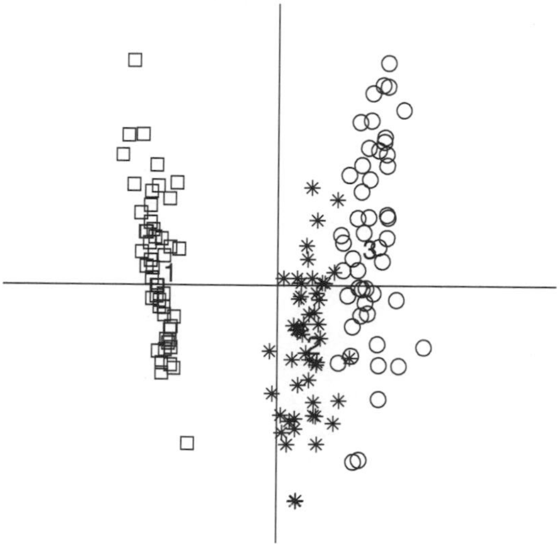

Figure 4.1 *CDA of Fisher's (1936) iris data.*

Consequently, we have

$$(\tilde{\mathbf{G}}'\tilde{\mathbf{G}})^{-}\tilde{\mathbf{G}}'\tilde{\mathbf{H}} = (\mathbf{G}'\mathbf{G})^{-1}\mathbf{G}'\tilde{\mathbf{H}}, \tag{4.21}$$

where $\tilde{\mathbf{H}} = \mathbf{Q}_{1_n}\mathbf{H}$.

Fisher (1936) gives an example on discrimination of three species of iris (setosa, versicolor, and virginica) based on four predictor variables. Fifty instances each of the three species of iris were measured with respect to their sepal length and width, and petal length and width. CDA is applied to this data set to discriminate among the species. The results are shown in Figure 4.1. In the figure, squares indicate setosa, asterisks indicate versicolor, and circles virginica. Three integers indicate the centroids of the three species (1 = setosa; 2 = versicolor; 3 = virginica). Setosa is well discriminated from the other two by the first discriminant function indicated by the x-axis, while versicolor and virginica are only weakly discriminated on both discriminant functions.

When there are only two groups to discriminate ($q = 2$), CDA reduces to multiple regression analysis by representing the contrast between the two groups in a single criterion variable (Fisher 1936). CDA also reduces to a form of constrained CA (Section 4.7) when cases within groups have identical predictor values.

4.5 Multidimensional Scaling (MDS)

Multidimensional scaling (MDS) refers to a class of techniques for analysis of proximity data between "stimuli." MDS represents a set of stimuli as points in a multidimensional space in such a way that similar stimuli are located close together, while dissimilar stimuli are located far apart (Takane 2007a; Takane et al. 2009). Consider, as an example, a road map. It is relatively straightforward to measure distances between cities. However, the reverse operation, that of recovering a map (or more specifically, finding relative locations of the cities) based on a set of intercity distances, is not as easy. The role of MDS is to perform this reverse operation. As such, MDS has not been considered a mainstream multivariate analysis (MVA) technique. We take up this technique in this section because the theory of Euclidean distances underlying MDS is a common thread running through many other MVA techniques.

A variety of MDS techniques have been developed depending on the kind of proximity data analyzed, the form of functional relationships between the observed data and the distance model, the type of fitting criteria, and so on. See Borg and Groenen (2005) for an overview. In recent years, however, a form of MDS called nonmetric MDS has been very popular (Kruskal 1964a, b; Shepard 1962) among practitioners because of its flexibility. This type of MDS is capable of finding a stimulus configuration under the assumption that observed proximity data are only monotonically related to underlying distances (see Section 5.6). In this section, however, we primarily focus on the simplest kinds of MDS mostly amenable to matrix analysis (Takane 2004).

The reverse operation mentioned above is particularly simple when a set of error-free Euclidean distances is given as data. Suppose that a set of m stimuli is represented in an r dimensional Euclidean space. Let $\mathbf{V} = [v_{ia}]$ $(i = 1, \cdots, m;\ a = 1, \cdots, r)$ represent the matrix of stimulus coordinates. Then the order m matrix of squared Euclidean distances between stimuli can be expressed as

$$\mathbf{D}^{(2)} = \mathbf{1}_m\mathbf{1}_m'\text{diag}(\mathbf{V}\mathbf{V}') - 2\mathbf{V}\mathbf{V}' + \text{diag}(\mathbf{V}\mathbf{V}')\mathbf{1}_m\mathbf{1}_m', \tag{4.22}$$

where $\mathbf{1}_m$ is the m-component vector of ones. Define

$$\mathbf{S} = -\frac{1}{2}\mathbf{Q}_{1_m}\mathbf{D}^{(2)}\mathbf{Q}_{1_m}, \tag{4.23}$$

where $\mathbf{Q}_{1_m} = \mathbf{I}_m - \mathbf{1}_m\mathbf{1}_m'/m$ is the centering operator of order m. This transformation is called the Young–Householder (1938) transformation. Then

$$\mathbf{S} = \mathbf{Q}_{1_m}\mathbf{V}\mathbf{V}'\mathbf{Q}_{1_m} = \mathbf{V}\mathbf{V}', \tag{4.24}$$

where it is assumed that $\mathbf{Q}_{1_m}\mathbf{V} = \mathbf{V}$. (The origin of the space is placed at the centroid of the stimulus configuration.) The relation above indicates that $\mathbf{V}$ can be obtained as a square root factor of $\mathbf{S}$ (typically, by the eigendecomposition), and that $\mathbf{S}$ must be a *psd* matrix of rank r.

Observed data typically contain sizable amounts of measurement errors, for which no exact reverse operation (such as the above) exists. In this case we look for $\mathbf{V}$ such that $\mathbf{V}\mathbf{V}'$ best approximates (in the LS sense) an observed $\mathbf{S}$ derived from observed proximity (dissimilarity) data by the Young–Householder transformation (Torgerson 1958). Let $\Delta^{(2)}$ denote the matrix of squared dissimilarity data, and define $\hat{\mathbf{S}} = (-1/2)\mathbf{Q}_{1_m}\Delta^{(2)}\mathbf{Q}_{1_m}$. We look for $\mathbf{V}$ that minimizes $\text{SS}(\hat{\mathbf{S}} - \mathbf{V}\mathbf{V}')$. Such a

$\mathbf{V}$ can be found by the eigendecomposition of $\hat{\mathbf{S}}$, assuming that there are at least r positive eigenvalues. Due to measurement errors, $\hat{\mathbf{S}}$ is usually not a *psd* matrix.

In the discussion above, it is assumed that there is only one set of stimuli, and that dissimilarity data are observed between pairs of stimuli in this set. In some cases, however, dissimilarity data are obtained between two distinct sets of "stimuli." Such an instance arises when a group of n subjects make preference judgments on a set of m stimuli. The preference data are regarded as representing similarities between actual stimuli and subjects' ideal stimuli. It is assumed that the stimuli which are more similar to subject's ideal stimulus are more preferred by the subject. MDS, in this case, represents both actual and ideal stimuli as points in a joint multidimensional space in such a way that their mutual distances are as inversely related to the observed preference data as possible. This variant of MDS is called unfolding analysis (Coombs 1964), and the model associated with it is called an ideal point model.

Let $\mathbf{U}$ denote the $n \times r$ matrix of coordinates of subjects' ideal points. The matrix of squared Euclidean distances between them can then be expressed as

$$\mathbf{D}^{(2)} = \mathrm{diag}(\mathbf{U}\mathbf{U}')\mathbf{1}_n\mathbf{1}_m' - 2\mathbf{U}\mathbf{V}' + \mathbf{1}_n\mathbf{1}_m'\mathrm{diag}(\mathbf{V}\mathbf{V}'). \tag{4.25}$$

Note that this matrix is now an $n \times m$ matrix and is a function of both $\mathbf{V}$ and $\mathbf{U}$. By applying transformations analogous to the Young–Householder transformation, we obtain

$$\mathbf{Z} = -\frac{1}{2}\mathbf{Q}_{1_n}\mathbf{D}^{(2)}\mathbf{Q}_{1_m} = \mathbf{Q}_{1_n}\mathbf{U}\mathbf{V}', \tag{4.26}$$

where we assume that $\mathbf{Q}_{1_m}\mathbf{V} = \mathbf{V}$. Note that the orders of centering matrices applied to the left-hand and right-hand sides of $\mathbf{D}^{(2)}$ are usually different ($n \neq m$). Note also that we may place either the centroid of $\mathbf{V}$ or that of $\mathbf{U}$ at the origin of the space, but not both. By the SVD of $\mathbf{Z}$, we find $\mathbf{Z} = \mathbf{U}^*\mathbf{V}^{*'}$, where $\mathbf{U}^*$ and $\mathbf{V}^*$ are $n \times r$ and $m \times r$ columnwise nonsingular matrices. These matrices are related to $\mathbf{V}$ and $\mathbf{U}$ by $\mathbf{V} = \mathbf{V}^*\mathbf{T}$, and $\mathbf{U} = (\mathbf{U}^* + \mathbf{1}_m\mathbf{u}_0')\mathbf{T}^{-1}$ for some nonsingular matrix $\mathbf{T}$ of order r, and some off-set vector $\mathbf{u}_0'$. (This vector locates the centroid of $\mathbf{U}$.) Schönemann (1970) developed a series of lengthy formulas to find $\mathbf{T}$ and $\mathbf{u}_0'$ explicitly by relating these parameters to the original preference data. However, these formulas are found to be very sensitive to errors in the observed data. Gold (1973) proposed a remedy for Schönemann's procedure by adding nonlinear terms in the estimation. More recently, Adachi (2009) proposed, in a somewhat different context, to reparameterize $\mathbf{T}$ by its SVD, say, $\mathbf{T} = \mathbf{A}\mathbf{D}\mathbf{B}'$, and estimate $\mathbf{A}$, $\mathbf{B}$ and $\mathbf{D}$ rather than $\mathbf{T}$ directly. Note that $\mathbf{A}$ and $\mathbf{B}$ are fully orthogonal, $\mathbf{D}$ is *pd* diagonal, and $\mathbf{T}^{-1} = \mathbf{B}\mathbf{D}^{-1}\mathbf{A}'$. Adachi's method may potentially be useful in estimating parameters in the unfolding model. It may also be noted that a fairly reliable iterative algorithm has been developed recently for nonmetric unfolding analysis (Busing et al. 2005).

So far in this section, it has been assumed that there is only a single matrix of dissimilarities to be analyzed. In some cases, however, dissimilarity data may be collected on the same set of stimuli from multiple subjects, giving rise to $K > 1$ square dissimilarity matrices. MDS designed to analyze such data is called individual differences (ID) MDS. One useful ID-MDS technique, called INDSCAL (Carroll and Chang 1970), represents both common and distinct aspects of such data by the weighted Euclidean distance model. Let $\mathbf{V}$ represent the matrix of stimulus coordinates common across all subjects, and let $\mathbf{W}_k$ $(k = 1, \cdots, K)$ denote the *nnd* diagonal matrix of order r representing the weights allocated by subject k on r dimensions. Let $\mathbf{D}_k^{(2)}$ denote the matrix of squared Euclidean distances between m stimuli for subject k. Then,

$$\mathbf{D}_k^{(2)} = \mathbf{1}_m\mathbf{1}'_m\text{diag}(\mathbf{V}\mathbf{W}_k\mathbf{V}') - 2\mathbf{V}\mathbf{W}_k\mathbf{V}' + \text{diag}(\mathbf{V}\mathbf{W}_k\mathbf{V}')\mathbf{1}_m\mathbf{1}'_m. \tag{4.27}$$

This model aims to explain common aspects of K dissimilarity matrices by the common stimulus configuration $\mathbf{V}$, and the differences between them by the weights attached to different dimensions by different subjects. To remove scale indeterminacy in the model, it may be assumed that $\text{diag}(\mathbf{V}'\mathbf{V}) = \mathbf{I}_r$. One remarkable feature of the weighted Euclidean distance model is that for $K > 1$ it has no rotational indeterminacy problem inherent in the simple (unweighted) Euclidean distance model.

We apply the Young–Householder transformation to $\mathbf{D}_k^{(2)}$ to obtain $\mathbf{S}_k \equiv (-1/2)\mathbf{Q}_{1_m}\mathbf{D}_k^{(2)}\mathbf{Q}_{1_m} = \mathbf{V}\mathbf{W}_k\mathbf{V}'$, where, as before, it is assumed that $\mathbf{Q}_{1_m}\mathbf{V} = \mathbf{V}$. Let

$$\bar{\mathbf{S}} = \sum_{k=1}^{K} \mathbf{S}_k/K = \mathbf{V}(\sum_{k=1}^{K} \mathbf{W}_k/K)\mathbf{V}' = \mathbf{V}\mathbf{V}', \tag{4.28}$$

where it is temporarily assumed that $\sum_{k=1}^{K} \mathbf{W}_k/K = \mathbf{I}$. (We later rescale $\mathbf{V}$ and $\mathbf{W}_k$ to satisfy the usual scaling convention noted above.) By the eigendecomposition of $\bar{\mathbf{S}}$, we obtain $\mathbf{V}^*$ such that $\bar{\mathbf{S}} = \mathbf{V}^*\mathbf{V}^{*'}$. Then, $\mathbf{V} = \mathbf{V}^*\mathbf{T}$ for some orthogonal matrix $\mathbf{T}$ of order r. Such a $\mathbf{T}$ can be obtained as follows: Let $\mathbf{S}^*$ denote a linear combination of $\mathbf{S}_k$ (different from $\bar{\mathbf{S}}$ above). That is,

$$\mathbf{S}^* = \mathbf{V}(\sum_{k=1}^{K} \mathbf{W}_k e_k)\mathbf{V}' = \mathbf{V}\bar{\mathbf{W}}\mathbf{V}', \tag{4.29}$$

where $\bar{\mathbf{W}} = \sum_{k=1}^{K} \mathbf{W}_k e_k$. Then

$$\begin{aligned}\bar{\mathbf{W}} &= (\mathbf{V}'\mathbf{V})^{-1}\mathbf{V}'\mathbf{S}^*\mathbf{V}(\mathbf{V}'\mathbf{V})^{-1} \\ &= \mathbf{T}'(\mathbf{V}^{*'}\mathbf{V}^*)^{-1}\mathbf{V}^{*'}\mathbf{S}^*\mathbf{V}^*(\mathbf{V}^{*'}\mathbf{V}^*)^{-1}\mathbf{T}.\end{aligned} \tag{4.30}$$

Define $\mathbf{C} = (\mathbf{V}^{*'}\mathbf{V}^*)^{-1}\mathbf{V}^{*'}\mathbf{S}^*\mathbf{V}^*(\mathbf{V}^{*'}\mathbf{V}*)^{-1}$. Then $\mathbf{C} = \mathbf{T}\bar{\mathbf{W}}\mathbf{T}'$. This indicates that $\mathbf{T}$ and $\bar{\mathbf{W}}$ can be obtained by the eigendecomposition of $\mathbf{C}$. For this decomposition to be unique, however, there must be at least one linear combination of $\mathbf{W}_k$ such that the diagonal elements of $\bar{\mathbf{W}}$ are all distinct. (Often, any one of $\mathbf{S}_k$'s may be used, although $\bar{\mathbf{S}}$ in (4.28) clearly does not qualify.)

The procedure described above, however, strictly applies to only infallible data. A procedure, called SUMSCAL, that simultaneously diagonalizes $\mathbf{C}_k = (\mathbf{V}^{*'}\mathbf{V}^*)^{-1}\mathbf{V}^{*'} \times \hat{\mathbf{S}}_k\mathbf{V}^*(\mathbf{V}^{*'}\mathbf{V}*)^{-1}$ $(k = 1, \cdots, K)$ as much as possible, where $\hat{\mathbf{S}}_k$ is the observed counterpart of $\mathbf{S}_k$, has been developed by de Leeuw and Pruzansky (1978). This algorithm assumes that $\mathbf{V}^*$ is given and only estimates $\mathbf{T}$. A more general algorithm, called CANDECOMP, that fits the model $\mathbf{V}\mathbf{W}_k\mathbf{V}'$ to $\hat{\mathbf{S}}_k$ has been developed by Carroll and

Chang (1970). This algorithm estimates both $\mathbf{V}$ and $\mathbf{W}_k$ that minimize the LS criterion. In the original INDSCAL model, $\mathbf{V}$ is not necessarily assumed columnwise orthogonal. In most applications, however, $\mathbf{V}$ is found nearly orthogonal, and requiring the orthogonality facilitates the convergence of computational algorithms (Takane et al. 2010). The model with the orthogonality restriction on $\mathbf{V}$ is called the orthogonal INDSCAL. Flury (1988) proposed a model called common principal components (CPC), in which $\mathbf{V}$ was assumed square and (fully) orthogonal.

In a similar vein, a weighted inner product model $\mathbf{Z}_k = \mathbf{U}\mathbf{W}_k\mathbf{V}' + \mathbf{E}_k$ has been proposed for a set of K rectangular matrices $\mathbf{Z}_k$. An algorithm similar to the above has been developed to fit this model (Sands and Young 1980). Let $\bar{\mathbf{Z}} = \sum_{k=1}^K \mathbf{Z}_k/K = \mathbf{U}(\sum_{k=1}^K \mathbf{W}_k/K)\mathbf{V}' = \mathbf{U}\mathbf{V}'$, where it is assumed that $\sum_{k=1}^K \mathbf{W}_k/K = \mathbf{I}$. By a rank decomposition (e.g., SVD) of $\bar{\mathbf{Z}}$, we find $\bar{\mathbf{Z}} = \mathbf{U}^*\mathbf{V}^{*'}$, where $\mathbf{U}^*$ and $\mathbf{V}^*$ are related to $\mathbf{U}$ and $\mathbf{V}$ by $\mathbf{U} = \mathbf{U}^*\mathbf{T}$ and $\mathbf{V} = \mathbf{V}^*(\mathbf{T}^{-1})'$ for some nonsingular matrix $\mathbf{T}$. To find such a $\mathbf{T}$, we form a linear combination of $\mathbf{Z}_k$, i.e., $\mathbf{Z}^* = \mathbf{U}(\sum_{k=1}^K \mathbf{W}_k e_k)\mathbf{V}' = \mathbf{U}\bar{\mathbf{W}}\mathbf{V}'$. Then, $\bar{\mathbf{W}} = (\mathbf{U}'\mathbf{U})^{-1}\mathbf{U}'\mathbf{Z}^*\mathbf{V}(\mathbf{V}'\mathbf{V})^{-1} = \mathbf{T}^{-1}(\mathbf{U}^{*'}\mathbf{U}^*)^{-1}\mathbf{U}^{*'}\mathbf{Z}^*\mathbf{V}^*(\mathbf{V}^{*'}\mathbf{V}^*)^{-1}\mathbf{T}$. Define $\mathbf{C} = (\mathbf{U}^{*'}\mathbf{U}^*)^{-1}\mathbf{U}^{*'}\mathbf{Z}^*\mathbf{V}^*(\mathbf{V}^{*'}\mathbf{V}^*)^{-1}$. Then $\mathbf{C} = \mathbf{T}\bar{\mathbf{W}}\mathbf{T}^{-1}$. This indicates that $\mathbf{T}$ can be obtained by the eigendecomposition of $\mathbf{C}$. The only notable difference from the INDSCAL model above is that $\mathbf{C}$ in this case is usually asymmetric, and its eigendecomposition takes the form of $\mathbf{C} = \mathbf{T}\mathbf{\Lambda}\mathbf{T}^{-1}$, where $\mathbf{T}^{-1} \neq \mathbf{T}'$.

4.6 Correspondence Analysis (CA)

When both $\mathbf{G}$ and $\mathbf{H}$ consist of dummy-coded categorical variables, CANO specializes in correspondence analysis (CA) of a contingency table $\mathbf{F} = \mathbf{G}'\mathbf{H}$. Examples of analysis by CA have been given in Sections 1.3 and 1.4. Computationally, CA amounts to GSVD$(\mathbf{D}_R^{-1}\mathbf{F}\mathbf{D}_C^{-1})_{D_R,D_C}$, where $\mathbf{D}_R = \mathbf{G}'\mathbf{G}$ and $\mathbf{D}_C = \mathbf{H}'\mathbf{H}$ are diagonal matrices of row and column totals of $\mathbf{F}$, respectively.

There are a number of criteria other than those in CANO, from which CA can be derived. Below we describe a somewhat unconventional way of introducing CA based on the unfolding model discussed in the previous section. Observe that the elements of $\mathbf{F}$ represent similarities between rows and columns of the table. In unfolding analysis, rows and columns with large joint frequencies (= similarities) are located close together, while those with small joint frequencies are located far apart.

Let f_{ij} denote the ijth element of $\mathbf{F}$ ($R \times C$), and let d_{ij}^2 denote the squared Euclidean distance between the ith row and jth column of $\mathbf{F}$. Let $\mathbf{U}^*$ and $\mathbf{V}^*$ represent the matrices of coordinates of row and column points. We wish to find $\mathbf{U}^*$ and $\mathbf{V}^*$ that minimize

$$g(\mathbf{U}^*, \mathbf{V}^*) = \sum_{i,j} f_{ij} d_{ij}^2(\mathbf{U}^*, \mathbf{V}^*) \tag{4.31}$$

subject to some normalization restriction on $\mathbf{U}^*$ (or on $\mathbf{V}^*$). This criterion forces d_{ij}^2 corresponding to a large f_{ij} to be relatively small, and d_{ij}^2 corresponding to a small f_{ij} to be relatively large. Let $\mathbf{D}^{(2)}$ represent the matrix of d_{ij}^2 written as

$$\mathbf{D}^{(2)} = \text{diag}(\mathbf{U}^*\mathbf{U}^{*'})\mathbf{1}_R\mathbf{1}_C' - 2\mathbf{U}^*\mathbf{V}^{*'} + \mathbf{1}_R\mathbf{1}_C'\text{diag}(\mathbf{V}^*\mathbf{V}^{*'}), \tag{4.32}$$

where $\mathbf{1}_R$ and $\mathbf{1}_C$ are R- and C-component vectors of ones. This allows (4.31) to be rewritten as

$$g(\mathbf{U}^*, \mathbf{V}^*) = \text{tr}(\mathbf{U}^{*'}\mathbf{D}_R\mathbf{U}^*) - 2\text{tr}(\mathbf{U}^{*'}\mathbf{F}\mathbf{V}^*) + \text{tr}(\mathbf{V}^{*'}\mathbf{D}_C\mathbf{V}^*). \tag{4.33}$$

We minimize this criterion $\mathbf{U}^*$ and $\mathbf{V}^*$ under the normalization restriction that $\mathbf{U}^{*'}\mathbf{D}_R\mathbf{U}^* = \mathbf{I}$. By minimizing (4.33) with respect to $\mathbf{V}^*$ for fixed $\mathbf{U}^*$, we obtain

$$\hat{\mathbf{V}}^* = \mathbf{D}_C^{-1}\mathbf{F}'\mathbf{U}^*. \tag{4.34}$$

By substituting this into (4.33), we find

$$\begin{aligned} g^*(\mathbf{U}^*) &= g(\mathbf{U}^*, \hat{\mathbf{V}}^*) \\ &= \min_{V^*|U^*} g(\mathbf{U}^*, \mathbf{V}^*) = \text{tr}(\mathbf{U}^{*'}\mathbf{D}_R\mathbf{U}^*) - \text{tr}(\mathbf{U}^{*'}\mathbf{F}\mathbf{D}_C^{-1}\mathbf{F}'\mathbf{U}^*). \end{aligned} \tag{4.35}$$

Observe that the first term in (4.35) is a constant, and so the whole criterion can be minimized by maximizing the second term, which in turn can be obtained by the generalized eigenequation of $\mathbf{D}_R^{-1}\mathbf{F}\mathbf{D}_C^{-1}\mathbf{F}'\mathbf{D}_R^{-1}$ with respect to $\mathbf{D}_R$, or equivalently, by $\text{GSVD}(\mathbf{D}_R^{-1}\mathbf{F}\mathbf{D}_C^{-1})_{D_R,D_C}$.

The generalized eigenequation or GSVD given above produces a constant eigenvector or singular vector (vectors with constant elements) corresponding to a unit eigenvalue (or a unit singular value). This is called the trivial solution, and is not of empirical interest. We may *a priori* eliminate the trivial solution by

$$\mathbf{F}^* = \mathbf{F} - \mathbf{D}_R\mathbf{1}_R\mathbf{1}_C'\mathbf{D}_C/n, \tag{4.36}$$

where n is the total sample size. (This is equal to Eq. (4.1).) We obtain GSVD of $\mathbf{F}^*$ instead of $\mathbf{F}$. This leads to $\text{GSVD}(\mathbf{Z})_{D_R,D_C}$, where

$$\mathbf{Z} = \mathbf{D}_R^{-1}\mathbf{F}^*\mathbf{D}_C^{-1} = \mathbf{D}_R^{-1}\mathbf{F}\mathbf{D}_C^{-1} - \mathbf{1}_R\mathbf{1}_C'/n. \tag{4.37}$$

(This is equivalent to $\mathbf{Z}$ in Eq. (1.5).) Let $\mathbf{Q}_{1_R/D_R} = \mathbf{I}_R - \mathbf{1}_R\mathbf{1}_R'\mathbf{D}_R/n$ and $\mathbf{Q}_{1_C/D_C} = \mathbf{I}_C - \mathbf{1}_C\mathbf{1}_C'\mathbf{D}_C/n$. Then, (4.36) can be rewritten as

$$\mathbf{F}^* = \mathbf{Q}_{1_R/D_R}'\mathbf{F}\mathbf{Q}_{1_C/D_C} = \mathbf{Q}_{1_R/D_R}'\mathbf{F} = \mathbf{F}\mathbf{Q}_{1_C/D_C}. \tag{4.38}$$

Pearson's (1900) chi-square statistic for testing the independence between rows and columns of a contingency table can be expressed as

$$\begin{aligned} \chi^2_{(R-1)(C-1)} &= n\text{SS}(\mathbf{D}_R^{-1}\mathbf{F}^*\mathbf{D}_C^{-1})_{D_R,D_C} \\ &= n\text{SS}(\mathbf{D}_R^{-1}\mathbf{Q}_{1_R/D_R}'\mathbf{F}\mathbf{D}_C^{-1})_{D_R,D_C} \\ &= n\text{SS}(\mathbf{Q}_{1_R/D_R}\mathbf{D}_R^{-1}\mathbf{F}\mathbf{D}_C^{-1})_{D_R,D_C}. \end{aligned} \tag{4.39}$$

This statistic asymptotically follows the chi-square distribution with $(R-1)(C-1)$ df under the hypothesis that there is no association between the rows and columns of $\mathbf{F}$.

Note 4.2 CA has many different names. This is due to the fact that it has been independently and repeatedly (re)discovered in many different disciplines. For example, it has been called the method of reciprocal averages (Horst 1935), Fisher's (1940) additive scoring, the second kind of Hayashi's (1952) quantification method, optimal scaling (Bock 1960), correspondence analysis (Benzécri 1973; Greenacre 1984), dual scaling (Nishisato 1980), etc. Its origin is generally attributed to Hirschfeld (1935), Fisher (1940), Maung (1941a, b), and Guttman (1941). According to de Leeuw (1983), however, the original idea can be traced as far back as Pearson (1904, 1906).

4.7 Constrained CA

Rows and columns of a contingency table are often accompanied by auxiliary information. Such information may be incorporated in CA as linear constraints. This is similar to other techniques (e.g., CANO) we have already discussed. However, there are two special circumstances in CA that "complicate" the matter. One concerns the *a priori* removal of the trivial solution discussed in the previous section. The other concerns the metric matrices $\mathbf{D}_R$ and $\mathbf{D}_C$, which are usually not constant diagonal. (A diagonal matrix is constant diagonal if it can be expressed in the form of $a\mathbf{I}$ for some $a \neq 0$.) Although the latter causes no problem in unconstrained CA, it matters in constrained CA.

As noted earlier, there are two "classes" of methods to incorporate external information. One is called the reparameterization method, and the other the null space method (Section 2.2.10). Canonical correspondence analysis (CCA) by ter Braak (1986) uses the former, while canonical analysis with linear constraints (CALC) by Böckenholt and Böckenholt (1990; Böckenholt and Takane 1994) uses the latter. The two methods are closely related, and both are subsumed as special cases of CPCA.

Let $\mathbf{U}^*$ and $\mathbf{V}^*$ represent matrices of row and column scores in CA. In CCA these scores are constrained by

$$\mathbf{U}^* = \mathbf{X}\tilde{\mathbf{U}}, \tag{4.40}$$

and

$$\mathbf{V}^* = \mathbf{Y}\tilde{\mathbf{V}} \tag{4.41}$$

for given row and column information matrices, $\mathbf{X}$ and $\mathbf{Y}$. These matrices may also be interpreted as predictor variables for rows and columns of a contingency table $\mathbf{F}$. Here, $\tilde{\mathbf{U}}$ and $\tilde{\mathbf{V}}$ are matrices of weights (regression coefficients) applied to $\mathbf{X}$ and $\mathbf{Y}$ to obtain $\mathbf{U}^*$ and $\mathbf{V}^*$. Obtaining $\mathbf{U}^*$ and $\mathbf{V}^*$ under these constraints amounts to $\text{GSVD}(\mathbf{Z})_{D_R,D_C}$, where

$$\mathbf{Z} = \mathbf{D}_R^{-1}\mathbf{P}'_{X/D_R}\mathbf{F}\mathbf{P}_{Y/\mathbf{D}_C}\mathbf{D}_C^{-1} = \mathbf{P}_{X/D_R}\mathbf{D}_R^{-1}\mathbf{F}\mathbf{D}_C^{-1}\mathbf{P}'_{Y/D_C}. \tag{4.42}$$

The CCA of $\mathbf{F}$ with the constraint matrices $\mathbf{X}$ and $\mathbf{Y}$ is denoted by $\text{CCA}(\mathbf{X},\mathbf{Y})$. ($\text{CCA}(\mathbf{X},\mathbf{I})$ may simply be denoted as $\text{CCA}(\mathbf{X})$.)

In CALC, on the other hand, the constraints are expressed as

$$\mathbf{R}'\mathbf{U}^* = \mathbf{O}, \tag{4.43}$$

and

$$\mathbf{C}'\mathbf{V}^* = \mathbf{O}, \tag{4.44}$$

where $\mathbf{R}$ and $\mathbf{C}$ are called row and column constraint matrices. Define

$$\mathbf{Q}_{R/D_R^{-1}} = \mathbf{I} - \mathbf{R}(\mathbf{R}'\mathbf{D}_R^{-1}\mathbf{R})^{-1}\mathbf{R}'\mathbf{D}_R^{-1}, \tag{4.45}$$

and

$$\mathbf{Q}_{C/D_C^{-1}} = \mathbf{I} - \mathbf{C}(\mathbf{C}'\mathbf{D}_C^{-1}\mathbf{C})^{-1}\mathbf{C}'\mathbf{D}_C^{-1}. \tag{4.46}$$

A solution for CALC is then found by $\text{GSVD}(\mathbf{Z})_{D_R,D_C}$, where

$$\mathbf{Z} = \mathbf{D}_R^{-1}\mathbf{Q}_{R/D_R^{-1}}\mathbf{F}\mathbf{Q}'_{C/D_C^{-1}}\mathbf{D}_C^{-1} = \mathbf{Q}'_{R/D_R^{-1}}\mathbf{D}_R^{-1}\mathbf{F}\mathbf{D}_C^{-1}\mathbf{Q}_{C/D_C^{-1}}. \tag{4.47}$$

In a manner analogous to CCA, CALC of $\mathbf{F}$ with the constraint matrices $\mathbf{R}$ and $\mathbf{C}$ is denoted as $\text{CALC}(\mathbf{R},\mathbf{C})$. ($\text{CALC}(\mathbf{R},\mathbf{O})$ may simply be denoted as $\text{CALC}(\mathbf{R})$.)

There are at least two interesting relationships between CCA and CALC (Takane et al. 1991): equivalence and complementarity. These relationships hold when $\mathbf{X}$ and $\mathbf{R}$, and $\mathbf{Y}$ and $\mathbf{C}$, assume special relationships with each other. Below we explain these relationships only for $\mathbf{X}$ and $\mathbf{R}$. (Those for $\mathbf{Y}$ and $\mathbf{C}$ are similar.) We first discuss equivalence, and then complementarity.

Let $\mathbf{X}$ be given such that

$$\mathbf{X}'\mathbf{D}_R\mathbf{1}_R = \mathbf{0}. \tag{4.48}$$

This condition eliminates the trivial solution from $\text{CCA}(\mathbf{X})$. Let $\mathbf{R}^*$ be such that $\text{Sp}(\mathbf{R}^*) = \text{Ker}([\mathbf{1}_R,\mathbf{X}])$. This implies that

$$\mathbf{X}'\mathbf{R}^* = \mathbf{O}, \tag{4.49}$$

and that

$$\mathbf{R}^{*'}\mathbf{1}_R = \mathbf{0}. \tag{4.50}$$

Define $\mathbf{R} = [\mathbf{D}_R\mathbf{1}_R, \mathbf{R}^*]$. The inclusion of $\mathbf{D}_R\mathbf{1}_R$ as part of $\mathbf{R}$ has the effect of eliminating the trivial solution from $\text{CALC}(\mathbf{R})$. By removing the subspace $(\text{Ker}(\mathbf{1}'_R\mathbf{D}_R))$, we obtain a slightly generalized version of Khatri's lemma (Section 2.2.8) stated as

$$\mathbf{Q}_{1_R/D_R} = \mathbf{P}_{[X,D_R^{-1}R^*]/D_R} = \mathbf{P}_{X/D_R} + \mathbf{P}_{D_R^{-1}R^*/D_R}. \tag{4.51}$$

Note that (4.49) implies that $\mathbf{X}$ and $\mathbf{D}_R^{-1}\mathbf{R}^*$ are $\mathbf{D}_R$-orthogonal (orthogonal with respect to $\mathbf{D}_R$). That is, the two projectors on the right-hand side of (4.51) are $\mathbf{D}_R$-orthogonal. Note also that (4.48) implies $\mathbf{Q}_{1_R/D_R}\mathbf{X} = \mathbf{X}$, and that (4.50) implies $\mathbf{Q}_{1_R/D_R}\mathbf{D}_R^{-1}\mathbf{R}^* = \mathbf{D}_R^{-1}\mathbf{R}^*$. That is, the subspaces spanned by the two projectors on the right-hand side of (4.51) are both subsumed under $\text{Sp}(\mathbf{Q}_{1_R/D_R})$. We also have

$$\mathbf{Q}_{1_R/D_R} = \mathbf{Q}'_{D_R 1_R/D_R^{-1}}, \tag{4.52}$$

and

$$\mathbf{P}_{D_R^{-1}R^*/D_R} = \mathbf{P}'_{R^*/D_R^{-1}}. \tag{4.53}$$

We can now rewrite the first term on the right-hand side of (4.51) as

$$\mathbf{P}_{X/D_R} = \mathbf{Q}'_{D_R 1_R/D_R^{-1}} - \mathbf{P}'_{R^*/D_R^{-1}} = \mathbf{Q}'_{[D_R 1_R, R^*]/D_R^{-1}}. \tag{4.54}$$

This indicates the equivalence between CCA($\mathbf{X}$) and CALC($\mathbf{R}$). Compare (4.42) and (4.47).

We can also rewrite the second term on the right-hand side of (4.51) as

$$\mathbf{P}_{D_R^{-1}R^*/D_R} = \mathbf{Q}_{1_R/D_R} - \mathbf{P}_{X/D_R} = \mathbf{Q}_{[1_R, X]/D_R} = \mathbf{Q}'_{D_R[1_R, X]/D_R^{-1}}. \tag{4.55}$$

This indicates the equivalence between CALC($\mathbf{D}_R[\mathbf{1}_R, \mathbf{X}]$) and CCA($\mathbf{D}_R^{-1}\mathbf{R}^*$). That is, when $\mathbf{R} = \mathbf{D}_R[\mathbf{1}_R, \mathbf{X}]$, CALC($\mathbf{R}$) is equivalent to CCA($\mathbf{D}_R^{-1}\mathbf{R}^*$). Compare (4.42) and (4.47).

We may write

$$\mathbf{Q}_{1_R/D_R} = \mathbf{P}_{X/D_R} + \mathbf{Q}'_{D_R[1_R, X]/D_R^{-1}} \tag{4.56}$$

$$= \mathbf{Q}'_{[D_R 1_R, R^*]/D_R^{-1}} + \mathbf{P}_{D_R^{-1}R^*/D_R}. \tag{4.57}$$

Note that (4.54) and (4.55) indicate that the terms of (4.56) and (4.57) are equal. These equations show the complementarity of CCA($\mathbf{X}$) $=$ CALC($[\mathbf{D}_R\mathbf{1}_R, \mathbf{R}^*]$) and CALC($\mathbf{D}_R[\mathbf{1}_R, \mathbf{X}]$) $=$ CCA($\mathbf{D}_R^{-1}\mathbf{R}^*$). Note that the role of $\mathbf{X}$ and that of $\mathbf{R}^*$ are not symmetric due to $\mathbf{D}_R$, which is usually not constant diagonal. As will be explained later, this introduces some complication when we try to define $\mathbf{X}$ and $\mathbf{R}^*$ from more fundamental quantities that play symmetric roles.

It should be clear by now how CCA and CALC are related to CPCA. Let $\mathbf{Z}^* = \mathbf{K}^{-1}\mathbf{F}\mathbf{L}^{-1}$, and $\mathbf{K} = \mathbf{D}_R$ and $\mathbf{L} = \mathbf{D}_C$ in CPCA. To eliminate the trivial solution, we analyze $\mathbf{Z} = \mathbf{Q}_{1_R/D_R}\mathbf{Z}^*$ instead of $\mathbf{Z}^*$. By decomposing $\mathbf{Z}$ according to the external information $\mathbf{G}$ in CPCA, we find

$$\begin{aligned}\mathbf{Z} &= \mathbf{P}_{G/D_R}\mathbf{Q}_{1_R/D_R}\mathbf{Z}^* + \mathbf{Q}_{G/D_R}\mathbf{Q}_{1_R/D_R}\mathbf{Z}^* \\ &= \mathbf{P}_{G/D_R}\mathbf{Z}^* + (\mathbf{Q}_{1_R/D_R} - \mathbf{P}_{G/D_R})\mathbf{Z}^* \\ &= \mathbf{P}_{G/D_R}\mathbf{Z}^* + \mathbf{Q}'_{D_R[1_R, G]/D_R^{-1}}\mathbf{Z}^*.\end{aligned} \tag{4.58}$$

GSVD of the first term in (4.58) with the metric matrices $\mathbf{D}_R$ and $\mathbf{D}_C$ is equivalent to CCA($\mathbf{X}$) if we set $\mathbf{G} = \mathbf{X}$. (The matrix $\mathbf{G}$ must satisfy the condition required of $\mathbf{X}$ in (4.48).) GSVD of the second term is equivalent to CALC($\mathbf{D}_R[\mathbf{1}_R, \mathbf{G}]$) which, as shown above, analyzes the complementary part to the first term.

Corresponding to the decompositions of the data matrix, Pearson's chi-square statistic defined in (4.39), representing the total association between rows and columns of a contingency table, can also be decomposed into the sum of part chi-squares, each representing a particular kind of association between them. Let

$$\mathbf{A}^* = \mathbf{Q}_{1_R/D_R}\mathbf{D}_R^{-1}\mathbf{F}\mathbf{D}_C^{-1} \tag{4.59}$$

(this is equivalent to **Z** in (4.37),

$$\mathbf{B}^* = \mathbf{P}_{X/D_R}\mathbf{D}_R^{-1}\mathbf{F}\mathbf{D}_C^{-1}, \tag{4.60}$$

and

$$\mathbf{E}^* = \mathbf{P}_{D_R^{-1}R^*/D_R}\mathbf{D}_R^{-1}\mathbf{F}\mathbf{D}_C^{-1}, \tag{4.61}$$

and let $r = \text{rank}(\mathbf{X})$ and $\text{rank}(\mathbf{R}^*) = R - r - 1$. Define

$$\chi^2_{(R-1)(C-1)} = n\text{SS}(\mathbf{A}^*)_{D_R,D_C} \tag{4.62}$$

(this is equivalent to (4.39)),

$$\chi^2_{r(C-1)} = n\text{SS}(\mathbf{B}^*)_{D_R,D_C}, \tag{4.63}$$

and

$$\chi^2_{(R-r-1)(C-1)} = n\text{SS}(\mathbf{E}^*)_{D_R,D_C}. \tag{4.64}$$

Then

$$\chi^2_{(R-1)(C-1)} = \chi^2_{r(C-1)} + \chi^2_{(R-r-1)(C-1)}, \tag{4.65}$$

where $\chi^2_{r(C-1)}$ asymptotically follows the chi-square distribution with $r(C-1)$ df under the hypothesis that the portion of variabilities among the rows of a contingency table predictable by **X** has no association with the columns, while $\chi^2_{(R-r-1)(C-1)}$ asymptotically follows the chi-square distribution with $(R-r-1)(C-1)$ df under the hypothesis that the portion of variabilities among the rows left unaccounted for by **X** has no association with the columns.

Here is an example of analysis by CCA and CALC, through which we demonstrate how **X** and $\mathbf{R}^*$ may be constructed in practice. The data reported in Table 4.1 have been used repeatedly by many researchers (Böckenholt and Böckenholt 1990; Haberman 1979; Takane et al. 1991). The table cross-tabulates 3,181 respondents by religion, education, and attitude toward abortion. There are three categories in religion: northern protestant (np), southern protestant (sp), and catholic (ct), three categories in years of education (1 = no more than 8 years, 2 = between 9 and 12 years, and 3 = at least 13 years), and three categories in attitude: positive (ps), neutral (nt), and negative (ng). The religion and years of education were interactively coded to form nine categories labeled as np1, np2, np3, sp1, sp2, sp3, ct1, ct2, and ct3 representing the rows of the table.

Unconstrained CA was first applied to this data set. The results are reported in Column (1) of Table 4.2. The total association between rows and columns is found to be 257.5 in chi-square with 16 df, indicating a strong association between them. The unidimensional solution from CA accounts for 94.7% of the total chi-square, indicating that a majority of the total association is captured by this component. The column scores are given in standard coordinates, while the row scores in principal coordinates (i.e., the right singular vector is multiplied by the corresponding singular value).

The unconstrained CA has revealed that the row scores are fairly linear over the

Table 4.1 *Attitudes toward nonclinical abortion: Classification by respondent's religion and education**.

Religion	Label	Education (Years)	Attitude Positive	Neutral	Negative
Northern	np1	≤ 8	49	46	115
Protestant	np2	9-12	293	140	277
	np3	≥ 13	244	66	100
Southern	sp1	≤ 8	27	34	117
Protestant	sp2	9-12	134	98	167
	sp3	≥ 13	138	38	73
Catholic	ct1	≤ 8	25	40	88
	ct2	9-12	172	193	312
	ct3	≥ 13	93	57	135

*Reproduced from Table 6.14 of Haberman (1979) with permission from Elsevier.

Table 4.2 *Results of analysis of the data in Table 4.1.*

		(1) CA	(2) Only **X**	(3) Only **Y**	(4) Both	(5) #Only **X**
Chi-square		257.5	244.1	.235.8	234.2	232.9
df		(16)	(8)	(8)	(4)	(8)
% Chi-square		100.0%	94.8%	91.6%	90.9%	90.0%
Column	ps	.352	$*$.350	.324	.323	.343
scores	nt	−.066	−.062	.020	.020	−.086
	ng	−.260	−.261	−.284	−.283	−.253
	np1	−1.033	−1.121	−1.021	−1.107	−.823
	np2	.349	.338	.362	.344	.303
	np3	1.724	1.797	1.712	1.794	1.472
Row	sp1	−1.749	−1.620	−1.806	−1.594	−2.195
scores	sp2	−.154	−.161	−.070	−.143	−.007
	sp3	1.378	1.299	1.342	1.307	1.548
	ca1	−1.495	−1.299	−1.421	−1.295	−1.748
	ca2	−.655	−.764	−.710	−.780	−.362
	ca3	−.334	−.228	−.335	−.266	−.554
%SS accounted		94.7%	94.0%	91.6%	90.9%	89.9%

*Column scores are in standard coordinates, and row scores in principal coordinates.
#Eliminating the effect of $\mathbf{T}_2$ (to be explained later in the text).

levels of education within each of the three religious groups. Furthermore, this linear trend is similar for the first two religious groups (np and sp), although it is rather different for the third group (ct). There also are quite large differences in attitude toward abortion among the three religious groups. The column scores are also fairly linear. These features of the data are captured by the following contrast matrices:

$$\mathbf{T}_1 = \begin{bmatrix} -1 & 0 & 1 & -1 & 0 & 1 & 0 & 0 & 0 \\ 0 & 0 & 0 & 0 & 0 & 0 & -1 & 0 & 1 \\ 1 & 1 & 1 & -1 & -1 & -1 & 0 & 0 & 0 \\ 1 & 1 & 1 & 1 & 1 & 1 & -2 & -2 & -2 \end{bmatrix}', \tag{4.66}$$

and

$$\mathbf{S}_1 = [-1, 0, 1]'. \tag{4.67}$$

The first column of $\mathbf{T}_1$ specifies an identical linear trend over education for np and sp, while the second column a linear trend for ct that may be distinct from the first two religious groups. The last two columns of $\mathbf{T}_1$ jointly specify the main effects of the religious groups. We then define

$$\mathbf{X} = \mathbf{Q}_{1_R/D_R}\mathbf{T}_1, \tag{4.68}$$

and

$$\mathbf{Y} = \mathbf{Q}_{1_C/D_C}\mathbf{S}_1. \tag{4.69}$$

The premultiplication of $\mathbf{T}_1$ and $\mathbf{S}_1$ by $\mathbf{Q}_{1_R/D_R}$ and $\mathbf{Q}_{1_C/D_C}$ eliminates the trivial solution.

Three analyses were conducted by CCA using $\mathbf{X}$ and $\mathbf{Y}$ defined above. Column (2) of Table 4.2 reports the results obtained when only the row side constraint $\mathbf{X}$ is incorporated. Column (3) shows the results obtained when only the column side constraint $\mathbf{Y}$ is incorporated. Column (4) gives the results when both $\mathbf{X}$ and $\mathbf{Y}$ are incorporated. The row-constrained CCA captures 94.8% of the total chi-square, and the unidimensional solution captures almost all (94.7%) of the chi-square captured by the row-constrained CCA. The column-constrained CCA, on the other hand, captures 91.6% of the total chi-square. Only the unidimensional solution can be derived in this case since $\mathbf{Y}$ has only one column. When both row and column constraints are imposed, 90.9% of the total chi-square can still be explained. For the same reason as above, only the unidimensional solution can be obtained. It may be concluded that the constraints defined above indeed capture most of the prevailing tendencies in the data set.

What matrices of $\mathbf{R}$ and $\mathbf{C}$ in CALC would give equivalent results to CCA above? They are usually not unique. The ones given below are mere instances of infinitely many possibilities. However, columns of these matrices are chosen in such a way that they have specific empirical meanings. Define

$$\mathbf{R}^* \equiv \mathbf{T}_2 = \begin{bmatrix} 1 & -2 & 1 & 0 & 0 & 0 & 0 & 0 & 0 \\ 0 & 0 & 0 & 1 & -2 & 1 & 0 & 0 & 0 \\ 0 & 0 & 0 & 0 & 0 & 0 & 1 & -2 & 1 \\ 1 & -1 & 0 & -1 & 1 & 0 & 0 & 0 & 0 \end{bmatrix}', \tag{4.70}$$

and

$$\mathbf{C}^* \equiv \mathbf{S}_2 = [1, -2, 1]'. \tag{4.71}$$

We then define

$$\mathbf{R} = [\mathbf{D}_R\mathbf{1}_R, \mathbf{R}^*], \tag{4.72}$$

and

$$\mathbf{C} = [\mathbf{D}_C\mathbf{1}_C, \mathbf{C}^*] \tag{4.73}$$

in CALC$(\mathbf{R},\mathbf{C})$. As stated earlier, the inclusion of $\mathbf{D}_R\mathbf{1}_R$ and $\mathbf{D}_C\mathbf{1}_C$ in $\mathbf{R}$ and $\mathbf{C}$ eliminate the trivial solution. Note that $\mathbf{T}_1'\mathbf{T}_2 = \mathbf{O}$ and $\mathbf{S}_1'\mathbf{S}_2 = 0$. Each of the first three columns of $\mathbf{R}^*$ specifies a separate quadratic trend over education within each of the three religious groups. The fourth column specifies the interaction between the first two religious groups and the first two levels of education. Requiring $\mathbf{R}^{*'}\mathbf{U}^* = \mathbf{O}$ implies that there are no such effects as those represented by $\mathbf{R}^*$. That is, only the effects represented by $\mathbf{X}$ may exist. Similarly, the single column in $\mathbf{C}^*$ specifies a quadratic trend over the three categories of attitude, and $\mathbf{C}^{*'}\mathbf{V}^* = \mathbf{O}$ implies that there is no such trend; only a linear trend may exist.

We now address how $\mathbf{X}$ and $\mathbf{R}^*$ that yield complementary results may be constructed. We discuss this only for row-constraints. Column constraints work similarly. We also discuss this only for CCA, since all CALC problems can be turned into equivalent CCA problems. Recall that CCA$(\mathbf{X})$ and CCA$(\mathbf{D}_R^{-1}\mathbf{R}^*)$ provide complementary results if $\mathbf{X}$ and $\mathbf{R}^*$ satisfy (4.49), (4.48), and (4.50). Suppose that $\mathbf{X}$ is as given by (4.68). Then $\mathbf{R}^*$ given in (4.70) satisfies all of these conditions. This means that CCA$(\mathbf{Q}_{1_R/D_R}\mathbf{T}_1)$ and CCA$(\mathbf{D}_R^{-1}\mathbf{T}_2)$ give complementary results. Note, however, that both $\mathbf{T}_1$ and $\mathbf{T}_2$ are matrices of contrast vectors (both orthogonal to $\mathbf{1}_R$), and so they algebraically play symmetric roles. What happens if we interchange $\mathbf{T}_1$ and $\mathbf{T}_2$ in CCA above? We obtain CCA$(\mathbf{Q}_{1_R/D_R}\mathbf{T}_2)$ and CCA$(\mathbf{D}_R^{-1}\mathbf{T}_1)$. It can be easily verified that these two CCAs also give complementary results. But then what is the relationship between CCA$(\mathbf{Q}_{1_R/D_R}\mathbf{T}_1)$ and CCA$(\mathbf{D}_R^{-1}\mathbf{T}_1)$, and that between CCA$(\mathbf{Q}_{1_R/D_R}\mathbf{T}_2)$ and CCA$(\mathbf{D}_R^{-1}\mathbf{T}_2)$? To answer this question, it is essential to understand partial CCA (PCCA) proposed by ter Braak (1988).

In PCCA, the effects of nuisance variables (covariates) are eliminated from $\mathbf{X}$. Let $\mathbf{M}$ denote the covariates. We partial out the effect of $\mathbf{M}$ from $\mathbf{X}$ by

$$\mathbf{X}^* = (\mathbf{I} - \mathbf{M}(\mathbf{M}'\mathbf{D}_R\mathbf{M})^{-1}\mathbf{M}'\mathbf{D}_R)\mathbf{X} = \mathbf{Q}_{M/D_R}\mathbf{X}, \tag{4.74}$$

and then use $\mathbf{X}^*$ in CCA. This leads to:

$$\text{GSVD}(\mathbf{P}_{X^*/D_R}\mathbf{D}_R^{-1}\mathbf{F}\mathbf{D}_C^{-1})_{D_R,D_C}. \tag{4.75}$$

Note 4.3 PCCA is different from partial CA (Yanai 1988), which analyzes

$$\tilde{\mathbf{F}} = \mathbf{G}'\mathbf{Q}_N\mathbf{H}, \tag{4.76}$$

obtained by partialling out the effect of $\mathbf{N}$ from the dummy-coded categorical variables $\mathbf{G}$ and $\mathbf{H}$. This is thus a special case of partial CANO. Note that this analysis requires case (subject) level covariates.

Let $\mathbf{X} = \mathbf{T}_1$ and $\mathbf{M} = [\mathbf{1}_R, \mathbf{T}_2]$ in PCCA. Note first that

$$\begin{aligned}\mathbf{Q}'_{M/D_R}\mathbf{D}_R &= \mathbf{T}_1(\mathbf{T}'_1\mathbf{D}_R^{-1}\mathbf{T}_1)^{-1}\mathbf{T}'_1\\ &= \mathbf{D}_R\mathbf{Q}_{M/D_R} = \mathbf{Q}'_{M/D_R}\mathbf{D}_R\mathbf{Q}_{M/D_R}.\end{aligned} \tag{4.77}$$

The first equality in (4.77) holds due to Khatri's lemma (Section 2.2.8). We then have

$$\begin{aligned}\mathbf{P}_{X^*/D_R} &= \mathbf{P}_{Q_{M/D_R}X/D_R}\\ &= \mathbf{D}_R^{-1}\mathbf{D}_R\mathbf{Q}_{M/D_R}\mathbf{X}(\mathbf{X}'\mathbf{Q}'_{M/D_R}\mathbf{D}_R\mathbf{X})^{-1}\mathbf{X}'\mathbf{Q}'_{M/D_R}\mathbf{D}_R\\ &= \mathbf{D}_R^{-1}\mathbf{T}_1(\mathbf{T}'_1\mathbf{D}_R^{-1}\mathbf{T}_1)^{-1}\mathbf{T}'_1\mathbf{T}_1(\mathbf{T}'_1\mathbf{T}_1(\mathbf{T}'_1\mathbf{D}_R^{-1}\mathbf{T}_1)^{-1}\mathbf{T}'_1\mathbf{T}_1)^{-1}\\ &\qquad \times\mathbf{T}'_1\mathbf{T}_1(\mathbf{T}'_1\mathbf{D}_R^{-1}\mathbf{T}_1)^{-1}\mathbf{T}'_1\\ &= \mathbf{D}_R^{-1}\mathbf{T}_1(\mathbf{T}'_1\mathbf{D}_R^{-1}\mathbf{T}_1)^{-1}\mathbf{T}'_1 = \mathbf{P}_{D_R^{-1}T_1/D_R}.\end{aligned} \tag{4.78}$$

That is, $\mathbf{D}_R^{-1}\mathbf{T}_1$ represents the effect of $\mathbf{T}_1$ eliminating the effect of $\mathbf{T}_2$. (To be more exact, it is $\mathbf{T}_1$ eliminating $[\mathbf{1}_R, \mathbf{T}_2]$, but $\mathbf{1}_R$ is only for eliminating the trivial solution.) This means that CCA$(\mathbf{D}_R^{-1}\mathbf{T}_1)$ analyzes the part of the association in a contingency table that can be accounted for by the effect of $\mathbf{T}_1$ after eliminating the effect of $\mathbf{T}_2$. CCA$(\mathbf{Q}_{1_R/D_R}\mathbf{T}_1)$, on the other hand, analyzes the effect of $\mathbf{T}_1$ ignoring $\mathbf{T}_2$. The effect of $\mathbf{T}_1$ eliminating $\mathbf{T}_2$ pertains to the unique effect of $\mathbf{T}_1$, while the effect of $\mathbf{T}_1$ ignoring $\mathbf{T}_2$ pertains to the total effect of $\mathbf{T}_1$, disregarding the possibility that some portion of it may also be explained by $\mathbf{T}_2$. The effect of $\mathbf{T}_2$ eliminating $\mathbf{T}_1$ $(\mathbf{D}_R^{-1}\mathbf{T}_2)$ and the effect of $\mathbf{T}_2$ ignoring $\mathbf{T}_1$ $(\mathbf{Q}_{1_R/D_R}\mathbf{T}_2)$ are similarly interpreted. Column (5) of Table 4.2 gives the results of CCA$(\mathbf{D}_R^{-1}\mathbf{T}_1)$. They are slightly different from those obtained by CCA$(\mathbf{Q}_{1_R/D_R}\mathbf{T}_1)$ given in Column (2). (In this example, the distinction between them is not of much empirical interest. See the next section for a more illuminating example of the distinction.)

Van der Heijden and his collaborators (van der Heijden and de Leeuw 1985; van der Heijden and Meijerink 1989; van der Heijden et al. 1989) proposed CA of residuals from certain log-linear models. Their method imposes constraints of the form $(\mathbf{D}_R[\mathbf{1}_R, \mathbf{T}_j])'\mathbf{U}^* = \mathbf{O}$ $(j = 1, 2)$, called the zero average restrictions on $\mathbf{U}^*$. This leads to a special case of CALC, CALC$(\mathbf{D}_R[\mathbf{1}_R, \mathbf{T}_i])$ $(i = 1, 2 \neq j)$, which is equivalent to CCA$(\mathbf{D}_R^{-1}\mathbf{T}_j)$. That is, the zero average restrictions pertain to the effects of eliminating. This is in contrast with the conventional CALC, i.e., CALC$([\mathbf{D}_R\mathbf{1}_R, \mathbf{T}_j])$, which is equivalent to CCA$(\mathbf{Q}_{1_R/D_R}\mathbf{T}_i)$ $(i \neq j)$ pertaining to the effect of ignoring (Takane and Jung 2009b).

4.8 Nonsymmetric CA (NSCA)

Just as CANO (Section 4.3) specializes in (symmetric) CA (SCA; Section 4.6) when both sets of variables consist of dummy-coded indicator variables, RA (Section 4.2) reduces to nonsymmetric CA (NSCA; Lauro and D'Ambra 1984) in such situations. In NSCA, one set of the categorical variables is taken as predictor variables, and the other as criterion variables. We assume that rows of a contingency table represent predictive categories, and columns represent criterion categories. The predictability of the rows on the columns can be expressed as

$$\mathbf{A} = \mathbf{Q}_{1_R/D_R}\mathbf{D}_R^{-1}\mathbf{F} = \mathbf{D}_R^{-1}\mathbf{F} - \mathbf{1}_R\mathbf{1}_C'\mathbf{D}_C/n. \tag{4.79}$$

Recall that $\mathbf{F}$ indicates an R by C contingency table, $\mathbf{D}_R$ and $\mathbf{D}_C$ denote the diagonal matrices of row and column totals of $\mathbf{F}$, $\mathbf{1}_R$ and $\mathbf{1}_C$ are R- and C-component vectors of ones, and $n = \mathbf{1}_R'\mathbf{F}\mathbf{1}_C$ is the total sample size. The first term on the right-hand side of (4.79) indicates conditional probabilities of columns given rows, i.e., $p_{ij}/p_{i.}$, where p_{ij} is the joint probability of row i and column j, and $p_{i.}$ is the marginal probability of row i. The second term, on the other hand, indicates the marginal probabilities of columns, i.e., $p_{.j}$, spread across all rows. The difference between $p_{ij}/p_{i.}$ and $p_{.j}$ indicates how well row i predicts column j. The larger this value, the higher the predictability. Notice that the first term in (4.79) is equal to $\hat{\mathbf{B}} = (\mathbf{G}'\mathbf{G})^{-1}\mathbf{G}'\mathbf{H}$ in RA, while the second term is similar to the adjustment needed to eliminate the trivial solution in SCA. We apply GSVD$(\mathbf{A})_{D_R,I}$ to find a low dimensional representation of the rows and columns in NSCA.

The matrix $\mathbf{A}$ in (4.79) may be contrasted with the analogous matrix in SCA defined as

$$n\mathbf{A}^* = n\mathbf{Q}_{1_R/D_R}\mathbf{D}_R^{-1}\mathbf{F}\mathbf{D}_C^{-1} = n\mathbf{D}_R^{-1}\mathbf{F}\mathbf{D}_C^{-1} - \mathbf{1}_R\mathbf{1}_C. \tag{4.80}$$

Notice that this matrix is n times $\mathbf{A}^*$ defined in (4.59) (and n times $\mathbf{Z}$ defined in (4.37)). The ijth element of $n\mathbf{A}^*$ is equal to $p_{ij}/p_{i.}p_{.j} - 1$, indicating the strength of association between row i and column j. It is zero if i and j are independent ($p_{ij} = p_{i.}p_{.j}$). If it is positive, i and j have a positive association, and if it is negative, they have a negative association. The difference between (4.79) and (4.80) is rather trivial. If we postmultiply (4.79) by $(\mathbf{D}_C/n)^{-1}$ (the diagonal matrix of reciprocals of marginal probabilities of columns), we obtain (4.80).

NSCA has many features similar to those of SCA. In SCA, linear constraints may be incorporated if such information is available. This is called constrained CA (the previous section). Something similar can also be done in NSCA. In NSCA, however, the constraints can be incorporated only in row (predictive) categories, while in SCA, it can be done for both rows and columns. In SCA, Pearson's chi-square statistic (measuring the total association between rows and columns of a contingency table) is decomposed according to the constraints imposed. In NSCA, the CATANOVA C statistic (Light and Margolin 1971) measuring the total predictability of rows on columns is analogously decomposed.

Let

$$\text{TSS} = 1 - \text{tr}(\mathbf{D}_C^2)/n^2 \tag{4.81}$$

denote the total between-column variability in $\mathbf{F}$, and let

$$\text{BSS}_A = \frac{1}{n}\text{SS}(\mathbf{A})_{D_R,I} = \text{tr}(\mathbf{A}'\mathbf{D}_R\mathbf{A})/n \tag{4.82}$$

denote the portion of TSS that can be predicted by the rows of $\mathbf{F}$. Then the Goodman-Kruskal (1954) τ index is defined by

$$\tau_A = \text{BSS}_A/\text{TSS}. \tag{4.83}$$

This index measures the overall predictability of rows on columns. The significance of the τ_A can be tested by the CATANOVA C statistic (Light and Margolin 1971) defined by

$$C_A = (n-1)(C-1)\tau_A. \tag{4.84}$$

This statistic (similarly to Pearson's chi-square statistic) asymptotically follows the chi-square distribution with $(R-1)(C-1)$ df under the hypothesis that the rows have no predictive power on the columns.

Nonsymmetric versions of constrained CA (CCA and CALC) can be readily derived. Let $\mathbf{T}_1$ and $\mathbf{T}_2$ represent matrices of constraints to be imposed on the rows of $\mathbf{F}$ analogous to (4.66) and (4.70). Recall that these matrices consist of contrasts among the rows that satisfy $\mathbf{T}_1'\mathbf{T}_2 = \mathbf{O}$ and $\text{Sp}(\mathbf{T}_1, \mathbf{T}_2]) = \text{Ker}(\mathbf{1}_R')$. Let $\mathbf{X} = \mathbf{Q}_{1_R/D_R}\mathbf{T}_1$, and define

$$\mathbf{B} = \mathbf{P}_{X/D_R}\mathbf{D}_R^{-1}\mathbf{F}. \tag{4.85}$$

This matrix is analogous to $\mathbf{B}^*$ defined in (4.60). We also define BSS_B, τ_B, and C_B analogously to (4.82), (4.83), and (4.84). That is,

$$\text{BSS}_B = \frac{1}{n}\text{SS}(\mathbf{B})_{D_R,I} = \text{tr}(\mathbf{B}'\mathbf{D}_R\mathbf{B})/n, \tag{4.86}$$

$$\tau_B = \text{BSS}_B/\text{TSS}, \tag{4.87}$$

and

$$C_B = (n-1)(C-1)\tau_B. \tag{4.88}$$

The BSS_B indicates the portion of BSS_A that can be accounted for by $\mathbf{T}_1$ ignoring $\mathbf{T}_2$, τ_B the portion of τ_A that can be explained by $\mathbf{T}_1$. The significance of τ_B can be tested using C_B, which has the asymptotic chi-square distribution with $r_B(C-1)$ df, where $r_B = \text{rank}(\mathbf{T}_1)$ under the hypothesis that $\mathbf{T}_1$ has no predictability on columns.

Let $\mathbf{R}^* = \mathbf{T}_2$, and define

$$\mathbf{E} = \mathbf{P}_{D_R^{-1}R^*/D_R}\mathbf{D}_R^{-1}\mathbf{F}. \tag{4.89}$$

This matrix is analogous to $\mathbf{E}^*$ defined in (4.61). Define also BSS_E, τ_E, and C_E analogously to the above. That is,

$$\text{BSS}_E = \frac{1}{n}\text{SS}(\mathbf{E})_{D_R,I} = \text{tr}(\mathbf{E}'\mathbf{D}_R\mathbf{E})/n, \tag{4.90}$$

$$\tau_E = \text{BSS}_E/\text{TSS}, \tag{4.91}$$

and

$$C_E = (n-1)(C-1)\tau_E. \tag{4.92}$$

The τ_E indicates the portion of τ_A that can be explained by $\mathbf{T}_2$ eliminating $\mathbf{T}_1$, and its significance can be tested using C_E, which asymptotically follows the chi-square distribution with $r_E(C-1)$ df, where $r_E = \text{rank}(\mathbf{T}_2)$, under the hypothesis that $\mathbf{T}_2$ has no predictability on columns. We have $\tau_A = \tau_B + \tau_C$, and $C_A = C_B + C_E$. That is, the overall predictability of rows on columns is decomposed into two parts, the predictability of $\mathbf{T}_1$ ignoring $\mathbf{T}_2$ and that of $\mathbf{T}_2$ eliminating $\mathbf{T}_1$ (Takane and Jung 2009b). By interchanging the role of $\mathbf{T}_1$ and $\mathbf{T}_2$ above, we obtain the effect of $\mathbf{T}_2$ ignoring $\mathbf{T}_1$ and that of $\mathbf{T}_1$ eliminating $\mathbf{T}_2$. See the previous section for the distinction between the effects of ignoring and eliminating.

Reduced rank approximations of $\mathbf{A}$ in (4.79), $\mathbf{B}$ in (4.85), and $\mathbf{E}$ in (4.89) are possible. They are all obtained by GSVD of these matrices with the row metric matrix $\mathbf{D}_R$ (and the identity column metric matrix).

There is an interesting phenomenon called Simpson's paradox in the analysis of contingency tables. This paradox stipulates possible reversals of the relationship between two variables depending on whether the effect of a third variable is taken into account (eliminated) or not (ignored). As an example, look at Table 4.3, which cross-tabulates 800 patients in terms of treatment (medication or control) and prognosis (recovered or nonrecovered). The table indicates the recovery rate is slightly better for the medication group. However, Table 4.4 shows a different picture. This table breaks down the previous table by gender (male or female). Observe that for both males and females the control group has a better recovery rate than the medication group. This somewhat paradoxical phenomenon stems from the fact that disproportionately many males, whose recovery rate is generally higher than females, are assigned to the medication group. Takane and Jung (2009b) used the tests of ignoring and eliminating to characterize this paradox. The former corresponds with the test of the treatment effect on prognosis in Table 4.3 (ignoring gender), while the latter with the treatment effect eliminating gender in Table 4.4. The value of the C statistic for the former is $C = .50$ with 1 df (nonsignificant), while for the latter $C = 13.49$ with 1 df (significant). This is as expected above.

Table 4.3 *Cross-classification of 800 patients in terms of treatment (medication or control) and prognosis (recovery or nonrecovery)**.

	Prognosis		
Treatment	Recovery	Nonrecovery	Recovery rate
Medication	200	200	.5
Control	190	210	.475

*Reproduced from Table 1 of Takane and Jung (2009b) with permission from Springer.

Takane and Jung (2009a) also incorporated a ridge type of regularized estimation in NSCA. They found that this kind of estimation can yield estimates of parameters in NSCA which are more reliable than the conventional method. See the article cited above for details.

4.9 Multiple-Set CANO (GCANO)

Multiple-set canonical correlation analysis (GCANO) relates more than two sets of variables. It can be considered a natural extension of the usual (2-set) CANO into multiple ($K \geq 2$) sets. GCANO has no direct relationship with CPCA except that it

Table 4.4 *The breakdown of Table 4.3 by gender (males and females)**.

Gender	Treatment	Prognosis Recovery	Prognosis Nonrecovery	Recovery rate
Male	Medication	180	120	.6
	Control	70	30	.7
Female	Medication	20	80	.2
	Control	120	180	.4

*Reproduced from Table 2 of Takane and Jung (2009b) with permission from Springer.

computationally reduces to an SVD or GSVD of a certain matrix. It also reduces to a PCA of columnwise standardized data when each data set in GCANO consists of a single continuous variable. The objective of GCANO is usually different from that of PCA, however. Whereas PCA extracts components which are most representative of a single data matrix, GCANO extracts components that maximally account for the relationships (associations) among multiple sets of variables.

GCANO may also be viewed as a method for integrating information from multiple sources (Takane and Oshima-Takane 2002). For example, information about a stimulus object may come through multimodal sensory channels, e.g., visual, auditory, and tactile. The information coming through multiple pathways must be integrated to enable an identification judgment about the object. GCANO mimics this information integration mechanism.

As another example, let us look at Table 4.5 (Takane et al. 2008), in which three expert judges evaluated six brands of wine according to several criteria. These criteria may be different across the judges. We may ask: (1) What are the most discriminating factors among the six brands of wine that are commonly used by the three judges? (2) Where are those wines positioned in terms of those factors? These questions are best answered by GCANO, which extracts a set of components most representative of all three judges to characterize the wines.

Table 4.5 *Wine tasting data used in Abdi and Valentin (2007).*

	Oak-	Expert 1			Expert 2				Expert 3		
wine	type	fruit	wood	coffee	red fruit	roast	vanilla	wood	fruit	butter	wood
1	1	1	6	7	2	5	7	6	3	6	7
2	2	5	3	2	4	4	4	2	4	4	3
3	2	6	1	1	5	2	1	1	7	1	1
4	2	7	1	2	7	2	1	2	2	2	2
5	1	2	5	4	3	5	6	5	2	6	6
6	1	3	4	4	3	5	4	5	1	7	5

Let $\mathbf{G}_k$ $(k = 1, \cdots, K)$ denote the n-case by p_k-variable data matrix for the kth set.

We assume that $\mathbf{G}_k$ is columnwise centered. Let $\mathbf{G}$ denote the $n \times p$ $(= \sum_k p_k)$ row block matrix, $\mathbf{G} = [\mathbf{G}_1, \cdots, \mathbf{G}_K]$. Let $\mathbf{V}^*$ denote the $p \times r$ weight matrix applied to $\mathbf{G}$ to derive canonical variates (scores). Let $\mathbf{V}^*$ be partitioned in the same way as $\mathbf{G}$ is partitioned, namely $\mathbf{V}^* = [\mathbf{V}_1^{*'}, \cdots, \mathbf{V}_K^{*'}]'$, where $\mathbf{V}_k^*$ $(k = 1, \cdots, K)$ is a $p_k \times r$ matrix. In GCANO, we look for $\mathbf{V}^*$ which maximizes

$$\phi(\mathbf{V}^*) = \mathrm{SS}(\mathbf{G}\mathbf{V}^*) = \mathrm{tr}(\mathbf{V}^{*'}\mathbf{G}'\mathbf{G}\mathbf{V}^*) \tag{4.93}$$

subject to the restriction that $\mathbf{V}^{*'}\mathbf{C}\mathbf{V}^* = \mathbf{I}$, where $\mathbf{C}$ is the block diagonal matrix with $\mathbf{C}_k = \mathbf{G}_k'\mathbf{G}_k$ as the kth diagonal block. This leads to the following generalized eigenequation:

$$\mathbf{G}'\mathbf{G}\mathbf{V}^* = \mathbf{C}\mathbf{V}^*\Lambda, \tag{4.94}$$

where Λ is the diagonal matrix of the r largest generalized eigenvalues of $\mathbf{G}'\mathbf{G}$ with respect to $\mathbf{C}$ arranged in descending order, and $\mathbf{V}^*$ is the matrix of the corresponding generalized eigenvectors. Once the generalized eigenequation is solved, the matrix of canonical variates $\mathbf{U}$ can be calculated by $\mathbf{U} = \mathbf{G}\mathbf{V}^*\Lambda^{-1/2}$.

There is another popular criterion, called a homogeneity criterion, used to derive GCANO (Gifi 1990; see also Meredith (1964) and Carroll (1968)). It is stated as

$$\psi(\mathbf{U}, \mathbf{B}) = \sum_{k=1}^{K} \mathrm{SS}(\mathbf{U} - \mathbf{G}_k\mathbf{B}_k), \tag{4.95}$$

where $\mathbf{B}_k$ is the $p_k \times r$ matrix of weights on $\mathbf{G}_k$. (The matrix $\mathbf{B}$ is a column block matrix with $\mathbf{B}_k$ as the kth block.) We minimize the criterion with respect to $\mathbf{B}$ and $\mathbf{U}$ subject to the normalization restriction $\mathbf{U}'\mathbf{U} = \mathbf{I}_r$. This requires the formation of linear combinations of $\mathbf{G}_k$ that are as homogeneous as possible. Minimizing the above criterion with respect to $\mathbf{B}_k$ for fixed $\mathbf{U}$ gives

$$\hat{\mathbf{B}}_k = (\mathbf{G}_k'\mathbf{G}_k)^{-1}\mathbf{G}_k'\mathbf{U}. \tag{4.96}$$

By substituting this back into (4.95), we obtain

$$\begin{aligned}\psi^*(\mathbf{U}) \equiv \psi(\mathbf{U}, \hat{\mathbf{B}}) &= \min_{B|U} \psi(\mathbf{U}, \mathbf{B}) \\ &= \sum_{k=1}^{K} \mathrm{SS}(\mathbf{U} - \mathbf{P}_{G_k}\mathbf{U}) = \mathrm{tr}(\mathbf{U}'(\sum_{k=1}^{K} \mathbf{Q}_{G_k})\mathbf{U}),\end{aligned} \tag{4.97}$$

where $\mathbf{Q}_{G_k} = \mathbf{I}_m - \mathbf{G}_k(\mathbf{G}_k'\mathbf{G}_k)^{-1}\mathbf{G}_k'$. Let $\mathbf{P} = \sum_{k=1}^{K} \mathbf{P}_{G_k}$, where $\mathbf{P}_{G_k} = \mathbf{G}_k(\mathbf{G}_k'\mathbf{G}_k)^{-1}\mathbf{G}_k'$. Minimizing (4.97) with respect to $\mathbf{U}$ under the restriction that $\mathbf{U}'\mathbf{U} = \mathbf{I}$ is equivalent to maximizing $g(\mathbf{U}) = \mathrm{tr}(\mathbf{U}'\mathbf{P}\mathbf{U})$ under the same restriction. This maximization problem leads to the following eigenequation:

$$\mathbf{G}\mathbf{C}^{-1}\mathbf{G}'\mathbf{U} = \mathbf{U}\Lambda \tag{4.98}$$

to be solved. Note that $\mathbf{P} = \mathbf{G}\mathbf{C}^{-1}\mathbf{G}'$. Once $\mathbf{U}$ is obtained, $\mathbf{B}$ is obtained by (4.96), and $\mathbf{V}^*$ in (4.94) is found by $\mathbf{V}^* = \mathbf{B}\Lambda^{1/2} = \mathbf{C}^{-1}\mathbf{G}'\mathbf{U}\Lambda^{1/2}$.

The two formulations of the GCANO problem above give equivalent results. We can transform one eigenequation into the other. By premultiplying (4.94) by $\mathbf{GC}^{-1}$ and setting $\mathbf{U} = \mathbf{GV}^*\Lambda^{-1}$, we obtain (4.98). Conversely, by premultiplying (4.98) by $\mathbf{G}'$ and setting $\mathbf{V}^* = \mathbf{C}^{-1}\mathbf{G}'\mathbf{U}\Lambda$, we obtain (4.94). Both eigenequations are closely related to GSVD$(\mathbf{GC}^{-1})_{I,C}$. Solving (4.94) is more economical than solving (4.98) when $p < n$, while the opposite is true when $n < p$. The choice should be obvious in practical situations.

GCANO reduces to the usual (two-set) CANO when $K = 2$. This can be shown as follows (ten Berge 1979): Let $\mathbf{A} = [\mathbf{G}_1^*, \mathbf{G}_2^*] = \mathbf{GC}^{-1/2}$, where $\mathbf{G}_j* = \mathbf{G}_j(\mathbf{G}_j'\mathbf{G}_j)^{-1/2}$ for $j = 1, 2$. Then,

$$\mathbf{AA}' = \mathbf{GC}^{-1}\mathbf{G}' = \mathbf{P}_{G_1} + \mathbf{P}_{G_2}, \tag{4.99}$$

and

$$\mathbf{A}'\mathbf{A} = \mathbf{C}^{-1/2}\mathbf{G}'\mathbf{GC}^{-1/2} = \begin{bmatrix} \mathbf{I} & \mathbf{G}_1^{*\prime}\mathbf{G}_2^* \\ \mathbf{G}_2^{*\prime}\mathbf{G}_1^* & \mathbf{I} \end{bmatrix}. \tag{4.100}$$

Let $\mathbf{WDV}'$ represent the SVD of $\mathbf{G}_1^{*\prime}\mathbf{G}_2^*$. It can be easily verified that the matrix of normalized eigenvectors and the diagonal matrix of eigenvalues of $\mathbf{A}'\mathbf{A}$ are given by

$$\tilde{\mathbf{W}} = \frac{1}{\sqrt{2}}\begin{bmatrix} \mathbf{W} & \mathbf{W} \\ \mathbf{V} & -\mathbf{V} \end{bmatrix}, \tag{4.101}$$

and

$$\tilde{\mathbf{D}} = \begin{bmatrix} \mathbf{I}+\mathbf{D} & \mathbf{O} \\ \mathbf{O} & \mathbf{I}-\mathbf{D} \end{bmatrix}, \tag{4.102}$$

respectively. That is,

$$\mathbf{A}'\mathbf{A}\tilde{\mathbf{W}} = \tilde{\mathbf{W}}\tilde{\mathbf{D}}. \tag{4.103}$$

(This is analogous to $\mathbf{C}^{-1/2}\mathbf{G}'\mathbf{GC}^{-1/2}\mathbf{C}^{1/2}\mathbf{V}^* = \mathbf{C}^{1/2}\mathbf{V}^*\Lambda$ or $\mathbf{G}'\mathbf{GV}^* = \mathbf{CV}^*\Lambda$.) By premultiplying both sides of (4.103) by $\mathbf{A}$, we obtain

$$(\mathbf{AA}')\mathbf{A}\tilde{\mathbf{W}} = \mathbf{A}\tilde{\mathbf{W}}\tilde{\mathbf{D}}, \tag{4.104}$$

which is nothing but the eigenequation for $\mathbf{AA}'$. (This is equivalent to (4.98).) The matrix of normalized eigenvectors of $\mathbf{AA}'$ is given by $\mathbf{A}\tilde{\mathbf{W}}\tilde{\mathbf{D}}^{-1/2} = \mathbf{GV}^*\Lambda^{-1/2}$.

Takane et al. (2008) incorporated a ridge type of regularized estimation in GCANO (RGCANO) and compared the results with those obtained by nonregularized GCANO. RGCANO obtains $\mathbf{V}^*$ that maximizes $\text{tr}(\mathbf{V}^{*\prime}(\mathbf{G}'\mathbf{G} + \delta\mathbf{J}_p)\mathbf{V}^*)$ subject to $\mathbf{V}{*}'(\mathbf{C} + \delta\mathbf{J}_p)\mathbf{V}^* = \mathbf{I}$, where δ is the regularization parameter, $\mathbf{J}_p$ is a block diagonal matrix with $\mathbf{G}_k'(\mathbf{G}_k\mathbf{G}_k')^-\mathbf{G}_k$ (the orthogonal projector onto the row space of $\mathbf{G}_k$) as the kth diagonal block. (Note that this matrix is equal to an identity matrix if the $\mathbf{G}_k$'s are all columnwise nonsingular.) An example of regularized MCA, a special case of GCANO, will be given in the following section.

Hwang et al. (2012) extended GCANO to functional data (Ramsay and Silverman 2005), and Hwang et al. (2013) integrated GCANO and PCA in a unified framework. This method obtains the eigendecomposition of $\mathbf{R} = \alpha\mathbf{GC}^{-1}\mathbf{G}' + \beta\mathbf{GG}'$, where α and β modulate the importance of the two constituents. A larger value of α gives more importance to the GCANO part (the first term), while a larger value of β gives more importance to the PCA part (the second term).

Takane and Oshima-Takane (2002) developed a nonlinear information integrator by combining GCANO and multilayered neural network (NN) models. This method nonlinearly transforms each set of input data by NN models in such a way that the resultant outputs from different sets of input data are as homogeneous as possible.

4.10 Multiple Correspondence Analysis (MCA)

Multiple correspondence analysis (MCA) is a popular technique for structural analysis of multivariate categorical data (Greenacre 1984; Lebart et al. 1984; Nishisato 1980). MCA assigns scores to rows (often representing subjects) and columns (often representing response categories of multiple-choice items) of a data matrix, yielding a graphical display of the rows and columns of the data matrix. The graphical display facilitates our intuitive understanding of the relationships between the response categories and subjects. MCA works for discrete categorical data much the same way as PCA does for continuous multivariate data, and it is often referred to as PCA of categorical data.

There are a number of alternative criteria from which MCA can be derived. One is to use the multidimensional scaling (MDS; Section 4.5) idea in a way similar to simple CA (Section 4.6). In MDS, we represent both rows (subjects) and columns (response categories) of a data matrix in a multidimensional Euclidean space in such a way that those response categories chosen by particular subjects are located close to the subjects, while those categories not chosen by those subjects are located far from them.

Let $\mathbf{U}$ and $\mathbf{V}^*$ represent the matrices of coordinates of subjects and response categories, respectively, and let $\mathbf{G} = [\mathbf{G}_1, \cdots, \mathbf{G}_K]$ represent the data matrix, where each $\mathbf{G}_k = [g_{ik(j)}]$ is an indicator matrix such that

$$g_{ik(j)} = \begin{cases} 1 & \text{if subject } i \text{ chooses category } j \text{ in item } k, \\ 0 & \text{otherwise.} \end{cases} \tag{4.105}$$

We would like to find $\mathbf{U}$ and $\mathbf{V}^*$ that minimize

$$\phi(\mathbf{U}, \mathbf{V}^*) = \sum_i \sum_{k,j} g_{ik(j)} d_{ik(j)}^2(\mathbf{U}, \mathbf{V}^*), \tag{4.106}$$

subject to an orthonormalization restriction on $\mathbf{U}$. Here, $d_{ik(j)}^2$ is the squared Euclidean distance between the ith subject and the jth response category in item k. This criterion is analogous to (4.31) and leads to a solution similar to that in CA. Specifically, we obtain $\text{GSVD}(\mathbf{D}_R^{-1}\mathbf{G}\mathbf{D}_C^{-1})_{D_R,D_C}$, where $\mathbf{D}_R$ and $\mathbf{D}_C$ are the diagonal matrix of row and column totals of $\mathbf{G}$. Note that $\mathbf{D}_R$ is often constant diagonal, in which case the GSVD problem above reduces to $\text{GSVD}(\mathbf{G}\mathbf{D}_C^{-1})_{I,\mathbf{D}_C}$. The trivial solution may be removed *a priori* by columnwise centering $\mathbf{G}$.

Another way to derive MCA is via GCANO (the previous section). As will be seen below, GCANO reduces to MCA with one minor qualification when all K data sets consist of indicator variables. In GCANO we columnwise center the data matrix $\mathbf{G}$, which will eliminate the trivial solution in MCA. In MCA, however, each $\mathbf{G}_k$ is a indicator matrix, and the columnwise centering of $\mathbf{G}_k$ reduces the rank of the resultant matrix by one. This means that the diagonal blocks in the block diagonal matrix $\mathbf{C}$ are all rank deficient for which no regular inverse exists. Fortunately, it has been shown (Takane et al. 2008) that the original (columnwise uncentered) $\mathbf{G}$ can be used to calculate $\mathbf{C}$ without affecting the solution in any essential way.

Hwang and Takane (2002a) developed constrained MCA, in which additional constraints are imposed on response categories. Takane and Hwang (2006) incorporated a ridge type of regularized estimation (Section 2.2.8) into MCA. Using Greenacre's car data (Example 1.3), they demonstrated a dramatic gain in efficiency of estimates by regularized MCA over the conventional estimation method. It was found particularly beneficial for response categories with small observed frequencies which tend to be poorly estimated in small samples.

One interesting application of MCA is analysis of sorting data (Takane 1980a). In sorting data, a group of subjects are asked to sort a set of stimuli into several groups in terms of their similarities. The sorting data can be considered a special kind of multiple-choice data, where the rows represent the set of stimuli and the columns the sorting groups generated by the subjects. Suppose, for example, that five stimuli are sorted into three clusters by subject k, say stimuli 1 and 4 into cluster 1, stimulus 2 into cluster 2, and stimuli 3 and 5 into cluster 3. Then we form $\mathbf{G}_k$ as follows:

$$\mathbf{G}_k = \begin{bmatrix} 1 & 0 & 0 \\ 0 & 1 & 0 \\ 0 & 0 & 1 \\ 1 & 0 & 0 \\ 0 & 0 & 1 \end{bmatrix}.$$

A collection of $\mathbf{G}_k$ serves as input data to MCA.

Here is an example (Takane 1980a). Ten university students were asked to sort a set of 29 "Have" words into as many groups as they like in terms of their similarity in meaning. The 29 Have words were: 1 = Accept, 2 = Beg, 3 = Belong, 4 = Borrow, 5 = Bring, 6 = Buy, 7 = Earn, 8 = Find, 9 = Gain, 10 = Get, 11 = Get rid of, 12 = Give, 13 = Have, 14 = Hold, 15 = Keep, 16 = Lack, 17 = Lend, 18 = Lose, 19 = Need, 20 = Offer, 21 = Own, 22 = Receive, 23 = Return, 24 = Save, 25 = Sell, 26 = Steal, 27 = Take, 28 = Use, and 29 = Want. For comparison, both nonregularized and regularized MCA were applied to the data. The 29-fold cross validation (Section 5.3) found that the optimal value of δ was 1. Figure 4.2 displays the two-dimensional stimulus configurations obtained by the two kinds of MCA along with 95% confidence regions obtained by the bootstrap procedure (Section 5.2). The configurations are similar in both cases. We find words such as 13 = Have, 21 = Own, 3 = Belong, 15 = Keep, etc. on the left-hand side of the configurations, while 12 = Give, 25 = Sell, 11 = Get rid of, 18 = Lose, etc. on the right-hand side. At the bottom, we see 16 = Lack, 19 = Need, and 29 = Want, while at the top, we find 22 = Receive, 10 = Get, 7 = Earn, 8 = Find, etc. We may interpret dimension 1 (the horizontal direction) as a contrast between a stable state of possession on the left and an unstable state of possession

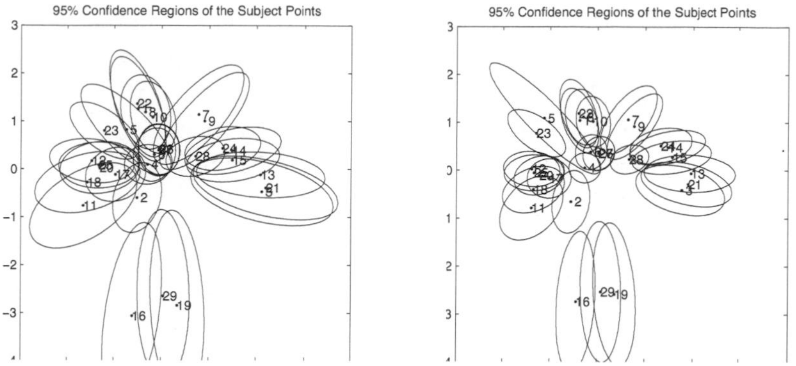

Figure 4.2 *Two-dimensional stimulus configuration for 29 have words obtained by MCA along with 95% confidence regions: Nonregularized left, regularized right. (Reproduced from Figures 6 and 7 of Takane et al. (2008) with permission from the Psychometric Society.)*

on the right, while dimension 2 (the vertical direction) as a contrast between a stable state of nonpossession at the bottom and an unstable state of the same at the top. Although the configurations are similar, confidence regions are almost uniformly smaller in the regularized case, indicating that more reliable estimates are obtained by the regularized estimation method than the nonregularized counterpart.

4.11 Vector Preference Models

Vector preference models are often used to represent individual differences in preference (Bechtel 1976; Carroll 1972; Slater 1960; Tucker 1959). In these models, a set of stimuli are represented as points, while subjects' preference tendencies are represented by vectors in a multidimensional space. Subjects' preferences for the stimuli are predicted by the projections of the stimulus points onto the subject vectors. The length of the projection vectors (from the origin) indicates how well the subjects' preferences are represented by particular directions in the space. Multidimensional ideal point models (unfolding analysis) discussed in Section 4.5 are used for similar purposes. However, they represent both stimuli and subjects as points in a multidimensional space, and distances between them are assumed to be inversely related to preference judgments.

Vector preference models are also used when pairs of stimuli are presented to subjects who are asked to judge the extent to which they prefer one stimulus to the other. In this case, the comparative preference judgments are predicted by the differences in the projections of the two stimuli onto subject vectors. Let $\mathbf{Z}$ denote an n-subject by m-stimulus-pairs comparative preference data matrix ($m = p(p-1)/2$, where p is the number of stimuli), and assume that p stimulus points and n subject vectors are represented in an r dimensional space, whose coordinates are given by $\mathbf{U}$ and $\mathbf{V}$ (or $\mathbf{V}^*$), respectively. (We omitted subscript r on $\mathbf{U}$ and $\mathbf{V}$ to indicate the number of columns in these matrices explicitly.) Then three representative vector preference models are stated as follows.

(i) The BTC model (Bechtel et al. 1971):

$$\mathbf{Z} = \mathbf{V}^*\mathbf{U}'\mathbf{A}' + \mathbf{1}_n\mathbf{c}' + \mathbf{E}. \tag{4.107}$$

(ii) The THL model (Heiser and de Leeuw 1981; Takane 1980b):

$$\mathbf{Z} = (\mathbf{1}_n\mathbf{f}' + \mathbf{V}\mathbf{U}')\mathbf{A}' + \mathbf{E}. \tag{4.108}$$

(iii) The wandering vector model (WVM; De Soete and Carroll 1983):

$$\mathbf{Z} = (\mathbf{1}_n\mathbf{v}' + \mathbf{V})\mathbf{U}'\mathbf{A}' + \mathbf{E}. \tag{4.109}$$

Here $\mathbf{1}_n$ is the n-component vector of ones, $\mathbf{A}$ is an $m \times p$ design matrix for pair comparison (see Note 4.5 below), $\mathbf{E}$ indicates a matrix of residual terms, $\mathbf{c}$ is the m-component vector of pairwise unscalabilities, and $\mathbf{f}$ represents the mean preference values of the p stimuli, and $\mathbf{v}$ indicates subjects' mean preference vector. The pairwise unscalabilities indicate the degree to which comparative preference judgments between two stimuli cannot be predicted by the difference between the preference values of the two stimuli. Note that (4.109) can be obtained by setting $\mathbf{f} = \mathbf{U}\mathbf{v}$ in (4.108).

Note 4.5 Rows of $\mathbf{A}$ represent stimulus pairs, and columns represent individual stimuli. If, for example, the first row of $\mathbf{Z}$ represents a comparative judgment between stimuli 1 and 2, and if a larger value indicates a larger preference for stimulus 1, then the first row of $\mathbf{A}$ has 1 in the first column, -1 in the second column, and 0s elsewhere. Suppose there are four stimuli ($p = 4$). Then, $\mathbf{A}$ may look like:

$$\mathbf{A} = \begin{bmatrix} 1 & -1 & 0 & 0 \\ 1 & 0 & -1 & 0 \\ 1 & 0 & 0 & -1 \\ 0 & 1 & -1 & 0 \\ 0 & 1 & 0 & -1 \\ 0 & 0 & 1 & -1 \end{bmatrix}$$

assuming that the first column of $\mathbf{Z}$ contains the comparative judgments between stimuli 1 vs 2, the second column between 1 and 3, and so on. The rows of $\mathbf{A}$ may be permuted depending on which rows of $\mathbf{Z}$ represent comparisons of which stimuli. They may also be reflected depending on whether a larger value in the corresponding column of $\mathbf{Z}$ indicates a favor for the first stimulus or the second stimulus in the pair. It can be easily verified that $\mathbf{A}'\mathbf{A} = 4\mathbf{Q}_{1_4}$ for $\mathbf{A}$ above. In general, $\mathbf{A}'\mathbf{A} = p\mathbf{Q}_{1_p}$, where $\mathbf{Q}_{1_p}$ is the centering matrix of order p (i.e., $\mathbf{Q}_{1_p} = \mathbf{I}_p - \mathbf{1}_p\mathbf{1}_p'/p$).

Estimates of parameters in the three models above are given as follows:

(i) Obtain $\mathbf{V}^*$ and $\mathbf{U}$ by SVD($\hat{\mathbf{M}}$), where $\hat{\mathbf{M}} = \mathbf{Z}\mathbf{A}(\mathbf{A}'\mathbf{A})^+ = \mathbf{Z}\mathbf{A}/p$, and the estimate of $\mathbf{c}$ by $\hat{\mathbf{c}}' = \mathbf{1}_n'\mathbf{Z}\mathbf{Q}_A/n$, where $\mathbf{Q}_A = \mathbf{I} - \mathbf{A}(\mathbf{A}'\mathbf{A})^-\mathbf{A}'$.

(ii) Define $\hat{\mathbf{M}}$ as in (i). Then, $\hat{\mathbf{f}}$ is obtained by $\hat{\mathbf{f}}' = \mathbf{1}_n'\hat{\mathbf{M}}/n$, and $\mathbf{V}$ and $\mathbf{U}$ by SVD($\mathbf{Q}_{1_n}\hat{\mathbf{M}}$), where $\mathbf{Q}_{1_n} = \mathbf{I} - \mathbf{1}_n\mathbf{1}_n'/n$.

(iii) Obtain $\mathbf{V}^*$, $\mathbf{U}$ and $\mathbf{V}$ as in (i), and set $\mathbf{V} = \mathbf{Q}_{1_n}\mathbf{V}^*$ and $\hat{\mathbf{v}}' = \mathbf{1}_n'\mathbf{V}^*/n$.

The estimates above can be directly obtained by minimizing $SS(\mathbf{E}) = \text{tr}(\mathbf{E}'\mathbf{E})$ in (4.107) through (4.109). However, these models are all regarded as special cases of CPCA, and so we may alternatively use the results of estimation already derived for CPCA by identifying which parameters in these models correspond to which parameters in CPCA. Let $\mathbf{G} = \mathbf{1}_n$, $\mathbf{H} = \mathbf{A}$, $\mathbf{K} = \mathbf{I}$, and $\mathbf{L} = \mathbf{I}$ in (3.25). Then, the equation can be rewritten as

$$\mathbf{Z} = \mathbf{P}_{1_n}\mathbf{Z}\mathbf{P}_A + \mathbf{Q}_{1_n}\mathbf{Z}\mathbf{P}_A + \mathbf{P}_{1_n}\mathbf{Z}\mathbf{Q}_A + \mathbf{Q}_{1_n}\mathbf{Z}\mathbf{Q}_A. \tag{4.110}$$

Here,

$$\mathbf{P}_{1_n}\mathbf{Z}\mathbf{P}_A = \mathbf{1}_n\hat{\mathbf{f}}'\mathbf{A}', \tag{4.111}$$

$$\mathbf{Q}_{1_n}\mathbf{Z}\mathbf{P}_A = \mathbf{Q}_{1_n}\hat{\mathbf{M}}\mathbf{A}', \tag{4.112}$$

and

$$\mathbf{P}_{1_n}\mathbf{Z}\mathbf{Q}_A = \mathbf{1}_n\hat{\mathbf{c}}'. \tag{4.113}$$

The BTC model does not split the first and second terms on the right-hand side of (4.110), and obtains SVD of $\hat{\mathbf{M}}$ in $\mathbf{Z}\mathbf{P}_A = \hat{\mathbf{M}}\mathbf{A}'$. The THL model obtains SVD of $\mathbf{Q}_{1_n}\hat{\mathbf{M}}$ in the second term, and does not split the third and fourth terms. The WVM obtains SVD of $\hat{\mathbf{M}}$, as in the BTC model, and splits $\mathbf{V}^*\mathbf{U}'$ into $\mathbf{1}_n'\mathbf{V}^*\mathbf{U}'$ and $\mathbf{Q}_{1_n}\mathbf{V}^*\mathbf{U}' = \mathbf{V}\mathbf{U}'$. The WVM does not split the third and fourth terms, as in the THL model. Since $\mathbf{A}'\mathbf{A} = p\mathbf{Q}_{1_p}$, and so $\hat{\mathbf{M}}\mathbf{Q}_{1_p} = \hat{\mathbf{M}}$, $\text{GSVD}(\hat{\mathbf{M}})_{I,A'A}$ reduces to $\text{SVD}(\hat{\mathbf{M}})$.

We give an example of application of the BTC model. The data used are pair-wise preference data on a set of nine stimuli. The stimuli are nine celebrities selected from three distinct categories (politicians, athletes, and entertainers): 1. Brian Mulroney (ex-prime minister of Canada), 2. Ronald Reagan (ex-president of the United States), 3. Margaret Thatcher (ex-prime minister of the United Kingdom), 4. Jacqueline Gareau (twice winner of the Boston Marathon in the women's division), 5. Wayne Gretzky (professional ice hockey player), 6. Steve Podborski (professional skier), 7. Paul Anka (male vocalist), 8. Tommy Hunter (country-western singer), and 9. Ann Murray (female vocalist). Subjects were 100 students at a large Canadian university, approximately one half of which were anglophones, while the other half were francophones and allophones (whose mother tongues are other than English or French). The stimuli were presented in pairs, and subjects were asked to indicate the extent to which they preferred to meet one over the other on 25-point rating scales.

In this analysis, we first obtained the rank-free estimate of $\mathbf{M}$, and then applied SVD to this matrix. Figure 4.3 displays the derived two-dimensional stimulus configuration. In the figure, integers from 1 to 9 indicate the points representing the nine stimuli. It may be observed that the nine stimuli are roughly clustered into three groups from which they were selected. One exception is 4 (Jacqueline Gareau) who is located closer to the entertainers' group (7, 8, and 9) despite the fact that she is an athlete. This is probably because she was relatively unknown among the subjects, who thought that she was an entertainer. The ten unlabeled arrows indicate preference vectors of the first ten subjects (out of 100 subjects in total). If we project stimulus points onto these vectors, we can obtain the best prediction of the preference values of the stimuli by these subjects.

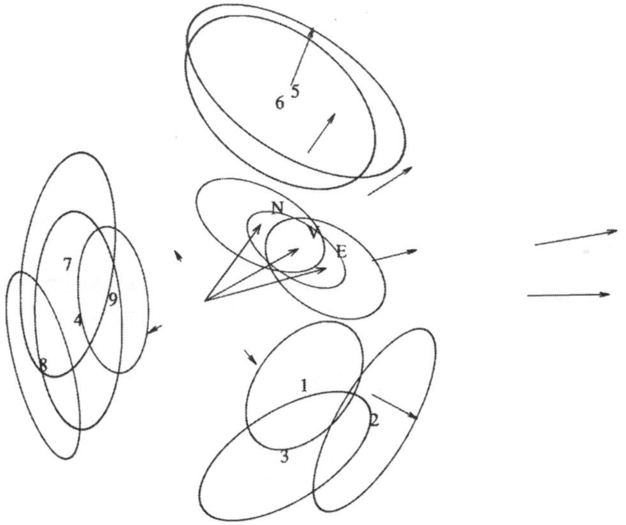

Figure 4.3 *Stimulus configuration and 95% confidence regions derived from the BTC model. (Reproduced from Figure 2 of Takane and Shibayama (1991) with permission from the Psychometric Society.)*

Ellipses surrounding the nine stimulus points indicate 95% confidence regions obtained by the bootstrap method (Efron 1979). In the bootstrap method (see Section 5.2 for more details), 100 bootstrap samples were created by repeated sampling of observations from the original data with replacement. Each data set was analyzed by CPCA to derive variance-covariance estimates of estimated parameters. Confidence regions were then drawn under the assumption of asymptotic normality of the estimates.

In the figure the vector labeled as V indicates the mean preference tendency across all subjects. It may be observed that on average athletes (5 and 6) are the most preferred, followed by politicians (1, 2 and 3), and then by entertainers (7, 8 and 9; 4 is exceptional). The vectors labeled E and N indicate mean preference tendencies of anglophone and nonanglophone students, respectively. Anglophone students tend to prefer politicians slightly more than nonanglophone students. (The three politicians are all anglophone.) Note, however, that there is a considerable overlap in the confidence regions corresponding to the two vectors, so there does not seem to be any statistically significant difference between the two groups.

4.12 Two-Way CANDELINC

Carroll et al. (1980) considered the minimization of

$$\phi(\mathbf{M}) = \mathrm{SS}(\mathbf{Z} - \mathbf{GMH}') \tag{4.114}$$

with respect to $\mathbf{M}$ subject to the reduced rank restriction on $\mathbf{M}$. The model $\mathbf{GMH}'$ is called the CANDELINC (canonical decomposition with linear constraints). Earlier, Carroll and Chang (1970) proposed a model and algorithm called CANDECOMP (canonical decomposition) for reduced rank approximations of multiway tables. CANDELINC is a special case of CANDECOMP in a dual sense. It is a special case in which the data are two-way arrays, in which case CANDECOMP reduces to an ALS algorithm for obtaining a few dominant components of a data matrix (Horst 1935; see also Daugavet (1968) and Section 4.18). It is also a special case of CANDECOMP in which the structural term in the model is constrained by the model matrices $\mathbf{G}$ and $\mathbf{H}$. The CANDELINC model is similar to the growth curve model with an additional rank restriction on $\mathbf{M}$ (the next section). It is also a special case of CPCA in which only the first term in CPCA is isolated (the second and third terms in CPCA are considered to be part of the disturbance terms) with identity metric matrices ($\mathbf{K} = \mathbf{I}$ and $\mathbf{L} = \mathbf{I}$).

When both $\mathbf{G}$ and $\mathbf{H}$ are columnwise orthogonal (i.e., $\mathbf{G}'\mathbf{G} = \mathbf{I}$ and $\mathbf{H}'\mathbf{H} = \mathbf{I}$), (4.114) can be rewritten as

$$\phi(\mathbf{M}) = \mathrm{SS}(\hat{\mathbf{M}} - \mathbf{M}) + \mathrm{SS}(\mathbf{Z}) - \mathrm{SS}(\hat{\mathbf{M}}), \tag{4.115}$$

where

$$\hat{\mathbf{M}} = \mathbf{G}'\mathbf{ZH}. \tag{4.116}$$

Note 4.6 Let $\mathbf{U}$ and $\mathbf{V}$ be columnwise orthogonal matrices. Then

$$\mathrm{SS}(\mathbf{UAV}') = \mathrm{SS}(\mathbf{A}). \tag{4.117}$$

For a given $\mathbf{Z}$, $\mathbf{G}$, and $\mathbf{H}$, $\mathrm{SS}(\mathbf{Z}) - \mathrm{SS}(\hat{\mathbf{M}})$ is a constant, so the minimum of $\phi(\mathbf{M})$ is attained by the minimum of $\mathrm{SS}(\hat{\mathbf{M}} - \mathbf{M})$, which is found by $\mathrm{SVD}(\hat{\mathbf{M}})$.

If $\mathbf{G}$ and $\mathbf{H}$ are not columnwise orthogonal, they are orthogonalized by the following procedure: Let $\mathbf{G} = \mathbf{F}_G\mathbf{R}_G'$ and $\mathbf{H} = \mathbf{F}_H\mathbf{R}_H'$ be the compact QR decompositions of $\mathbf{G}$ and $\mathbf{H}$. Then

$$\phi(\mathbf{M}) = \mathrm{SS}(\hat{\mathbf{J}} - \mathbf{J}) + \mathrm{SS}(\mathbf{Z}) - \mathrm{SS}(\hat{\mathbf{J}}), \tag{4.118}$$

where

$$\hat{\mathbf{J}} = \mathbf{F}_G'\mathbf{Z}\mathbf{F}_H, \tag{4.119}$$

and

$$\mathbf{J} = \mathbf{R}_G'\mathbf{M}\mathbf{R}_H. \tag{4.120}$$

Again, $\mathrm{SS}(\mathbf{Z}) - \mathrm{SS}(\hat{\mathbf{J}})$ is a constant, so the minimum of $\phi(\mathbf{M})$ is attained by the minimum of $\mathrm{SS}(\hat{\mathbf{J}}-\mathbf{J})$ with respect to $\mathbf{M}$, which is found by $\mathrm{GSVD}(\hat{\mathbf{M}})_{G'G,H'H}$, where

$$\hat{\mathbf{M}} = (\mathbf{G}'\mathbf{G})^{-}\mathbf{G}'\mathbf{Z}\mathbf{H}(\mathbf{H}'\mathbf{H})^{-} = (\mathbf{R}_G')^{-}\hat{\mathbf{J}}\mathbf{R}_H^{-}, \tag{4.121}$$

$$\mathbf{G}'\mathbf{G} = \mathbf{R}_G\mathbf{R}_G', \tag{4.122}$$

and

$$\mathbf{H}'\mathbf{H} = \mathbf{R}_H\mathbf{R}_H'. \tag{4.123}$$

Although (4.118) has been derived as a special case of (4.115), a more general decomposition of $\phi(\mathbf{M})$ also exists that more directly shows the relationship between the minimization of (4.118) and $\mathrm{GSVD}(\hat{\mathbf{M}})_{G'G,H'H}$ (Takane and Hunter 2001), namely

$$\begin{aligned}\phi(\mathbf{M}) &= \mathrm{SS}(\mathbf{P}_G\mathbf{Z}\mathbf{P}_H - \mathbf{G}\mathbf{M}\mathbf{H}') + \mathrm{SS}(\mathbf{Z} - \mathbf{P}_G\mathbf{Z}\mathbf{P}_H)\\ &= \mathrm{SS}(\mathbf{G}(\hat{\mathbf{M}}-\mathbf{M})\mathbf{H}') + \mathrm{SS}(\mathbf{Z}) - \mathrm{SS}(\mathbf{G}\hat{\mathbf{M}}\mathbf{H}').\end{aligned} \tag{4.124}$$

The decomposition above can be generalized to:

$$\begin{aligned}\phi(\mathbf{M})_{K,L} &= \mathrm{SS}(\mathbf{Z} - \mathbf{G}\mathbf{M}\mathbf{H}')_{K,L}\\ &= \mathrm{SS}(\mathbf{P}_{G/K}\mathbf{Z}\mathbf{P}_{H/L}' - \mathbf{G}\mathbf{M}\mathbf{H}')_{K,L}\\ &\qquad + \mathrm{SS}(\mathbf{Z} - \mathbf{P}_{G/K}\mathbf{Z}\mathbf{P}_{H/L}')_{K,L}\end{aligned} \tag{4.125}$$

by introducing nonidentity metric matrices $\mathbf{K}$ and $\mathbf{L}$. Since the second term on the right-hand side of (4.125) is a constant, $\phi(\mathbf{M})_{K,L}$ can be minimized by minimizing the first term, which in turn can be achieved by $\mathrm{GSVD}(\mathbf{P}_{G/K}\mathbf{Z}\mathbf{P}_{H/L}')_{K,L}$. This is identical to the analysis given in Section 3.2.2. The second term on the right-hand side of (4.125) represents the sum of squares (SS) of $\mathbf{Z}$ left unaccounted for by $\mathbf{G}$ and $\mathbf{H}$, while the first term represents the SS that cannot be explained by the specified number of components of $\mathbf{P}_{G/K}\mathbf{Z}\mathbf{P}_{H/L}'$, the portion of $\mathbf{Z}$ that can be explained by $\mathbf{G}$ and $\mathbf{H}$.

4.13 Growth Curve Models (GCM)

Growth curve models (GCM; Potthoff and Roy 1964), also known as GMANOVA (generalized multivariate analysis of variance), provide useful methods for analyzing patterns of change in repeated measurements, and investigating how such patterns are related to various characteristics of subjects. Let $\mathbf{Z}$ denote an n by m data matrix, where n indicates the number of subjects, and m the number of time points at which the measurements are taken. Let $\mathbf{G}$ (n by p) and $\mathbf{H}$ (m by q) represent design matrices (predictor variables) for subjects and measurement occasions, respectively. The matrix $\mathbf{G}$ may be a matrix of dummy variables indicating treatment groups, and $\mathbf{H}$ a matrix of orthogonal polynomials of certain order describing basic trends in measurements taken over time. For the moment, we assume that both $\mathbf{G}$ and $\mathbf{H}$ are columnwise nonsingular. The conventional growth curve model (hereafter, simply GCM) postulates

$$\mathbf{Z} = \mathbf{G}\mathbf{M}\mathbf{H}' + \mathbf{E}^*, \tag{4.126}$$

where $\mathbf{M}$ is the p by q matrix of regression coefficients, and $\mathbf{E}^*$ is the matrix of disturbance terms. We assume that $\mathbf{e}^* = \text{vec}(\mathbf{E}^*)$ follows the nm-variate normal distribution with the mean vector $\mathbf{0}$ and the variance-covariance matrix $\Sigma^* \otimes \mathbf{I}_n$. We write this as

$$\mathbf{e}^* \sim \mathcal{N}(\mathbf{0}, \Sigma^* \otimes \mathbf{I}_n). \tag{4.127}$$

Note that this model is essentially the same as the one in the previous section (2-way CANDELINC) without the rank restriction on $\mathbf{M}$.

Khatri (1966; see also Grizzle and Allen (1969) and Siotani et al. (1985)) found the maximum likelihood estimate (MLE) of $\mathbf{M}$ under the distributional assumption given above. The derivation is rather intriguing, and is reproduced below. We first postmultiply (4.126) by

$$\mathbf{T} = [\mathbf{T}_1, \mathbf{T}_2], \tag{4.128}$$

where $\mathbf{T}_1$ is such that

$$\mathbf{H}'\mathbf{T}_1 = \mathbf{I}, \tag{4.129}$$

and $\mathbf{T}_2$ is such that $\text{Sp}(\mathbf{T}_2) = \text{Ker}(\mathbf{H}')$, which implies $\mathbf{H}'\mathbf{T}_2 = \mathbf{O}$. We find

$$\mathbf{Y} = [\mathbf{Y}_1, \mathbf{Y}_2] = \mathbf{ZT} = [\mathbf{GM}, \mathbf{O}] + \mathbf{E}, \tag{4.130}$$

where $\mathbf{E} = \mathbf{E}^*\mathbf{T}$. This means that $\text{Ex}[\mathbf{Y}] = [\mathbf{GM}, \mathbf{O}]$, and $\text{Var}[\text{vec}(\mathbf{Y})] = \text{Var}[\text{vec}(\mathbf{E})] = (\mathbf{T}'\Sigma^*\mathbf{T}) \otimes \mathbf{I}_n$, where Ex and Var are the expectation and variance-covariance operators. Let $\Sigma = \mathbf{T}'\Sigma^*\mathbf{T}$ be partitioned as

$$\Sigma = \begin{bmatrix} \Sigma_{11} & \Sigma_{12} \\ \Sigma_{21} & \Sigma_{22} \end{bmatrix} \tag{4.131}$$

conformably to the partition of $\mathbf{Y}$. Then, the conditional distribution of $\mathbf{Y}_1$ given $\mathbf{Y}_2$ is given by

$$\mathbf{Y}_1 | \mathbf{Y}_2 \sim \mathcal{N}(\mathbf{GM} + \mathbf{Y}_2\mathbf{B}, \Sigma_{11.2} \otimes \mathbf{I}_n), \tag{4.132}$$

where $\mathbf{B} = \Sigma_{22}^{-1}\Sigma_{21}$, and $\Sigma_{11.2} = \Sigma_{11} - \Sigma_{12}\Sigma_{22}^{-1}\Sigma_{21}$.

Note 4.7. Recall that if

$$\begin{pmatrix} \mathbf{y}_1 \\ \mathbf{y}_2 \end{pmatrix} \sim \mathcal{N}\left(\begin{pmatrix} \mu_1 \\ \mu_2 \end{pmatrix}, \begin{bmatrix} \Sigma_{11} & \Sigma_{12} \\ \Sigma_{21} & \Sigma_{22} \end{bmatrix}\right), \tag{4.133}$$

$$\mathbf{y}_1 | \mathbf{y}_2 \sim \mathcal{N}(\mu_1 + \Sigma_{12}\Sigma_{22}^{-1}(\mathbf{y}_2 - \mu_2), \Sigma_{11} - \Sigma_{12}\Sigma_{22}^{-1}\Sigma_{21}), \tag{4.134}$$

and

$$\mathbf{y}_2 \sim \mathcal{N}(\mu_2, \Sigma_{22}). \tag{4.135}$$

The parameters in the conditional mean can be obtained by fitting the following multivariate multiple regression model:

$$\mathbf{Y}_1 = \mathbf{GM} + \mathbf{Y}_2\mathbf{B} + \tilde{\mathbf{E}}, \tag{4.136}$$

where $\tilde{\mathbf{E}}$ is the matrix of disturbance terms having the same variances-covariances as $\mathbf{Y}_1|\mathbf{Y}_2$. This is called a covariance adjustment (Rao 1967).

By Seber's trick (Section 2.2.10), the OLSE (which is identical to the conditional MLE in this case) of $\mathbf{M}$ and $\mathbf{B}$ can be obtained by:

$$\hat{\mathbf{M}} = (\mathbf{G}'\mathbf{G})^{-1}\mathbf{G}'(\mathbf{Y}_1 - \mathbf{Y}_2\hat{\mathbf{B}}), \tag{4.137}$$

and

$$\hat{\mathbf{B}} = (\mathbf{Y}_2'\mathbf{Q}_G\mathbf{Y}_2)^{-1}\mathbf{Y}_2'\mathbf{Q}_G\mathbf{Y}_1 = (\mathbf{T}_2'\mathbf{S}\mathbf{T}_2)^{-1}\mathbf{T}_2'\mathbf{S}\mathbf{T}_1, \tag{4.138}$$

where $\mathbf{Q}_G = \mathbf{I}_n - \mathbf{G}(\mathbf{G}'\mathbf{G})^{-1}\mathbf{G}'$, and

$$\mathbf{S} = \mathbf{Z}'\mathbf{Q}_G\mathbf{Z}. \tag{4.139}$$

Note that these estimates are not affected by $\Sigma_{11.2}$, the conditional covariances among the columns of $\mathbf{Y}_1$ given $\mathbf{Y}_2$. The estimate of $\mathbf{M}$ in (4.137) can be simplified as

$$\begin{aligned}\hat{\mathbf{M}} &= (\mathbf{G}'\mathbf{G})^{-1}\mathbf{G}'\mathbf{Z}(\mathbf{T}_1 - \mathbf{T}_2\hat{\mathbf{B}}) \\ &= (\mathbf{G}'\mathbf{G})^{-1}\mathbf{G}'\mathbf{Z}\mathbf{Q}_{T_2/S}\mathbf{T}_1 &(4.140)\\ &= (\mathbf{G}'\mathbf{G})^{-1}\mathbf{G}'\mathbf{Z}\mathbf{P}'_{H/S^{-1}}\mathbf{T}_1 &(4.141)\\ &= (\mathbf{G}'\mathbf{G})^{-1}\mathbf{G}'\mathbf{Z}\mathbf{S}^{-1}\mathbf{H}(\mathbf{H}'\mathbf{S}^{-1}\mathbf{H})^{-1}. &(4.142)\end{aligned}$$

Note that $\mathbf{H}'\mathbf{T}_1 = \mathbf{I}$. To obtain (4.141) from (4.140), we used Khatri's lemma (Section 2.2.8). Equation (4.142) gives the conditional MLE of $\mathbf{M}$. However, since the marginal distribution of $\mathbf{Y}_2$ is unrelated to $\mathbf{M}$, it is also the unconditional MLE of $\mathbf{M}$. When $\mathbf{G}'\mathbf{G}$ and/or $\mathbf{H}'\mathbf{S}^{-1}\mathbf{H}$ are singular in the equations above, their inverses can be replaced by their respective g-inverses. Then, $\hat{\mathbf{M}}$ is not unique, but $\mathbf{G}\hat{\mathbf{M}}\mathbf{H} = \mathbf{P}_G\mathbf{Z}\mathbf{P}'_{H/S^{-1}}$ is uniquely determined.

The estimate of $\mathbf{M}$ in (4.142) is similar to the following estimate of $\mathbf{M}$ given by Potthoff and Roy (1964):

$$\hat{\mathbf{M}} = (\mathbf{G}'\mathbf{G})^{-1}\mathbf{G}'\mathbf{Z}\mathbf{W}^{-1}\mathbf{H}(\mathbf{H}'\mathbf{W}^{-1}\mathbf{H})^{-1} \tag{4.143}$$

for an arbitrary symmetric *pd* matrix $\mathbf{W}$. The important difference is that whereas in (4.143), Potthoff and Roy (1964) could not single out any particular matrix as best for $\mathbf{W}$ in (4.143), whereas it is determined as $\mathbf{S}^{-1}$ in (4.142).

It is straightforward to impose a rank restriction rank$(\mathbf{M}) \leq \min(p,q)$ on model (4.126), as has been done by Reinsel and Velu (1998; see also the previous section on two-way CANDELINC), and further allow a ridge type of RLS estimation (Section 2.2.12) with or without a rank constraint, as has been done by Takane et al. (2010). All these developments can be considered special cases of CPCA in which we isolate only the first term in CPCA.

In GMANOVA, however, the main emphasis has been on tests of hypotheses about $\mathbf{M}$ rather than obtaining the reduced rank estimate of $\mathbf{M}$. This leads to a kind of higher order structure in CPCA discussed in Section 3.3.2. A linear hypothesis about $\mathbf{M}$ can generally be stated as

$$\mathbf{R}'\mathbf{M}\mathbf{C} = \mathbf{O}, \tag{4.144}$$

where $\mathbf{R}$ and $\mathbf{C}$ are given constraint matrices. We assume without loss of generality that $\text{Sp}(\mathbf{R}) \subset \text{Sp}(\mathbf{G}')$, and similarly $\text{Sp}(\mathbf{C}) \subset \text{Sp}(\mathbf{H}')$. These conditions are automatically satisfied if $\mathbf{G}$ and $\mathbf{H}$ have full column ranks. An OLSE of $\mathbf{M}$ under the above hypothesis can be derived as follows: Let $\mathbf{X}$ and $\mathbf{Y}$ be such that $\mathbf{R}'\mathbf{X} = \mathbf{O}$ and $\text{Sp}[\mathbf{R}, \mathbf{X}] = \text{Sp}(\mathbf{G}')$, and $\mathbf{C}'\mathbf{Y} = \mathbf{O}$ and $\text{Sp}[\mathbf{C}, \mathbf{Y}] = \text{Sp}(\mathbf{H}')$. (These conditions reduce to $\text{Sp}(\mathbf{X}) = \text{Ker}(\mathbf{R}')$ and $\text{Sp}(\mathbf{Y}) = \text{Ker}(\mathbf{C}')$, respectively, when $\mathbf{G}$ and $\mathbf{H}$ have full column ranks.) Then, $\mathbf{M}$ in (4.144) can be reparameterized as

$$\mathbf{M} = \mathbf{X}\mathbf{M}_{XY}\mathbf{Y}' + \mathbf{M}_Y\mathbf{Y}' + \mathbf{X}\mathbf{M}_X, \tag{4.145}$$

where $\mathbf{M}_{XY}$, $\mathbf{M}_Y$ and $\mathbf{M}_X$ are matrices of unknown parameters. This representation of $\mathbf{M}$ is not unique. For identification, we assume

$$\mathbf{X}'\mathbf{G}'\mathbf{K}\mathbf{G}\mathbf{M}_Y = \mathbf{O}, \tag{4.146}$$

(where $\mathbf{K} = \mathbf{I}$ in GMANOVA), and

$$\mathbf{Y}'\mathbf{H}'\mathbf{L}\mathbf{H}\mathbf{M}_X' = \mathbf{O}, \tag{4.147}$$

(where $\mathbf{L} = \mathbf{S}^{-1}$ in GMANOVA). These constraints are similar to (3.9) and (3.10). Putting (4.145) in model (4.126), we obtain

$$\mathbf{Z} = \mathbf{G}\mathbf{X}\mathbf{M}_{XY}\mathbf{Y}'\mathbf{H}' + \mathbf{G}\mathbf{M}_Y\mathbf{Y}'\mathbf{H}' + \mathbf{G}\mathbf{X}\mathbf{M}_X\mathbf{H}' + \mathbf{E}^*. \tag{4.148}$$

LS estimates of $\mathbf{M}_{XY}$, $\mathbf{M}_Y$ and $\mathbf{M}_X$ under the metric matrices $\mathbf{K}$ and $\mathbf{L}$ subject to (4.146) and (4.147) are given by

$$\hat{\mathbf{M}}_{XY} = (\mathbf{X}'\mathbf{G}'\mathbf{K}\mathbf{G}\mathbf{X})^-\mathbf{X}'\mathbf{G}'\mathbf{K}\mathbf{Z}\mathbf{L}\mathbf{H}\mathbf{Y}(\mathbf{Y}'\mathbf{H}'\mathbf{L}\mathbf{H}\mathbf{Y})^-, \tag{4.149}$$

$$\hat{\mathbf{M}}_Y = \mathbf{P}_{G(G'KG)^-R/K}\mathbf{Z}\mathbf{H}\mathbf{L}\mathbf{Y}(\mathbf{Y}'\mathbf{H}'\mathbf{L}\mathbf{H}\mathbf{Y})^-, \tag{4.150}$$

and

$$\hat{\mathbf{M}}_X = (\mathbf{X}'\mathbf{G}'\mathbf{K}\mathbf{G}\mathbf{X})^-\mathbf{X}'\mathbf{G}\mathbf{K}\mathbf{Z}\mathbf{P}'_{H(H'LH)^-C/L}. \tag{4.151}$$

These are analogous to (3.13), (3.15), and (3.16). Putting (4.149) through (4.151) into (4.148) leads to

$$\mathbf{Z} = \mathbf{P}_{GX/K}\mathbf{Z}\mathbf{P}'_{HY/L} + \mathbf{P}_{G(G'KG)^-R/K}\mathbf{Z}\mathbf{P}'_{HY/L} + \mathbf{P}_{GX/K}\mathbf{Z}\mathbf{P}'_{H(H'LH)^-C/L} + \hat{\mathbf{E}}^*, \tag{4.152}$$

where $\hat{\mathbf{E}}^*$ is defined as $\mathbf{Z}$ minus the sum of the first three terms in (4.152).

There are four terms in the decomposition above. These four terms may be more easily conceptualized as follows. The column space of $\mathbf{Z}$ is split into three mutually orthogonal subspaces (in the metric of $\mathbf{K}$) with the associated projectors, $\mathbf{P}_{GX/K}$, $\mathbf{P}_{G(G'KG)^-R/K}$ and $\mathbf{Q}_{G/K}$. Similarly, the row space of $\mathbf{Z}$ is split into three orthogonal subspaces (in the metric of $\mathbf{L}$) with the associated projectors of $\mathbf{P}_{HY/L}$, $\mathbf{P}_{H(H'LH)^-C/L}$, and $\mathbf{Q}_{H/L}$. By combining the two partitions, we obtain the nine-term partition listed in Table 4.6. The first three terms in (4.152) correspond with (a), (b) and (d) in the

Table 4.6 *Decomposition in GMANOVA.*

Decomposition of Sp($\mathbf{Z}$)	Decomposition of Sp($\mathbf{Z}'$) $\mathbf{P}_{HY/L}$	$\mathbf{P}_{H(H'LH)^-C/L}$	$\mathbf{Q}_{H/L}$
$\mathbf{P}_{GX/K}$	(a)	(b)	(c)
$\mathbf{P}_{G(G'KG)^-R/K}$	(d)	(e)	(f)
$\mathbf{Q}_{G/K}$	(g)	(h)	(i)

table. The fourth term in (4.152), $\hat{\mathbf{E}}^*$, represents the sum of all the remaining terms ((c), (e), (f), (g), (h), and (i)) in the table. See von Rosen (1995) for a similar decomposition.

The hypothesis (4.145) may be generalized to (Rao 1985):

$$\mathbf{M} = \mathbf{X}\mathbf{M}_{XY}\mathbf{Y}' + \mathbf{M}_Y\mathbf{Y}' + \mathbf{X}\mathbf{M}_X + \mathbf{E}_r^*, \tag{4.153}$$

where the additional term $\mathbf{E}_r^*$ is assumed to have a prescribed rank r, and is such that

$$\mathbf{X}'\mathbf{G}'\mathbf{K}\mathbf{G}\mathbf{E}_r^* = \mathbf{O}, \tag{4.154}$$

and

$$\mathbf{E}_r^*\mathbf{H}'\mathbf{L}\mathbf{H}\mathbf{Y} = \mathbf{O}. \tag{4.155}$$

Under (4.153), OLSEs of $\mathbf{M}_{XY}$, $\mathbf{M}_Y$, and $\mathbf{M}_X$ remain the same as in (4.149), (4.150) and (4.151). The estimate of $\mathbf{E}_r^*$, on the other hand, can be obtained as follows: Let $\tilde{\mathbf{E}}_r^*$ be such that

$$\mathbf{G}\tilde{\mathbf{E}}_r^*\mathbf{H}' = \mathbf{P}_{G(G'KG)^-R/K}\mathbf{Z}\mathbf{P}'_{H(H'LH)^-C/L}, \tag{4.156}$$

which is the rank-free OLSE of $\mathbf{G}\mathbf{E}_r^*\mathbf{H}'$. This corresponds with term (e) in Table 4.6. The low rank approximation of $\mathbf{G}\tilde{\mathbf{E}}_r^*\mathbf{H}'$ is obtained by the GSVD$(\mathbf{G}\tilde{\mathbf{E}}_r^*\mathbf{H}')_{K,L}$. Let $\hat{\mathbf{W}}$ represent the low rank approximation of $\mathbf{G}\hat{\mathbf{E}}_r^*\mathbf{H}'$. Then, the low rank approximation $\hat{\mathbf{E}}_r^*$ of $\mathbf{E}_r^*$ is found by

$$\hat{\mathbf{E}}_r^* = (\mathbf{G}'\mathbf{K}\mathbf{G})^-\mathbf{G}'\mathbf{K}\hat{\mathbf{W}}\mathbf{L}\mathbf{H}(\mathbf{H}'\mathbf{L}\mathbf{H})^-, \tag{4.157}$$

or by the GSVD of

$$\tilde{\mathbf{E}}_r^* = (\mathbf{G}'\mathbf{K}\mathbf{G})^-\mathbf{R}(\mathbf{R}'(\mathbf{G}'\mathbf{K}\mathbf{G})^-\mathbf{R})^-\mathbf{R}'(\mathbf{G}'\mathbf{K}\mathbf{G})^-\mathbf{G}'\mathbf{K}\mathbf{Z}\times \\ \mathbf{L}\mathbf{H}(\mathbf{H}'\mathbf{L}\mathbf{H})^-\mathbf{C}(\mathbf{C}'(\mathbf{H}'\mathbf{L}\mathbf{H})^-\mathbf{C})^-\mathbf{C}'(\mathbf{H}'\mathbf{L}\mathbf{H})^- \tag{4.158}$$

with the metric matrices $\mathbf{G}'\mathbf{K}\mathbf{G}$ and $\mathbf{H}'\mathbf{L}\mathbf{H}$.

4.14 Extended Growth Curve Models (ExGCM)

The conventional GCM discussed in the previous section has been generalized to the extended GCM (ExGCM), which has more than one structural term like $\mathbf{G}\mathbf{M}\mathbf{H}'$. The model may be written as

$$\mathbf{Z} = \sum_{j=1}^{J}\mathbf{G}_j\mathbf{M}_j\mathbf{H}_j' + \mathbf{E}^*, \tag{4.159}$$

where $\mathbf{G}_j$, $\mathbf{H}_j$, and $\mathbf{M}_j$ $(j = 1, \cdots, J)$ are analogous to $\mathbf{G}$, $\mathbf{H}$, and $\mathbf{M}$ in (4.126), and $\mathbf{E}^*$ is, as before, the matrix of disturbance terms. The general form of ExGCM given above was first introduced by Verbyla and Venables (1988), who also developed an iterative procedure for finding MLEs of $\mathbf{M}_j$'s under (4.127). No closed-form solution exists for the MLE of the $\mathbf{M}_j$'s unless Σ^* is known or simplifies to $\Sigma^* = \sigma^2 \mathbf{I}_m$. A model similar to (4.159) has been proposed by Takane et al. (1995). Their model, called DCDD (different constraints on different dimensions), imposes a reduced rank restriction on the $\mathbf{M}_j$'s, and will be discussed in Chapter 6 along with algorithms to fit the model.

In this section we primarily focus on a special case of (4.159) called the Banken model (von Rosen 1989), in which it is assumed that

$$\text{Sp}(\mathbf{G}_j) \supset \text{Sp}(\mathbf{G}_{j+1}) \tag{4.160}$$

for $j = 1, \cdots, J-1$, and that the $\mathbf{H}_j$'s are disjoint (i.e., $\text{Sp}(\mathbf{H}_j) \cap \text{Sp}(\mathbf{H}_k) = \{\mathbf{0}\}$) for $j \neq k$. Under this assumption, a closed-form solution for the MLE of the $\mathbf{M}_j$'s has been derived by von Rosen (1989). Fujikoshi and Satoh (1996) considered the condition that

$$\text{Sp}(\mathbf{H}_j) \subset \text{Sp}(\mathbf{H}_{j+1}) \tag{4.161}$$

for $j = 1, \cdots, J-1$. This condition looks different from (4.160). However, it turns out that the two conditions are equivalent, as will be shown below for $J = 2$. Let the Banken model be written as

$$\mathbf{Z} = \mathbf{G}_1\mathbf{M}_1\mathbf{H}_1' + \mathbf{G}_2\mathbf{M}_2\mathbf{H}_2' + \mathbf{E}^* \tag{4.162}$$

for $J = 2$. Let $\tilde{\mathbf{G}}_1$ be such that $\text{Sp}(\tilde{\mathbf{G}}_1) \oplus \text{Sp}(\mathbf{G}_2) = \text{Sp}(\mathbf{G}_1)$. Then, (4.162) can be rewritten as

$$\begin{aligned} \mathbf{Z} &= [\tilde{\mathbf{G}}_1, \mathbf{G}_2] \begin{bmatrix} \mathbf{M}_{11}^* \\ \mathbf{M}_{21}^* \end{bmatrix} \mathbf{H}_1' + \mathbf{G}_2\mathbf{M}_2\mathbf{H}_2' + \mathbf{E}^* \\ &= \tilde{\mathbf{G}}_1\mathbf{M}_{11}^*\mathbf{H}_1' + \mathbf{G}_2[\mathbf{M}_{21}^*, \mathbf{M}_2] \begin{bmatrix} \mathbf{H}_1' \\ \mathbf{H}_2' \end{bmatrix} + \mathbf{E}^*. \end{aligned} \tag{4.163}$$

If we reset $[\mathbf{H}_1, \mathbf{H}_2]$ as $\mathbf{H}_2$, it holds that $\text{Sp}(\mathbf{H}_2) \supset \text{Sp}(\mathbf{H}_1)$. Now $\tilde{\mathbf{G}}_1$ and $\mathbf{G}_2$ are disjoint. Note that the proof can easily be extended to $J > 2$.

The constraint in the Banken model, (4.160) or (4.161), may look restrictive. However, as the following example shows, a large number of empirically interesting models are subsumed under the Banken model. Suppose that there are two groups of subjects, say male and female, whose weights are measured over time. Suppose further that the female group exhibits both linear and quadratic trends in weight change, while the male group shows only a linear trend. We may then set $\mathbf{G}_1$ indicating both male and female groups, while $\mathbf{G}_2$ for the female group only (i.e., the portion of $\mathbf{G}_1$ pertaining to the female group only). We also set $\mathbf{H}_1$ and $\mathbf{H}_2$ to represent the linear and quadratic trends separately. (Alternatively, we may set $\mathbf{G}_1$ and $\mathbf{G}_2$ to indicate the two groups of subjects separately, and set $\mathbf{H}_1$ to represent the linear trend only, while $\mathbf{H}_2$ for both trends.)

Verbyla and Venables (1988) proposed an iterative method to obtain MLEs of parameters in the general extended GCM. They used a transformational approach similar to Khatri's solution for $J = 1$ discussed in the previous section. Their iterative method is known to converge in one iteration for the Banken model. Each step of their algorithm consists of a simple OLS estimation of multiple regression equations.

Let $\mathbf{H} = [\mathbf{H}_1, \cdots, \mathbf{H}_J]$, and similarly to Khatri's solution in the previous section, let

$$\mathbf{T} = [\mathbf{T}_1, \mathbf{T}_2], \tag{4.164}$$

where $\mathbf{T}_1$ is such that $\mathbf{H}'\mathbf{T}_1 = \mathbf{I}$, and $\mathbf{T}_2$ is such that $\mathrm{Sp}(\mathbf{T}_2) = \mathrm{Ker}(\mathbf{H}')$. Define

$$\mathbf{Y} = \mathbf{Z}\mathbf{T}. \tag{4.165}$$

Then,

$$\begin{aligned} \mathbf{Y} = [\mathbf{Y}_1, \mathbf{Y}_2] &= [\mathbf{Y}_{11}, \cdots, \mathbf{Y}_{1J}, \mathbf{Y}_2] \\ &= [\mathbf{G}_1\mathbf{M}_1, \cdots, \mathbf{G}_J\mathbf{M}_J, \mathbf{O}] + \mathbf{E}, \end{aligned} \tag{4.166}$$

where $\mathbf{E} = \mathbf{E}^*\mathbf{T}$. The conditional expectation of $\mathbf{Y}_1$ given $\mathbf{Y}_2$ is then given by

$$\begin{aligned} \mathrm{Ex}[\mathbf{Y}_1|\mathbf{Y}_2] &= [\mathbf{G}_1\mathbf{M}_1, \cdots, \mathbf{G}_J\mathbf{M}_J] + \mathbf{Y}_2[\mathbf{B}_1, \cdots, \mathbf{B}_J], \\ &= [\mathbf{X}_1\mathbf{A}_1, \cdots, \mathbf{X}_J\mathbf{A}_J], \end{aligned} \tag{4.167}$$

where $\mathbf{X}_j = [\mathbf{G}_j, \mathbf{Y}_2]$, and $\mathbf{A}'_j = [\mathbf{M}'_j, \mathbf{B}'_j]'$ for $j = 1, \cdots, J$. Verbyla and Venables (1988) update $\mathbf{A}_j$ $(j = 1, \cdots, J)$ conditionally on the other $\mathbf{A}_k$'s $(k \neq j)$ in a manner similar to an ALS (alternating LS) algorithm. The estimate of $\mathbf{A}_j$ given the other $\mathbf{A}_k$'s is obtained from the conditional expectation of $\mathbf{Y}_{1j}$ given $\mathbf{Y}_{1k}$ $(k \neq j)$, namely

$$\mathrm{Ex}[\mathbf{Y}_{1j}|\mathbf{Y}_{1k}, k \neq j] = \mathbf{X}_j\mathbf{A}_j + \sum_{k \neq j}(\mathbf{Y}_{1k} - \mathbf{X}_k\mathbf{A}_k)\mathbf{C}_{kj}. \tag{4.168}$$

The estimates of $\mathbf{A}_j$ and $\mathbf{C}_{kj}$ $(k \neq j)$ above can be obtained by regressing $\mathbf{Y}_{1j}$ onto $[\mathbf{X}_j, \mathbf{U}_{(-j)}]$, where $\mathbf{U}_{(-j)}$ is the row block matrix formed by $\mathbf{Y}_{1k} - \mathbf{X}_k\mathbf{A}_k$ $(k \neq j)$ arranged side by side. In the case of general ExGCM, this updating scheme is applied cyclically over j and repeatedly until convergence. The initial $\mathbf{A}_j$ may be obtained simply by regressing $\mathbf{Y}_{1j}$ onto $\mathbf{X}_j$.

In the case of the Banken model, the iterative procedure described above converges in one iteration due to the special relationship among the $\mathbf{G}_j$'s introduced in (4.160). We start with $j = J$, and obtain the OLSE of $\mathbf{A}_J$ by regressing $\mathbf{Y}_{1J}$ onto $\mathbf{X}_J\mathbf{A}_J$. We do not need to include $\mathbf{U}_{(-J)}$ in this regression because it is orthogonal to $\mathbf{X}_J$, and so has no effect on the estimate of $\mathbf{A}_J$. We then estimate $\mathbf{A}_{J-1}$ by regressing $\mathbf{Y}_{J-1}$ onto $[\mathbf{X}_{J-1}, \mathbf{Y}_J - \mathbf{X}_J\hat{\mathbf{A}}_J]$, where $\hat{\mathbf{A}}_J$ is the estimate of $\mathbf{A}_J$ obtained in the previous step. Note that we appended the residuals from the previous regression, i.e., $\mathbf{Y}_J - \mathbf{X}_J\hat{\mathbf{A}}_J$, to $\mathbf{X}_{J-1}$ in this regression. In the next step, we further append $\mathbf{Y}_{J-1} - \mathbf{X}_{J-1}\hat{\mathbf{A}}_{J-1}$. This process is repeated until $j = 1$.

Von Rosen (1989) derived MLEs of $\mathbf{M}_j$'s and Σ for the Banken model more directly. Below we summarize his solution. For simplicity, we assume that both $\mathbf{G}_j$'s and $\mathbf{H}_j$'s are columnwise nonsingular, although von Rosen (1989) deals with more general cases. Let

$$\mathbf{R}_r = \prod_{k=0}^{r-1} \mathbf{T}_k, \quad \text{where} \quad \mathbf{T}_0 = \mathbf{I} \tag{4.169}$$

for $r = 1, \cdots, J+1$, where J is the number of structural terms in the model,

$$\mathbf{T}_i = \mathbf{I} - \mathbf{R}_i\mathbf{H}_i(\mathbf{H}_i'\mathbf{R}_i'\mathbf{S}_i^{-1}\mathbf{R}_i\mathbf{H}_i)^{-1}\mathbf{H}_i'\mathbf{R}_i'\mathbf{S}_i^{-1} \tag{4.170}$$

for $i = 1, \cdots, J$,

$$\mathbf{S}_i = \sum_{j=1}^{i} \mathbf{W}_j, \tag{4.171}$$

where

$$\mathbf{W}_j = \mathbf{R}_j\mathbf{Z}'\mathbf{P}_{G_{j-1}}\mathbf{Q}_{G_j}\mathbf{P}_{G_{j-1}}\mathbf{Z}\mathbf{R}_j' \tag{4.172}$$

with $\mathbf{G}_0 = \mathbf{I}$. Then,

$$\mathbf{G}_r\hat{\mathbf{M}}_r\mathbf{H}_r' = \mathbf{P}_{G_r}(\mathbf{Z} - \sum_{i=r+1}^{J} \mathbf{G}_i\hat{\mathbf{M}}_i\mathbf{H}_i')\mathbf{P}'_{R_rH_r/S_r^{-1}} \tag{4.173}$$

(from which $\hat{\mathbf{M}}_r$ can be easily recovered), and

$$\begin{aligned} n\hat{\Sigma}^* &= (\mathbf{Z} - \sum_{i=1}^{J} \mathbf{G}_i\hat{\mathbf{M}}_i\mathbf{H}_i')'(\mathbf{Z} - \sum_{i=1}^{J} \mathbf{G}_i\hat{\mathbf{M}}_i\mathbf{H}_i') \\ &= \mathbf{S}_J + \mathbf{R}_{J+1}\mathbf{Z}'\mathbf{P}_m\mathbf{Z}\mathbf{R}_{J+1}'. \end{aligned} \tag{4.174}$$

For the Banken model, Verbyla and Venables' solution and von Rosen's solution should give identical results, although they look completely different. Takane and Zhou (2012) showed the algebraic equivalence between the two for the special case of $J = 2$.

4.15 Seemingly Unrelated Regression (SUR)

Consider J regression equations,

$$\mathbf{z}_j = \mathbf{G}_j\mathbf{b}_j + \mathbf{e}_j, \qquad j = 1, \cdots, J, \tag{4.175}$$

where $\mathbf{z}_j$ is an n-component vector of observations on the criterion variable j, $\mathbf{G}_j$ an n by p_j matrix of predictor variables, $\mathbf{b}_j$ an p_j-component vector of regression coefficients, and $\mathbf{e}_j$ an n-component vector of disturbance terms. We assume that $\text{Ex}[\mathbf{e}_j] = \mathbf{0}$, and that

$$\text{Ex}[\mathbf{e}_i\mathbf{e}_j'] = \sigma_{ij}\mathbf{I}_n. \tag{4.176}$$

We would like to estimate $\mathbf{b}_j$ based on observed $\mathbf{z}_j$ and $\mathbf{G}_j$ ($j = 1, \cdots, J$). This problem is called seemingly unrelated regression (SUR) by Zellner (1962, 1963).

The problem is interesting because while it is possible to estimate $\mathbf{b}_j$ separately for each j by OLS, a joint estimation may provide more efficient estimates. Zellner (1962) showed that this is indeed the case when σ_{ij} is nonzero, and the $\mathbf{G}_j$'s are not identical across all j. For the joint estimation we first rewrite the J separate equations into a single equation:

$$\mathbf{z} = \mathbf{D}_G\mathbf{b} + \mathbf{e}, \tag{4.177}$$

where

$$\mathbf{z} = \begin{pmatrix} \mathbf{z}_1 \\ \vdots \\ \mathbf{z}_J \end{pmatrix}, \quad \mathbf{b} = \begin{pmatrix} \mathbf{b}_1 \\ \vdots \\ \mathbf{b}_J \end{pmatrix}, \quad \mathbf{e} = \begin{pmatrix} \mathbf{e}_1 \\ \vdots \\ \mathbf{e}_J \end{pmatrix},$$

and $\mathbf{D}_G$ is the block diagonal matrix with $\mathbf{G}_j$ as the jth diagonal block,

$$\mathbf{D}_G = \begin{bmatrix} \mathbf{G}_1 & \mathbf{O} & \cdots & \mathbf{O} \\ \mathbf{O} & \mathbf{G}_2 & \cdots & \mathbf{O} \\ \vdots & \vdots & \ddots & \vdots \\ \mathbf{O} & \mathbf{O} & \cdots & \mathbf{G}_J \end{bmatrix}. \tag{4.178}$$

Zellner (1962) proposed to use a two-step Aitken estimator (GLSE), in which a consistent estimator of $\Sigma = [\sigma_{ij}]$ in (4.176) is first obtained and then used in the GLSE of $\mathbf{b}$ as if it were the population covariance matrix. This gives

$$\hat{\mathbf{b}} = (\mathbf{D}_G'(\hat{\Sigma}^{-1} \otimes \mathbf{I}_n)\mathbf{D}_G)^{-1}\mathbf{D}_G'(\hat{\Sigma}^{-1} \otimes \mathbf{I}_n)\mathbf{z}. \tag{4.179}$$

There are two commonly used estimators of σ_{ij} in (4.176). Zellner (1962) suggested to use the restricted estimate of σ_{ij}. This is obtained by first obtaining the separate OLSE of $\mathbf{b}_j$ in (4.175), from which the so-called restricted residual vector $\underline{\hat{\mathbf{e}}}_j$ is calculated, and then the restricted estimate of σ_{ij} is calculated by

$$n\underline{\hat{\sigma}}_{ij} = \underline{\hat{\mathbf{e}}}_i'\underline{\hat{\mathbf{e}}}_j. \tag{4.180}$$

Zellner (1963) also proposed the unrestricted estimate of σ_{ij}. This is calculated from the unrestricted residual vector $\hat{\mathbf{e}}_j$ calculated by regressing each $\mathbf{z}_j$ on all available predictor variables $[\mathbf{G}_1, \cdots, \mathbf{G}_J]$. Then

$$n\hat{\sigma}_{ij} = \hat{\mathbf{e}}_i'\hat{\mathbf{e}}_j. \tag{4.181}$$

The restricted and unrestricted estimators of σ_{ij} are both consistent and asymptotically equivalent. Zellner (1963) showed, for the case in which $J = 2$ and $\mathbf{G}_1'\mathbf{G}_2 = \mathbf{O}$, that the restricted estimate of σ_{ij} leads to a more efficient estimate of $\mathbf{b}$ when n is small and $\rho_{ij} = \sigma_{ij}/(\sigma_{ii}\sigma_{jj})^{1/2}$ is small. Conniffe (1982), however, argued that the advantage of the restricted estimator was usually minor and was limited to a small range of n and ρ_{ij}, and recommended the use of the unrestricted estimate of σ_{ij}.

Interestingly, GCM discussed in Section 4.13 can be regarded as a special case of the SUR model (Stanek and Koch 1985; Reinsel and Velu 1998). Note first that (4.130) can be rewritten as

$$\mathbf{Y} = \mathbf{GMR}' + \mathbf{E}. \tag{4.182}$$

By vectorizing model (4.182), we obtain

$$\mathbf{y} = (\mathbf{R} \otimes \mathbf{G})\mathbf{m} + \mathbf{e}, \tag{4.183}$$

where $\mathbf{y} = \text{vec}(\mathbf{Y})$, $\mathbf{m} = \text{vec}(\mathbf{M})$, and $\mathbf{e} = \text{vec}(\mathbf{E}) \sim \mathcal{N}(\mathbf{0}, \Sigma \otimes \mathbf{I}_n)$, and

$$\mathbf{R} = \begin{bmatrix} \mathbf{I}_q \\ \mathbf{O}_{m-q,q} \end{bmatrix}, \tag{4.184}$$

so that

$$\mathbf{R} \otimes \mathbf{G} = \begin{bmatrix} \mathbf{I}_q \otimes \mathbf{G} \\ \mathbf{O}_{m-q,q} \otimes \mathbf{G} \end{bmatrix} = \begin{bmatrix} \mathbf{G} & \mathbf{O} & \cdots & \mathbf{O} \\ \mathbf{O} & \mathbf{G} & \cdots & \mathbf{O} \\ \vdots & \vdots & \ddots & \vdots \\ \mathbf{O} & \mathbf{O} & \cdots & \mathbf{G} \\ \mathbf{O} & \mathbf{O} & \cdots & \mathbf{O} \end{bmatrix}. \tag{4.185}$$

Note that (4.183) is precisely in the form of (4.177). The two-step Aitken estimator (GLSE) of $\mathbf{m}$ is obtained by

$$\begin{aligned} \hat{\mathbf{m}} &= ((\mathbf{R} \otimes \mathbf{G})'(\mathbf{A}^{-1} \otimes \mathbf{I}_n)(\mathbf{R} \otimes \mathbf{G}))^{-1}(\mathbf{R} \otimes \mathbf{G})'(\mathbf{A}^{-1} \otimes \mathbf{I}_n)\mathbf{y} \\ &- [(\mathbf{R}'\mathbf{A}^{-1}\mathbf{R})^{-1}\mathbf{R}'\mathbf{\Lambda}^{-1} \otimes (\mathbf{G}'\mathbf{G})^{-1}\mathbf{G}']\mathbf{y}, \end{aligned} \tag{4.186}$$

where

$$\mathbf{A} = \mathbf{Y}'\mathbf{Q}_G\mathbf{Y}. \tag{4.187}$$

This is proportional to the unrestricted estimate of Σ.

By rearranging the elements of $\hat{\mathbf{m}}$, we obtain

$$\begin{aligned} \hat{\mathbf{M}} &= (\mathbf{G}'\mathbf{G})^{-1}\mathbf{G}'\mathbf{Y}\mathbf{A}^{-1}\mathbf{R}(\mathbf{R}'\mathbf{A}^{-1}\mathbf{R})^{-1} \\ &= (\mathbf{G}'\mathbf{G})^{-1}\mathbf{G}'\mathbf{Z}\mathbf{T}\mathbf{A}^{-1}\mathbf{T}'(\mathbf{T}')^{-1}\mathbf{R} \times \\ &\qquad (\mathbf{R}'\mathbf{T}^{-1}\mathbf{T}\mathbf{A}^{-1}\mathbf{T}'(\mathbf{T}')^{-1}\mathbf{R})^{-1} \qquad (4.188) \\ &= (\mathbf{G}'\mathbf{G})^{-1}\mathbf{G}'\mathbf{Z}\mathbf{S}^{-1}\mathbf{H}(\mathbf{H}'\mathbf{S}^{-1}\mathbf{H})^{-1}, \qquad (4.189) \end{aligned}$$

where $\mathbf{T}$ is as introduced in (4.128), and $\mathbf{S}$ is as defined in (4.139). Note that

$$\mathbf{A} = \mathbf{T}'\mathbf{S}\mathbf{T}, \tag{4.190}$$

so that

$$\mathbf{A}^{-1} = \mathbf{T}^{-1}\mathbf{S}^{-1}(\mathbf{T}')^{-1}, \tag{4.191}$$

and

$$\mathbf{S}^{-1} = \mathbf{T}\mathbf{A}^{-1}\mathbf{T}'. \tag{4.192}$$

Hence

$$(\mathbf{T}')^{-1}\mathbf{R} = [\mathbf{H}, **]\begin{bmatrix} \mathbf{I}_q \\ \mathbf{O}_{m-q,q} \end{bmatrix} = \mathbf{H}, \tag{4.193}$$

where $**$ is any matrix completing $(\mathbf{T}')^{-1}$, depending on the specific form of $\mathbf{T}_2$.

The relationship between SUR and GCM discussed above can be extended to SUR and the Banken model. Revankar (1974) discussed the two-step Aitken estimator (GLSE with the unrestricted estimate of σ_{ij} for SUR with $J = 2$ and $\mathbf{G}_1 = [\mathbf{G}_2, \mathbf{L}]$ for some $\mathbf{L}$ (i.e., $\mathbf{G}_2$ is nested within $\mathbf{G}_1$). He has shown that in this case the two-step Aitken estimator (GLSE) for $\mathbf{b}_2$ is identical to the direct OLSE of $\mathbf{b}_2$, that is,

$$\hat{\mathbf{b}}_2 = (\mathbf{G}_2'\mathbf{G}_2)^{-1}\mathbf{G}_2'\mathbf{z}_2, \tag{4.194}$$

and that the two-step Aitken estimator for $\mathbf{b}_1$ is given by

$$\hat{\mathbf{b}}_1 = (\mathbf{G}_1'\mathbf{G}_1)^{-1}\mathbf{G}_1'(\mathbf{z}_1 - \frac{\hat{\sigma}_{12}}{\hat{\sigma}_{22}}\mathbf{Q}_{G_2}\mathbf{z}_2), \tag{4.195}$$

$$= (\mathbf{G}_1'\mathbf{G}_1)^{-1}\mathbf{G}_1'(\mathbf{z}_1 - \frac{\hat{\underline{\mathbf{e}}}_2'\mathbf{Q}_{G_1}\mathbf{z}_1}{\hat{\underline{\mathbf{e}}}_2'\mathbf{Q}_{G_1}\hat{\underline{\mathbf{e}}}_2}\hat{\underline{\mathbf{e}}}_2), \tag{4.196}$$

where

$$\hat{\underline{\mathbf{e}}}_2 = \mathbf{Q}_{G_2}\mathbf{e}_2 = \mathbf{Q}_{G_2}\mathbf{z}_2 \tag{4.197}$$

is the restricted residual vector. Notice that we again used Seber's trick (Section 2.2.10) to derive (4.196) from (4.195). Note also that $\hat{\underline{\mathbf{e}}}_2'\mathbf{Q}_{G_1}\mathbf{z}_1 = \mathbf{e}_2'\mathbf{Q}_{G_2}\mathbf{Q}_{G_1}\mathbf{e}_1 = \mathbf{e}_2'\mathbf{Q}_{G_1}\mathbf{e}_1 = \hat{\mathbf{e}}_1'\hat{\mathbf{e}}_2 = n\hat{\sigma}_{12}$ (where $\hat{\mathbf{e}}_1 = \mathbf{Q}_{G_1}\mathbf{e}_1$ and $\hat{\mathbf{e}}_2 = \mathbf{Q}_{G_1}\mathbf{e}_2$ are unrestricted residual vectors), and $\hat{\underline{\mathbf{e}}}_2'\mathbf{Q}_{G_1}\hat{\underline{\mathbf{e}}}_2 = n\hat{\sigma}_{22}$. Note also that (4.197) can be derived as the direct OLS estimator of $\mathbf{b}_1$ in the following regression model (obtained by appending $\hat{\underline{\mathbf{e}}}_2$ to the original regression model for $\mathbf{z}_1$):

$$\mathbf{z}_1 = \mathbf{G}_1\mathbf{b}_1 + \hat{\underline{\mathbf{e}}}_2\theta + \mathbf{r}, \tag{4.198}$$

where θ is the regression coefficient for $\hat{\underline{\mathbf{e}}}_2$ and $\mathbf{r}$ is the vector of disturbance terms. An important point to notice is that by including $\hat{\underline{\mathbf{e}}}_2$ in (4.198) the efficiency of the estimate of $\mathbf{b}_1$ can be improved. That is, $\hat{\mathbf{b}}_1$ given in (4.196) is more efficient than the OLSE of $\mathbf{b}_1$ given as the first term of (4.195) and (4.196). This is called a covariance adjustment (Telser 1964; Conniffe 1982). This situation resembles that of the Banken model, in which $\mathbf{G}_2$ is assumed nested within $\mathbf{G}_1$.

4.16 Wedderburn–Guttman Decomposition

Let $\mathbf{Z}$ denote an $n \times m$ matrix, and let $\mathbf{M}$ and $\mathbf{N}$ be matrices such that $\mathbf{M}'\mathbf{Z}\mathbf{N}$ is square and nonsingular. It has been shown by Guttman (1944, 1952, 1957) that

$$\begin{aligned}\text{rank}(\mathbf{Z} - \mathbf{Z}\mathbf{N}(\mathbf{M}'\mathbf{Z}\mathbf{N})^{-1}\mathbf{M}'\mathbf{Z}) &= \text{rank}(\mathbf{Z}) - \text{rank}(\mathbf{Z}\mathbf{N}(\mathbf{M}'\mathbf{Z}\mathbf{N})^{-1}\mathbf{M}'\mathbf{Z}) \\ &= \text{rank}(\mathbf{Z}) - \text{rank}(\mathbf{M}'\mathbf{Z}\mathbf{N}).\end{aligned} \tag{4.199}$$

This is called the Wedderburn–Guttman (WG) theorem. This theorem has been used extensively in psychometrics to derive the rank of a residual matrix when the effects of certain components are extracted from the original data matrix $\mathbf{Z}$. Takane and Yanai (2005, 2007, 2009) generalized the theorem by identifying the necessary and sufficient condition under which the regular inverse of $\mathbf{M}'\mathbf{Z}\mathbf{N}$ in (4.199) may be replaced by a generalized inverse. It turns out that $\text{rank}(\mathbf{Z}-\mathbf{ZN}(\mathbf{M}'\mathbf{ZN})^{-}\mathbf{M}'\mathbf{Z}) = \text{rank}(\mathbf{Z}) - \text{rank}(\mathbf{M}'\mathbf{ZN})$ holds unconditionally. However, the rank subtractivity, i.e., $\text{rank}(\mathbf{Z}-\mathbf{ZN}(\mathbf{M}'\mathbf{ZN})^{-}\mathbf{M}'\mathbf{Z}) = \text{rank}(\mathbf{Z}) - \text{rank}(\mathbf{ZN}(\mathbf{M}'\mathbf{ZN})^{-}\mathbf{M}'\mathbf{Z})$, or equivalently $\text{rank}(\mathbf{ZN}(\mathbf{M}'\mathbf{ZN})^{-}\mathbf{M}'\mathbf{Z}) = \text{rank}(\mathbf{M}'\mathbf{ZN})$, requires a condition. Let $\mathbf{A} = \mathbf{N}(\mathbf{M}'\mathbf{ZN})^{-}\mathbf{M}'$. Then this condition is stated as (Takane and Yanai 2005)

$$\mathbf{ZAZAZ} = \mathbf{ZAZ}. \tag{4.200}$$

This condition along with the rank formula is called the extended WG theorem.

Note 4.8 There are several notable equivalent conditions to (4.200) (Takane and Yanai 2005): (i) $\mathbf{Z}^{-} \in \{(\mathbf{ZAZ})^{-}\}$, (ii) $\mathbf{Z}^{-} \in \{(\mathbf{Z}-\mathbf{ZAZ})^{-}\}$, (iii) $\mathbf{ZAZZ}^{-}$ is the projector onto $\text{Sp}(\mathbf{ZAZ})$ along $\text{Ker}(\mathbf{ZAZZ}^{-})$, and (iv) $\mathbf{Z}^{-}\mathbf{ZAZ}$ is the projector onto $\text{Sp}(\mathbf{Z}^{-}\mathbf{ZAZ})$ along $\text{Ker}(\mathbf{ZAZ})$. Note that there are many sufficient conditions, such as (v) $\mathbf{ZA}$ is the projector onto $\text{Sp}(\mathbf{ZA})$ along $\text{Ker}(\mathbf{ZA})$, and (vi) $\mathbf{AZ}$ is the projector onto $\text{Sp}(\mathbf{AZ})$ along $\text{Ker}(\mathbf{AZ})$, which all imply (i) - (iv) above.

The original and extended WG theorems imply the following decomposition of $\mathbf{Z}$:

$$\mathbf{Z} = \mathbf{ZN}(\mathbf{M}'\mathbf{ZN})^{-}\mathbf{M}'\mathbf{Z} + (\mathbf{Z}-\mathbf{ZN}(\mathbf{M}'\mathbf{ZN})^{-}\mathbf{M}'\mathbf{Z}), \tag{4.201}$$

which we call the Wedderburn–Guttman (WG) decomposition. In this section we elaborate on this decomposition, and discuss its relationship with decompositions given in Sections 2.2.11(iii) and 3.3.3.

We begin by rewriting the second term in (4.201) as a single matrix rather than a difference between two matrices. Let $\tilde{\mathbf{M}}$, and $\tilde{\mathbf{N}}$ be such that $\text{Sp}(\tilde{\mathbf{M}}) \subset \text{Sp}(\mathbf{Z})$, $\text{Sp}(\tilde{\mathbf{N}}) \subset \text{Sp}(\mathbf{Z}')$,

$$\text{rank}(\mathbf{M}'\mathbf{ZN}) + \text{rank}(\tilde{\mathbf{N}}'\mathbf{Z}^{-}\tilde{\mathbf{M}}) = \text{rank}(\mathbf{Z}), \tag{4.202}$$

where $\mathbf{Z}^{-}$ is a g-inverse of $\mathbf{Z}$,

$$\mathbf{M}'\mathbf{ZZ}^{-}\tilde{\mathbf{M}} = \mathbf{M}'\tilde{\mathbf{M}} = \mathbf{O}, \tag{4.203}$$

and

$$\tilde{\mathbf{N}}'\mathbf{Z}^{-}\mathbf{ZN} = \tilde{\mathbf{N}}'\mathbf{N} = \mathbf{O}. \tag{4.204}$$

Then the following decomposition holds:

$$\mathbf{Z} = \mathbf{ZN}(\mathbf{M}'\mathbf{ZN})^{-}\mathbf{M}'\mathbf{Z} + \tilde{\mathbf{M}}(\tilde{\mathbf{N}}'\mathbf{Z}^{-}\tilde{\mathbf{M}})^{-}\tilde{\mathbf{N}}'. \tag{4.205}$$

Let $\mathbf{Y}_1 = [\mathbf{N}, \mathbf{Z}^{-}\tilde{\mathbf{M}}]$, and $\mathbf{Y}_2 = [\mathbf{M}, (\mathbf{Z}^{-})'\tilde{\mathbf{N}}]$. Then we have

$$\mathbf{Y}_2'\mathbf{ZY}_1 = \begin{bmatrix} \mathbf{M}'\mathbf{ZN} & \mathbf{O} \\ \mathbf{O} & \tilde{\mathbf{N}}'\mathbf{Z}^{-}\tilde{\mathbf{M}} \end{bmatrix}, \tag{4.206}$$

and

$$\mathbf{ZY}_1(\mathbf{Y}_2'\mathbf{ZY}_1)^{-*}\mathbf{Y}_2'\mathbf{Z} = \mathbf{ZN}(\mathbf{M}'\mathbf{ZN})^{-}\mathbf{M}'\mathbf{Z} + \tilde{\mathbf{M}}(\tilde{\mathbf{N}}'\mathbf{Z}^{-}\tilde{\mathbf{M}})^{-}\tilde{\mathbf{N}}', \tag{4.207}$$

where

$$(\mathbf{Y}_2'\mathbf{Z}\mathbf{Y}_1)^{-*} = \begin{bmatrix} (\mathbf{M}'\mathbf{Z}\mathbf{N})^- & \mathbf{O} \\ \mathbf{O} & (\tilde{\mathbf{N}}'\mathbf{Z}^-\tilde{\mathbf{M}})^- \end{bmatrix}. \tag{4.208}$$

Clearly, $(\mathbf{Y}_2'\mathbf{Z}\mathbf{Y}_1)^{-*} \in \{(\mathbf{Y}_2'\mathbf{Z}\mathbf{Y}_1)^-\}$. From (4.206) and (4.202) we have

$$\text{rank}(\mathbf{Y}_2'\mathbf{X}\mathbf{Y}_1) = \text{rank}(\mathbf{M}'\mathbf{Z}\mathbf{N}) + \text{rank}(\tilde{\mathbf{N}}'\mathbf{Z}^-\tilde{\mathbf{M}}) = \text{rank}(\mathbf{Z}). \tag{4.209}$$

Decomposition (4.205) follows from Theorem 2.1 of Mitra (1968), which states that $\mathbf{Z}\mathbf{Y}_1(\mathbf{Y}_2'\mathbf{Z}\mathbf{Y}_1)^-\mathbf{Y}_2'\mathbf{Z} = \mathbf{Z}$ (i.e., $\mathbf{Y}_1(\mathbf{Y}_2'\mathbf{Z}\mathbf{Y}_1)^-\mathbf{Y}_2' \in \{\mathbf{Z}^-\}$) if and only if $\text{rank}(\mathbf{Y}_2'\mathbf{Z}\mathbf{Y}_1) = \text{rank}(\mathbf{Z})$.

Decomposition (4.205) may be considered as a rectangular version of Khatri's (1966) lemma (Section 2.2.8). In Khatri's original lemma, it was assumed that **Z** was symmetric and *pd*, and that **M** and **N** were identical to each other. In Khatri (1990), this condition was relaxed somewhat, but **Z** was still assumed square.

There are a few special cases of the decomposition above which are of interest to us. Let

$$\text{rank}(\mathbf{M}'\mathbf{Z}\mathbf{N}) = \text{rank}(\mathbf{Z}\mathbf{N}) = \text{rank}(\mathbf{M}'\mathbf{Z}\mathbf{Z}^-), \tag{4.210}$$

and

$$\text{rank}(\tilde{\mathbf{N}}'\mathbf{Z}^-\tilde{\mathbf{M}}) = \text{rank}(\tilde{\mathbf{M}}) = \text{rank}(\tilde{\mathbf{N}}'\mathbf{Z}^-). \tag{4.211}$$

Then,

$$\mathbf{Z}\mathbf{Z}^- = \mathbf{Z}\mathbf{N}(\mathbf{M}'\mathbf{Z}\mathbf{N})^-\mathbf{M}'\mathbf{Z}\mathbf{Z}^- + \tilde{\mathbf{M}}(\tilde{\mathbf{N}}'\mathbf{Z}^-\tilde{\mathbf{M}})^-\tilde{\mathbf{N}}'\mathbf{Z}^-, \tag{4.212}$$

where $\mathbf{Z}\mathbf{Z}^-$ is the projector onto Sp($\mathbf{Z}$) along Ker($\mathbf{Z}\mathbf{Z}^-$) (note that $\mathbf{Z}\mathbf{Z}^- = \mathbf{Z}(\mathbf{Z}\mathbf{Z}^-\mathbf{Z})^-\mathbf{Z}\mathbf{Z}^-$), the first term on the right-hand side of (4.212) is the projector onto Sp($\mathbf{Z}\mathbf{N}$) along Ker($\mathbf{M}'\mathbf{Z}\mathbf{Z}^-$), and the second term is the projector onto Sp($\tilde{\mathbf{M}}$) along Ker($\tilde{\mathbf{N}}'\mathbf{Z}^-$). In the notation adopted in (2.152) and (2.153), (4.212) may be written as

$$\mathbf{P}_{Z:(ZZ^-)'^{\perp}} = \mathbf{P}_{ZN:(M'ZZ^-)'^{\perp}} + \mathbf{P}_{\tilde{M}:(\tilde{N}'Z^-)'^{\perp}}. \tag{4.213}$$

The decomposition above follows by postmultiplying (4.205) by $\mathbf{Z}^-$. Conditions (4.210) and (4.211) ensure that the two terms on the right-hand side of (4.212) are the projectors with the prescribed onto and along spaces. Note that (4.213) is a special case of (iii-a") in (2.205). Note also that the two terms on the right-hand side of (4.212) (and of (4.213)) satisfy (2.94), the condition under which the sum of two projectors is also a projector.

Similarly, let

$$\text{rank}(\mathbf{M}'\mathbf{Z}\mathbf{N}) = \text{rank}(\mathbf{Z}^-\mathbf{Z}\mathbf{N}) = \text{rank}(\mathbf{M}'\mathbf{Z}), \tag{4.214}$$

and

$$\text{rank}(\tilde{\mathbf{N}}'\mathbf{Z}^-\tilde{\mathbf{M}}) = \text{rank}(\mathbf{Z}^-\tilde{\mathbf{M}}) = \text{rank}(\tilde{\mathbf{N}}). \tag{4.215}$$

Then,

$$\mathbf{Z}^-\mathbf{Z} = \mathbf{Z}^-\mathbf{Z}\mathbf{N}(\mathbf{M}'\mathbf{Z}\mathbf{N})^-\mathbf{M}'\mathbf{Z} + \mathbf{Z}^-\tilde{\mathbf{M}}(\tilde{\mathbf{N}}'\mathbf{Z}^-\tilde{\mathbf{M}})^-\tilde{\mathbf{N}}', \tag{4.216}$$

where $\mathbf{Z}^-\mathbf{Z}$ is the projector onto Sp($\mathbf{Z}^-\mathbf{Z}$) along Ker($\mathbf{Z}$) (note that $\mathbf{Z}^-\mathbf{Z} = \mathbf{Z}^-\mathbf{Z}(\mathbf{Z}\mathbf{Z}^-\mathbf{Z})^-\mathbf{Z}$), the first term on the right-hand side of (4.216) is the projector onto

$\text{Sp}(\mathbf{Z}^-\mathbf{Z}\mathbf{N})$ along $\text{Ker}(\mathbf{M}'\mathbf{Z})$, and the second term is the projector onto $\text{Sp}(\mathbf{Z}^-\tilde{\mathbf{M}})$ along $\text{Ker}(\tilde{\mathbf{N}})$. In the notation adopted in (2.152) and (2.153), (4.216) may be written as

$$\mathbf{P}_{Z^-Z:Z'^{\perp}} = \mathbf{P}_{Z^-ZN:(M'Z)'^{\perp}} + \mathbf{P}_{Z^-\tilde{M}:\tilde{N}^{\perp}}. \tag{4.217}$$

The decomposition above follows by premultiplying (4.205) by $\mathbf{Z}^-$. Conditions (4.214) and (4.215) ensure that the two terms on the right-hand side of (4.216) are the projectors with the prescribed onto and along spaces. Note that (4.217) is a special case of (iii-a") in (2.205). The two terms on the right-had side of (4.216) (and of (4.217)) satisfy (2.94), the condition under which the sum of two projectors is also a projector. Note that (4.210) and (4.214) are equivalent, as are (4.211) and (4.215). In (4.212), $\mathbf{M}'\mathbf{Z}\mathbf{Z}^-$ may be redefined as a new $\mathbf{M}'$, and in (4.216), $\mathbf{Z}^-\mathbf{Z}\mathbf{N}$ may be redefined as a new $\mathbf{N}$.

Let $\mathbf{Z}^{*'} = (\mathbf{Z}'\mathbf{Z})^-\mathbf{Z}' \in \{\mathbf{Z}^-\}$. Then, (4.205), (4.212), and (4.216) can be rewritten as:

$$\mathbf{Z} = \mathbf{Z}\mathbf{N}(\mathbf{M}'\mathbf{Z}\mathbf{N})^-\mathbf{M}'\mathbf{Z} + \tilde{\mathbf{M}}(\tilde{\mathbf{N}}'\mathbf{Z}^{*'}\tilde{\mathbf{M}})^-\tilde{\mathbf{N}}', \tag{4.218}$$

$$\begin{aligned}\mathbf{P}_Z &= \mathbf{Z}\mathbf{N}(\mathbf{M}'\mathbf{Z}\mathbf{N})^-\mathbf{M}'\mathbf{P}_Z + \tilde{\mathbf{M}}(\tilde{\mathbf{N}}'\mathbf{Z}^{*'}\tilde{\mathbf{M}})^-\tilde{\mathbf{N}}'\mathbf{Z}^{*'}, \\ &= \mathbf{P}_{ZN:(P_ZM)^{\perp}} + \mathbf{P}_{\tilde{M}:(Z^*\tilde{N})^{\perp}},\end{aligned} \tag{4.219}$$

and

$$\begin{aligned}\mathbf{P}_{Z'} &= \mathbf{P}_{Z'}\mathbf{N}(\mathbf{M}'\mathbf{Z}\mathbf{N})^-\mathbf{M}'\mathbf{Z} + \mathbf{Z}^{*'}\tilde{\mathbf{M}}(\tilde{\mathbf{N}}'\mathbf{Z}^{*'}\tilde{\mathbf{M}})^-\tilde{\mathbf{N}}', \\ &= \mathbf{P}_{P_{Z'}N:(Z'M)^{\perp}} + \mathbf{P}_{Z^{*'}\tilde{M}:\tilde{N}^{\perp}},\end{aligned} \tag{4.220}$$

respectively. Note that $\mathbf{P}_Z\mathbf{M} = \mathbf{M}$ if $\text{Sp}(\mathbf{M}) \subset \text{Sp}(\mathbf{Z})$, and $\mathbf{P}_{Z'}\mathbf{N} = \mathbf{N}$ if $\text{Sp}(\mathbf{N}) \subset \text{Sp}(\mathbf{Z}')$. Note also that (4.218) can be derived from (4.219) and (4.220) by $\mathbf{Z} = \mathbf{P}_Z\mathbf{Z} = \mathbf{Z}\mathbf{P}_{Z'}$. G-inverses of $\mathbf{Z}$ other than $\mathbf{Z}^{*'}$ defined above, e.g., $\mathbf{Z}^- = (\mathbf{Z}'\mathbf{K}\mathbf{Z})^-\mathbf{Z}'\mathbf{K}$, $\mathbf{Z}^- = (\mathbf{M}'\mathbf{Z})^-\mathbf{M}'$, etc. may also be used to derive other decompositions analogous to (4.219).

Set $\mathbf{M} = \mathbf{Z}\mathbf{H}$, $\mathbf{N} = \mathbf{H}$, $\tilde{\mathbf{M}} = \mathbf{Z}^*\tilde{\mathbf{H}}$, and $\tilde{\mathbf{N}} = \tilde{\mathbf{H}}$ in (4.219), where $\tilde{\mathbf{H}}$ is such that $\text{Sp}(\tilde{\mathbf{H}}) \dot{\oplus} \text{Sp}(\mathbf{H}) = \text{Sp}(\mathbf{Z}')$. Then we obtain

$$\begin{aligned}\mathbf{P}_Z &= \mathbf{Z}\mathbf{H}(\mathbf{H}'\mathbf{Z}'\mathbf{Z}\mathbf{H})^-\mathbf{H}'\mathbf{Z}' + \mathbf{Z}^*\tilde{\mathbf{H}}(\tilde{\mathbf{H}}'\mathbf{Z}^{*'}\mathbf{Z}^*\tilde{\mathbf{H}})^-\tilde{\mathbf{H}}\mathbf{Z}^{*'} \\ &= \mathbf{P}_{ZH} + \mathbf{P}_{Z^*\tilde{H}}.\end{aligned} \tag{4.221}$$

This is identical to decomposition (3.62) with $\mathbf{K} = \mathbf{I}$. Decompositions (3.63), (3.64), and (3.65) (with $\mathbf{K} = \mathbf{I}$) as well as other decompositions used in Section 3.3.3 can be similarly obtained from (4.219) by setting $\mathbf{M}$, $\mathbf{N}$, $\tilde{\mathbf{M}}$, and $\tilde{\mathbf{N}}$ appropriately.

4.17 Multilevel RA (MLRA)

Hierarchically structured data are often encountered in many scientific disciplines. In educational assessment studies, for example, students' performance in mathematics is measured in various schools. Such data are called hierarchically structured because students are nested within schools. Another example of such data arises in repeated measurement designs where some attributes of subjects are repeatedly measured over time.

Hierarchical (multilevel) linear models (HLM: Bock 1989; Bryk and Raudenbush 1992; Goldstein 1987; Hox 1995) are often used to analyze such data. Interpretations of parameters in such models, however, become increasingly more difficult as more levels, more predictor variables, and more criterion variables are accommodated (Takane and Hunter 2002). Takane and Zhou (2011) proposed multilevel redundancy analysis (MLRA) to facilitate interpretations in HLMs with multiple criterion variables. In a manner similar to CPCA, MLRA first decomposes variabilities in the criterion variables into several orthogonal parts, using predictor variables (external information) at different levels, and then applies SVD to the decomposed parts to find more parsimonious representations.

Let $\mathbf{Z}$ denote the n-subject by m-variable matrix of criterion variables. Let $\mathbf{1}_n$ denote the n-component vector of ones. Suppose that the subjects (the first level units) represent students who belong to one of J schools (the second level units) with n_j students in school j. Let $\mathbf{G}$ denote the n by J matrix of dummy variables indicating which students belong to which schools. Let

$$\mathbf{W}_0 = [\mathbf{w}_1, \cdots, \mathbf{w}_J]' \tag{4.222}$$

represent the J by q-variable matrix of school-level predictor variables. We define $\mathbf{W}_0^*$, the matrix of columnwise centered school-level predictor variables, by

$$\mathbf{W}_0^* = \mathbf{Q}_{1_J/G'G}\mathbf{W}_0, \tag{4.223}$$

where

$$\mathbf{Q}_{1_J/G'G} = \mathbf{I} - \mathbf{1}_J(\mathbf{1}_J'\mathbf{G}'\mathbf{G}\mathbf{1}_J)^{-1}\mathbf{1}_J'\mathbf{G}'\mathbf{G} \tag{4.224}$$

is the projector onto $\text{Ker}(\mathbf{1}_J'\mathbf{G}'\mathbf{G})$ along $\text{Sp}(\mathbf{1}_J)$.

Let $\mathbf{D}_{X^*}$ denote a block diagonal matrix with $\mathbf{X}_j^*$ as the jth diagonal block, where $\mathbf{X}_j^*$ is the n_j by t-variable matrix of student-level predictor variables. The matrix $\mathbf{D}_{X^*}$ represents interactions between schools and the student-level predictor variables. We assume that $\mathbf{X}_j^*$ is columnwise centered within each school. Let

$$\mathbf{J}_t = \mathbf{1}_J \otimes \mathbf{I}_t, \tag{4.225}$$

and define

$$\mathbf{X}^* = \mathbf{D}_{X^*}\mathbf{J}_t = \begin{bmatrix} \mathbf{X}_1^* \\ \vdots \\ \mathbf{X}_J^* \end{bmatrix}. \tag{4.226}$$

The matrix $\mathbf{X}^*$ represents the main effects of the student-level predictor variables. Let

$$\mathbf{W}_1 = [\mathbf{I}_t \otimes \mathbf{w}_1, \cdots, \mathbf{I}_t \otimes \mathbf{w}_J]' \tag{4.227}$$

and define $\mathbf{W}_1^*$ by

$$\mathbf{W}_1^* = \mathbf{Q}_{J_t/D_{XX}}\mathbf{W}_1, \tag{4.228}$$

where

$$\mathbf{Q}_{J_t/D_{XX}} = \mathbf{I} - \mathbf{J}_t(\mathbf{J}_t'\mathbf{D}_{XX}\mathbf{J}_t)^{-1}\mathbf{J}_t'\mathbf{D}_{XX} \tag{4.229}$$

is the projector onto $\text{Ker}(\mathbf{J}_t'\mathbf{D}_{XX})$ along $\text{Sp}(\mathbf{J}_t)$, and

$$\mathbf{D}_{XX} = \mathbf{D}_{X^*}'\mathbf{D}_{X^*}. \tag{4.230}$$

The full model may then be stated as

$$\begin{aligned}\mathbf{Z} = {} & \mathbf{1}_n\mathbf{c}_{00}' + \mathbf{G}\mathbf{W}_0^*\mathbf{C}_{01} + \mathbf{G}\mathbf{Q}_{[1_J,W_0^*]/G'G}\mathbf{U}_0 \\ & + \mathbf{X}^*\mathbf{C}_{10} + \mathbf{D}_{X^*}\mathbf{W}_1^*\mathbf{C}_{11} + \mathbf{D}_{X^*}\mathbf{Q}_{[J_t,W_1^*]/D_{XX}}\mathbf{U}_1 + \mathbf{E},\end{aligned} \tag{4.231}$$

where $\mathbf{c}_{00}'$, $\mathbf{C}_{01}$, $\mathbf{C}_{10}$, $\mathbf{C}_{11}$, $\mathbf{U}_0$ and $\mathbf{U}_1$ are matrices of regression coefficients, $\mathbf{Q}_{[1_J,W_0^*]/G'G}$ is the orthogonal projector onto $\text{Ker}([\mathbf{1}_J,\mathbf{W}_0^*]')$ in the metric of $\mathbf{G}'\mathbf{G}$, $\mathbf{Q}_{[J_t,W_1^*]/D_{XX}}$ is the orthogonal projector onto $\text{Ker}([\mathbf{J}_t,\mathbf{W}_1^*]')$ in the metric of $\mathbf{D}_{XX}$, and $\mathbf{E}$ is the matrix of disturbance terms.

The first term in (4.231) pertains to the grand means. The second and third terms pertain to the portions of the between-school effects that can and cannot be explained by the school-level predictor variables, respectively. The fourth term represents the portion of the within-school effects that can be explained by the (main) effects of the student-level predictor variables. The fifth term indicates the portion of the within-school effects that can be explained by the interactions between the school-level and student-level predictor variables. The sixth term pertains to the portion of the interactions between schools and the student-level predictor variables that cannot be explained by the main effects of the student-level predictor variables and the interactions between the school- and student-level predictor variables. Finally, the last term in the model represents residuals left unaccounted for by any systematic effects in the model.

There are several important special cases of the full model presented above. When no school-level predictor variables exist, neither terms 2 and 3 nor terms 5 and 6 can be isolated. In this case, the model reduces to a simple analysis of covariance model. When no student-level predictor variables exist, terms 4, 5, 6, and 7 cannot be separated. When neither the school-level nor student-level predictor variables exist, neither terms 2 and 3 nor terms 4, 5, 6, and 7 can be isolated. In this case, the model reduces to a simple one-way ANOVA model.

The seven terms in model (4.231) are all columnwise orthogonal. The orthogonality of the first three terms and the last four terms in (4.231) may be seen by noting that the former all pertain to subspaces of $\text{Sp}(\mathbf{G})$ (note that $\mathbf{1}_n = \mathbf{G}\mathbf{1}_J$), while the latter to $\text{Ker}(\mathbf{G}') = \text{Sp}(\mathbf{Q}_G)$. (Note in the latter that $\mathbf{Q}_G\mathbf{D}_{X^*} = \mathbf{D}_{X^*}$, and that $\mathbf{Q}_{D_{X^*}}\mathbf{Q}_G = \mathbf{Q}_G\mathbf{Q}_{D_{X^*}}$.) The orthogonality among the first three terms may be seen by noting that

$$\mathbf{G}\mathbf{W}_0^* = \mathbf{G}\mathbf{Q}_{1_J/G'G}\mathbf{W}_0 = \mathbf{Q}_{1_n}\mathbf{G}\mathbf{W}_0, \tag{4.232}$$

where $\mathbf{Q}_{1_n} = \mathbf{I} - \mathbf{1}_n\mathbf{1}_n'/n$, and that

$$\mathbf{G}\mathbf{Q}_{[1_J,W_0^*]/G'G} = \mathbf{Q}_{[1_n,GW_0^*]}\mathbf{G}, \tag{4.233}$$

where $\mathbf{Q}_{[1_n,GW_0^*]}$ is the orthogonal projector onto $\text{Ker}([\mathbf{1}_n,\mathbf{GW}_0^*]')$. The orthogonality among the last four terms may be seen by noting

$$\mathbf{D}_{X^*}\mathbf{W}_1^* = \mathbf{D}_{X^*}\mathbf{Q}_{J_t/D_{XX}}\mathbf{W}_1 = \mathbf{Q}_{X^*}\mathbf{D}_{X^*}\mathbf{W}_1, \tag{4.234}$$

where $\mathbf{Q}_{X^*} = \mathbf{I} - \mathbf{X}^*(\mathbf{X}^{*\prime}\mathbf{X}^*)^{-1}\mathbf{X}^{*\prime}$ is the orthogonal projector onto $\text{Ker}(\mathbf{X}^{*\prime})$,

$$\mathbf{D}_{X^*}\mathbf{Q}_{[J_t,W_1^*]/D_{XX}} = \mathbf{Q}_{[X^*,D_{X^*}W_1^*]}\mathbf{D}_{X^*}, \tag{4.235}$$

where $\mathbf{Q}_{[X^*,D_{X^*}W_1^*]}$ is the orthogonal projector onto $\text{Ker}([\mathbf{X}^*,\mathbf{D}_{X^*}\mathbf{W}_1^*]')$, and $\text{Sp}(\mathbf{Q}_{D_{X^*}}) = \text{Ker}(\mathbf{D}_{X^*}')$.

Due to the orthogonality, coefficients in model (4.231) can be separately estimated by OLS for each term. We have

$$\hat{\mathbf{c}}_{00} = \mathbf{1}_n'\mathbf{Z}/n, \tag{4.236}$$

$$\hat{\mathbf{C}}_{01} = (\mathbf{W}_0^{*\prime}\mathbf{G}'\mathbf{G}\mathbf{W}_0^*)^{-1}\mathbf{W}_0^{*\prime}\mathbf{G}'\mathbf{Z}, \tag{4.237}$$

$$\hat{\mathbf{U}}_0 = (\mathbf{Q}_{[1_J,W_0^*]/G'G}'\mathbf{G}'\mathbf{G}\mathbf{Q}_{[1_J,W_0^*]/G'G})^{-1}\mathbf{Q}_{[1_J,W_0^*]/G'G}'\mathbf{G}'\mathbf{Z}, \tag{4.238}$$

$$\hat{\mathbf{C}}_{10} = (\mathbf{X}^{*\prime}\mathbf{X}^*)^{-1}\mathbf{X}^{*\prime}\mathbf{Z}, \tag{4.239}$$

$$\hat{\mathbf{C}}_{11} = (\mathbf{W}_1^{*\prime}\mathbf{D}_{XX}\mathbf{W}_1^*)^{-1}\mathbf{W}_1^{*\prime}\mathbf{D}_{X^*}'\mathbf{Z}, \tag{4.240}$$

$$\hat{\mathbf{U}}_1 = (\mathbf{Q}_{[J_t,W_1^*]/D_{XX}}'\mathbf{D}_{XX}\mathbf{Q}_{[J_t,W_1^*]/D_{XX}})^{-1}\mathbf{Q}_{[J_t,W_1^*]/D_{XX}}'\mathbf{D}_{X^*}'\mathbf{Z}, \tag{4.241}$$

and finally,

$$\hat{\mathbf{E}} = \mathbf{Q}_{D_{X^*}}\mathbf{Q}_G\mathbf{Z}, \text{ where } \mathbf{Q}_G = \mathbf{I} - \mathbf{G}(\mathbf{G}'\mathbf{G})^{-1}\mathbf{G}'. \tag{4.242}$$

Putting the estimates of parameters given above into (4.231), we obtain the following decomposition of $\mathbf{Z}$:

$$\begin{aligned}\mathbf{Z} = {} & \mathbf{P}_{1_n}\mathbf{Z} + \mathbf{P}_{GW_0^*}\mathbf{Z} + \mathbf{P}_{GQ_{[1_J,W_0^*]/G'G}}\mathbf{Z} + \mathbf{P}_{X^*}\mathbf{Z} \\ & + \mathbf{P}_{D_{X^*}W_1^*}\mathbf{Z} + \mathbf{P}_{D_{X^*}Q_{[J_t,W_1^*]/D_{XX}}}\mathbf{Z} + \mathbf{Q}_{D_{X^*}}\mathbf{Q}_G\mathbf{Z}.\end{aligned} \tag{4.243}$$

Due to the orthogonality, the sum of squares (SS) of the elements of $\mathbf{Z}$ ($\text{SS}(\mathbf{Z}) = \text{tr}(\mathbf{Z}'\mathbf{Z})$) can also be uniquely decomposed into the sum of termwise sums of squares. Let SS_i denote the sum of squares (SS) of the elements of the ith term in (4.243). The SS_i $(i = 1, \cdots, 7)$ indicates the size of the effects of the ith term. In addition, we define

$$\text{SS}_T = \text{SS}(\mathbf{Z}) - \text{SS}_1, \tag{4.244}$$

$$\text{SS}_B = \text{SS}_2 + \text{SS}_3, \tag{4.245}$$

and

$$\text{SS}_W = \text{SS}_4 + \text{SS}_5 + \text{SS}_6 + \text{SS}_7. \tag{4.246}$$

The SS_T, SS_B, and SS_W stand for the total SS, the between-group (school) SS, and the within-group SS, respectively, in analogy to the univariate ANOVA situation. It holds that $\text{SS}_T = \text{SS}_B + \text{SS}_W$.

The columns of $\mathbf{Z}$ are often highly correlated and their variability can be more succinctly summarized by fewer components than the number of variables in $\mathbf{Z}$. Each of the estimated terms in the model (except term 1) may be subjected to rank reductions similarly to CPCA. Again, due to the orthogonality of the estimated terms, this can be done independently for each term. It is also possible to recombine some of the terms in the decomposition into one, which is then subjected to a rank reduction. Each rank reduction can be carried out by SVD or GSVD with a special metric matrix. For the second term in (4.243), for example, the former is denoted as $\text{SVD}(\mathbf{G}\mathbf{W}_0^*\hat{\mathbf{C}}_{01})$, while the latter by $\text{GSVD}(\hat{\mathbf{C}}_{01})_{W_0^{*\prime}G'GW_0^*}$, where the subscript indicates the left-hand side metric matrix. As has been shown earlier, these two SVDs are related in a simple manner. See Section 3.2, in particular Table 3.1.

Note 4.9 It is often assumed that $\mathbf{U}_0$ and $\mathbf{U}_1$ represent random rather than fixed-effects. Under the present set-up (4.231), however, the terms pertaining to these effects, i.e., $\mathbf{G}\mathbf{Q}_{[1_J,W_0^*]/G'G}\mathbf{U}_0$ and $\mathbf{D}_{X^*}\mathbf{Q}_{[J_t,W_1^*]/D_{XX}}\mathbf{U}_1$, are both orthogonal to all other effects in the model. The variance-covariance structure arising from this model satisfies the condition under which the OLSE and BLUE of the fixed-effects are identical. See Note 2.5. Note, however, that there is also a view that the effects of $\mathbf{U}_0$ and $\mathbf{U}_1$ should not be confined to the spaces orthogonal to the fixed-effects in the model.

4.18 Weighted Low Rank Approximations (WLRA)

Weighted low rank approximations (WLRA) allow PCA of a data matrix $\mathbf{Z}$ ($n \times m$) under an extremely flexible weighting scheme. Let $\mathbf{Z}_0$ denote a matrix of the same size as $\mathbf{Z}$ with a prescribed rank $r \leq \min(n,m)$. Let $\mathbf{z} = \text{vec}(\mathbf{Z})$ and $\mathbf{z}_0 = \text{vec}(\mathbf{Z}_0)$ (these are nm-component vectors), and define

$$f(\mathbf{z}_0) = (\mathbf{z} - \mathbf{z}_0)'\mathbf{W}(\mathbf{z} - \mathbf{z}_0), \tag{4.247}$$

where $\mathbf{W}$ is a given weight matrix of order nm (usually assumed symmetric and *pd*). We would like to minimize (4.247) with respect to $\mathbf{z}_0$ subject to the rank restriction stated above.

The minimization problem above does not have a closed-form solution except for the special case in which $\mathbf{W}$ factors out into the Kronecker product of row side and column side weight matrices, i.e., $\mathbf{W} = \mathbf{L} \otimes \mathbf{K}$, where $\mathbf{K}$ is a symmetric *pd* matrix of order n and $\mathbf{L}$ a symmetric *pd* matrix of order m. In this case the criterion above can be rewritten as

$$f(\mathbf{Z}_0) = \text{SS}(\mathbf{Z} - \mathbf{Z}_0)_{K,L}, \tag{4.248}$$

which can be minimized via $\text{GSVD}(\mathbf{Z})_{K,L}$. However, there are several iterative algorithms to minimize f in (4.247) under a general weight matrix (Manton et al. 2003; Markovsky et al. 2010; Schuermans et al. 2005; Wentzell et al. 1997). We discuss one of them below, which is based on an ALS (alternating LS) algorithm.

Under the rank restriction, the matrix $\mathbf{Z}_0$ can be reparameterized as

$$\mathbf{Z}_0 = \mathbf{B}\mathbf{A}', \tag{4.249}$$

where $\mathbf{B}$ and $\mathbf{A}$ are, respectively, $n \times r$ and $m \times r$ columnwise nonsingular matrices. (We may further require without loss of generality that $\mathbf{B}$ is columnwise orthogonal, although this is not essential.) Then $\mathbf{z}_0$ has the following two alternative expressions:

$$\mathbf{z}_0 = (\mathbf{A} \otimes \mathbf{I}_n)\text{vec}(\mathbf{B}) = \mathbf{A}^*\mathbf{b} \tag{4.250}$$
$$= (\mathbf{I}_m \otimes \mathbf{B})\text{vec}(\mathbf{A}') = \mathbf{B}^*\mathbf{a}, \tag{4.251}$$

where $\mathbf{A}^* = \mathbf{A} \otimes \mathbf{I}_n$, $\mathbf{B}^* = \mathbf{I}_m \otimes \mathbf{B}$, $\mathbf{b} = \text{vec}(\mathbf{B})$, and $\mathbf{a} = \text{vec}(\mathbf{A}')$. We alternately minimize (4.247) with respect to $\mathbf{b}$ given $\mathbf{a}$ and with respect to $\mathbf{a}$ given $\mathbf{b}$. The former is achieved by

$$\hat{\mathbf{b}} = (\mathbf{A}^{*\prime}\mathbf{W}\mathbf{A}^*)^{-1}\mathbf{A}^{*\prime}\mathbf{W}\mathbf{z}, \tag{4.252}$$

and the latter by

$$\hat{\mathbf{a}} = (\mathbf{B}^{*\prime}\mathbf{W}\mathbf{B}^*)^{-1}\mathbf{B}^{*\prime}\mathbf{W}\mathbf{z}. \tag{4.253}$$

We alternately apply the two updating formulas above until convergence. This algorithm is monotonically convergent. After convergence, we may calculate $\hat{\mathbf{Z}}_0 = \hat{\mathbf{B}}\hat{\mathbf{A}}'$ and apply SVD to $\hat{\mathbf{Z}}_0$ to derive matrices of component scores and loadings that reproduce $\hat{\mathbf{Z}}_0$.

Note that nm can be quite large, and each step in the algorithm above may take a considerable amount of computation time. Fortunately, it is rather rare to have a weight matrix $\mathbf{W}$ as general as in (4.247). For example, when $\mathbf{W}$ is diagonal, the algorithm simplifies considerably. Such a case arises when the elements in the data matrix are not correlated but differ only in their saliency. This is the situation treated by Gabriel and Zamir (1979) and Kiers (1997). Missing data elements may be indicated by a weight of 0, while observed data by 1 in the diagonal of $\mathbf{W}$. In such cases we may arrange the diagonal elements of $\mathbf{W}$ into an $n \times m$ matrix

$$\mathbf{W}^* = \begin{bmatrix} w_{11} & \cdots & w_{1m} \\ \vdots & \ddots & \vdots \\ w_{n1} & \cdots & w_{nm} \end{bmatrix} = [\mathbf{w}_1, \cdots, \mathbf{w}_m] = \begin{bmatrix} \mathbf{w}'_{(1)} \\ \vdots \\ \mathbf{w}'_{(n)} \end{bmatrix}, \tag{4.254}$$

where w_{ij} indicates the weight applied to the corresponding element z_{ij} in the data matrix. We also use columnwise and rowwise vector notation for $\mathbf{Z}$ and $\mathbf{Z}_0$, namely

$$\mathbf{Z} = [\mathbf{z}_1, \cdots, \mathbf{z}_m] = \begin{bmatrix} \mathbf{z}'_{(1)} \\ \vdots \\ \mathbf{z}'_{(n)} \end{bmatrix}, \tag{4.255}$$

and similarly,

$$\mathbf{Z}_0 = [\mathbf{z}_{01}, \cdots, \mathbf{z}_{0m}] = \begin{bmatrix} \mathbf{z}'_{(01)} \\ \vdots \\ \mathbf{z}'_{(0n)} \end{bmatrix}. \tag{4.256}$$

Let $\mathbf{a}_i$ $(i = 1, \cdots, m)$ denote the ith column vector of $\mathbf{A}'$, and let $\mathbf{b}_{(i)}$ $(i = 1, \cdots, n)$

denote the ith column vector of $\mathbf{B}'$. Let $\mathbf{D}_{w_i}$ denote the diagonal matrix with the elements of $\mathbf{w}_i$ as its diagonals, and let $\mathbf{D}_{w_{(i)}}$ be similarly defined. Then (4.247) can be rewritten as

$$f(\hat{\mathbf{Z}}) = \sum_i^m (\mathbf{z}_i - \mathbf{B}\mathbf{a}_i)'\mathbf{D}_{w_i}(\mathbf{z}_i - \mathbf{B}\mathbf{a}_i). \tag{4.257}$$

This means that f is columnwise separable, so that each $\mathbf{a}_i$ can be separately updated. That is,

$$\hat{\mathbf{a}}_i = (\mathbf{B}'\mathbf{D}_{w_i}\mathbf{B})^{-1}\mathbf{B}'\mathbf{D}_{w_i}\mathbf{z}_i. \tag{4.258}$$

The criterion (4.247) can also be rewritten as

$$\mathbf{f}(\hat{\mathbf{Z}}) = \sum_i^n (\mathbf{z}_{(i)} - \mathbf{A}\mathbf{b}_{(i)})'\mathbf{D}_{w_{(i)}}(\mathbf{z}_{(i)} - \mathbf{A}\mathbf{b}_{(i)}), \tag{4.259}$$

which leads to

$$\hat{\mathbf{b}}_{(i)} = (\mathbf{A}'\mathbf{D}_{w_{(i)}}\mathbf{A})^{-1}\mathbf{A}'\mathbf{D}_{w_{(i)}}\mathbf{z}_{(i)}. \tag{4.260}$$

We again alternate the two steps above until convergence.

When $\mathbf{W}$ is a block diagonal matrix such that

$$\mathbf{W} = \begin{bmatrix} \mathbf{W}_1 & \cdots & \mathbf{O} \\ \vdots & \ddots & \vdots \\ \mathbf{O} & \cdots & \mathbf{W}_m \end{bmatrix} \tag{4.261}$$

(the rows are correlated, but the columns are not) the situation is not as simple as above. Still, some simplification is possible. Each $\mathbf{a}_i$ can be updated by

$$\hat{\mathbf{a}}_i = (\mathbf{B}'\mathbf{W}_i\mathbf{B})^{-1}\mathbf{B}'\mathbf{W}_i\mathbf{z}_i, \tag{4.262}$$

but in order to update $\mathbf{b}$, (4.252) must be used. Essentially the same applies when the columns are correlated, but not the rows.

In the unweighted case (i.e., when the data are complete and no differential weights are applied) the above algorithm simplifies further to alternate applications of

$$\hat{\mathbf{B}} = \mathbf{Z}\mathbf{A}(\mathbf{A}'\mathbf{A})^{-1}, \tag{4.263}$$

and

$$\hat{\mathbf{A}}' = (\mathbf{B}'\mathbf{B})^{-1}\mathbf{B}'\mathbf{Z}. \tag{4.264}$$

These two steps are repeatedly applied until convergence. This is essentially Daugavet's (1968) algorithm referred to in connection with CANDECOMP (Section 4.12), which is a multiway extension of the former.

In this simplified case, however, it is more conventional (although not essential) to maintain columnwise orthogonality of $\mathbf{A}$ (or $\mathbf{B}$) in each iteration. This alters the algorithm slightly. In the modified algorithm, $\mathbf{A}$ is updated by first calculating

$$\mathbf{A}^* = \mathbf{Z}\mathbf{B}, \tag{4.265}$$

which is then factored into

$$\mathbf{A}^* = \mathbf{A}\mathbf{R}' \tag{4.266}$$

by the compact QR decomposition (Section 2.1.12). Note that $\mathbf{A}$ is columnwise orthogonal, and $\mathbf{R}'$ is upper triangular. This $\mathbf{A}$ is used in the next step to update $\mathbf{B}$ by

$$\mathbf{B} = \mathbf{Z}\mathbf{A}. \tag{4.267}$$

These two steps are repeated until convergence. It can be verified that $\mathbf{A}$ converges to the matrix of eigenvectors of $\mathbf{Z}'\mathbf{Z}$ corresponding to the r largest eigenvalues. By substituting (4.267) for $\mathbf{B}$ in (4.265), we obtain

$$\mathbf{A}\mathbf{R}' = \mathbf{Z}'\mathbf{Z}\mathbf{A}. \tag{4.268}$$

At convergence this equation must hold exactly. By premultiplying (4.268) by $\mathbf{A}'$, we obtain

$$\mathbf{R}' = \mathbf{A}'\mathbf{Z}'\mathbf{Z}\mathbf{A}, \tag{4.269}$$

which is upper diagonal and symmetric (*nnd*), and consequently diagonal. This indicates that (4.268) is the eigenequation for $\mathbf{Z}'\mathbf{Z}$ at convergence. When (4.268) is used as an updating equation for $\mathbf{A}$ (bypassing the updating of $\mathbf{B}$), it defines a simultaneous power iteration (e.g., Clint and Jennings 1970; see also Takane and Zhang (2009)) for obtaining the eigenvectors of $\mathbf{Z}'\mathbf{Z}$ corresponding to the r dominant eigenvalues.

4.19 Orthogonal Procrustes Rotation

We discussed an orthogonal approximation to a rectangular matrix in Section 2.3.4. Let $\mathbf{Z}$ and $\mathbf{G}$ be $n \times m$ ($n \geq m$) matrices. We consider minimizing

$$\phi(\mathbf{T}) = \mathrm{SS}(\mathbf{Z} - \mathbf{G}\mathbf{T}) \tag{4.270}$$

with respect to $\mathbf{T}$ subject to the constraint that $\mathbf{T}'\mathbf{T} = \mathbf{T}\mathbf{T}' = \mathbf{I}$. This is called the orthogonal Procrustes rotation problem. It is a slight extension of the minimization problem in (2.256). The criterion (4.270) can be expanded as

$$\phi(\mathbf{T}) = \mathrm{SS}(\mathbf{Z}) + \mathrm{SS}(\mathbf{G}\mathbf{T}) - 2\mathrm{tr}(\mathbf{Z}'\mathbf{G}\mathbf{T}) = c - 2\mathrm{tr}(\mathbf{Z}'\mathbf{G}\mathbf{T}), \tag{4.271}$$

since $\mathrm{SS}(\mathbf{G}\mathbf{T}) = \mathrm{SS}(\mathbf{G})$. Thus, minimizing $\phi(\mathbf{T})$ is equivalent to maximizing

$$f(\mathbf{T}) = \mathrm{tr}(\mathbf{Z}'\mathbf{G}\mathbf{T}) \tag{4.272}$$

under the same condition. This problem is isomorphic to the maximization of f in (2.241). The solution is obtained (Cliff 1966; Schönemann 1966) by

$$\mathbf{T} = \mathbf{U}\mathbf{V}' = \mathbf{G}'\mathbf{Z}(\mathbf{Z}'\mathbf{G}\mathbf{G}'\mathbf{Z})^{-1/2}, \tag{4.273}$$

where $\mathbf{G}'\mathbf{Z} = \mathbf{U}\mathbf{D}\mathbf{V}'$ is the SVD of $\mathbf{G}'\mathbf{Z}$. (See also ten Berge (1993).)

When the number of columns in $\mathbf{G}$ is larger than that of $\mathbf{Z}$, $\mathbf{T}$ is a tall matrix, for which $\mathbf{T}\mathbf{T}' = \mathbf{I}$ does not hold. There is no closed-form solution that minimizes (4.270) under this condition. However, iterative solutions exist (Green and Gower 1979; ten Berge and Knol 1984, Problem 3). This method first augments $\mathbf{Z}$ by appending an arbitrary matrix to make its size equal to that of $\mathbf{G}$, then solves the standard orthogonal rotation problem (with a square $\mathbf{T}$), as described above, and then updates the augmented part of $\mathbf{Z}$ by the augmented part of $\mathbf{GT}$. This process is repeated until convergence. This algorithm is monotonically convergent, since it is an ALS algorithm. The convergence is usually fairly quick, although the problem of convergence to suboptimal solutions exists, so the iterative process must be repeated several times to ensure global optimality. Kiers' (1990) and Kiers and ten Berge's (1992) majorization algorithms may also be used, which may be computationally more efficient than the above algorithm.

Note 4.10 When $\mathbf{T}$ is a tall matrix, the minimization of (4.270) and the maximization of (4.272) are not equivalent. There is a closed-form solution to the latter, which is essentially the same as above for a square $\mathbf{T}$ (Cliff 1966; ten Berge 1977). That is, we obtain the compact SVD of $\mathbf{G}'\mathbf{Z} = \mathbf{UDV}'$ and form a rectangular matrix $\mathbf{T} = \mathbf{UV}'$.

Likewise, there is no closed-form solution to the minimization of

$$\phi(\mathbf{T}) = \text{SS}(\mathbf{Z} - \mathbf{GTH}') \tag{4.274}$$

for a full orthogonal matrix $\mathbf{T}$. Again, iterative solutions exist (Koschat and Swayne 1991; Mooijaart and Commandeur 1990) for this case.

The minimization of (4.270) subject to the constraint that $\text{diag}(\mathbf{T}'\mathbf{T}) = \mathbf{I}$ is called the oblique Procrustes rotation problem. Again no closed-form solution exists for this problem. A quick and reliable algorithm developed by ten Berge and Nevels (1977) exists, however. See also Browne (1967) and Cramer (1974). One important characteristic of the problem is that the minimization problem can be split into separate minimization problems, that is, (4.270) in this case can be rewritten as

$$\phi(\mathbf{T}) = \sum_j \phi(\mathbf{t}_j) = \sum_j \text{SS}(\mathbf{z}_j - \mathbf{G}\mathbf{t}_j), \tag{4.275}$$

where $\mathbf{z}_j$ and $\mathbf{t}_j$ are the jth column vector of $\mathbf{Z}$ and $\mathbf{T}$, and each $\phi(\mathbf{t}_j)$ can be separately minimized with respect to $\mathbf{t}_j$ subject to the restriction that $\mathbf{t}_j'\mathbf{t}_j = 1$ for $j = 1, \cdots$. Note that the same algorithm can be used to minimize $\phi(\mathbf{t}_j)$ subject to $\mathbf{t}_j'\mathbf{K}\mathbf{t}_j = 1$, where $\mathbf{K}$ is an arbitrary *pd* matrix. Let $\mathbf{R}_k$ be such that $\mathbf{K} = \mathbf{R}_K\mathbf{R}_K'$ (a square root factor of $\mathbf{K}$), and define $\mathbf{t}_j^* = \mathbf{R}_K'\mathbf{t}_j$. The minimization problem can be reformulated as the minimization of

$$\phi(\mathbf{t}_j^*) = \text{SS}(\mathbf{z}_j - \mathbf{G}(\mathbf{R}_K')^{-1}\mathbf{t}_j^*) \tag{4.276}$$

subject to $\mathbf{t}_j^{*\prime}\mathbf{t}_j^* = 1$.

4.20 PCA of Image Data Matrices

Guttman (1953) proposed a PCA of image data matrices to simplify the estimation of unique variances in common factor analysis. The image data matrices replace each variable in the observed data matrix $\mathbf{Z}$ by its image, i.e., what can be predicted by the other $m-1$ variables in $\mathbf{Z}$. We assume that $\mathbf{Z}$ is columnwise centered and normalized in such a way that $\mathbf{Z}'\mathbf{Z} = \mathbf{R}$, where $\mathbf{R}$ is the correlation matrix. We also assume that $\mathbf{R}$ is nonsingular.

Let $\mathbf{Z}_I$ and $\mathbf{Z}_A$ denote the image and anti-image data matrices. Then

$$\mathbf{Z} = \mathbf{Z}_I + \mathbf{Z}_A, \tag{4.277}$$

where

$$\mathbf{Z}_I = \mathbf{Z}(\mathbf{I}_m - \mathbf{R}^{-1}\mathbf{D}), \tag{4.278}$$

and

$$\mathbf{Z}_A = \mathbf{Z}\mathbf{R}^{-1}\mathbf{D}, \tag{4.279}$$

where

$$\mathbf{D} = \text{diag}(\mathbf{Z}_A'\mathbf{Z}_A) = (\text{diag}(\mathbf{R}^{-1}))^{-1}. \tag{4.280}$$

The matrix $\mathbf{D}$ is diagonal with the variances of anti-image variables as its diagonal elements. (See below for derivations of $\mathbf{D}$, $\mathbf{Z}_I$, and $\mathbf{Z}_A$.) It follows that

$$\mathbf{C}_{II} = \mathbf{Z}_I'\mathbf{Z}_I = \mathbf{R} - 2\mathbf{D} + \mathbf{D}\mathbf{R}^{-1}\mathbf{D} = (\mathbf{R}-\mathbf{D})\mathbf{R}^{-1}(\mathbf{R}-\mathbf{D}), \tag{4.281}$$

$$\mathbf{C}_{AA} = \mathbf{Z}_A'\mathbf{Z}_A = \mathbf{D}\mathbf{R}^{-1}\mathbf{D}, \tag{4.282}$$

and

$$\mathbf{C}_{IA} = \mathbf{D} - \mathbf{D}\mathbf{R}^{-1}\mathbf{D}. \tag{4.283}$$

The matrices $\mathbf{C}_{II}$ and $\mathbf{C}_{AA}$ are respectively called the image and anti-image covariance matrices.

If we use $\mathbf{D}$ as estimates of the unique variances $\mathbf{U}^2$ in common factor analysis and scale $\mathbf{C}_{II}$ by $\mathbf{D}^{-1/2}$ from both sides, we obtain

$$\begin{aligned}\mathbf{C}_{II}^* &= \mathbf{D}^{-1/2}\mathbf{C}_{II}\mathbf{D}^{-1/2} = \mathbf{D}^{-1/2}(\mathbf{R} - 2\mathbf{D} + \mathbf{D}\mathbf{R}^{-1}\mathbf{D})\mathbf{D}^{-1/2} \\ &= \mathbf{D}^{-1/2}\mathbf{R}\mathbf{D}^{-1/2} - 2\mathbf{I} + \mathbf{D}^{1/2}\mathbf{R}^{-1}\mathbf{D}^{1/2}.\end{aligned} \tag{4.284}$$

It is well known that this matrix has the same set of eigenvectors as $\mathbf{D}^{-1/2}\mathbf{R}\mathbf{D}^{-1/2}$, and eigenvalues which are related in a simple manner. Let the ith eigenvalue of $\mathbf{R}^* = \mathbf{D}^{-1/2}\mathbf{R}\mathbf{D}^{-1/2}$ be denoted by s_i^2 and the corresponding eigenvalue of $\mathbf{C}_{II}^*$ by $\tilde{s}_i^2$. Then

$$\tilde{s}_i^2 = s_i^2 - 2 + 1/s_i^2. \tag{4.285}$$

Hence it suffices to obtain the eigendecomposition of $\mathbf{R}^*$ instead of $\mathbf{C}_{II}^*$, or the SVD of $\mathbf{Z}\mathbf{D}^{-1/2}$ instead of $\mathbf{Z}_I\mathbf{D}^{-1/2}$. (Note, however, that the order of eigenvalues (singular values) is likely to be different, so care should be taken in selecting the correct eigenvectors to be retained in the solution.)

We now show how (4.278), (4.279), and (4.280) can be derived (Yanai and Mukherjee 1987). Let $\mathbf{z}_j$ and $\mathbf{Z}_{(j)}$ denote the jth column vector of $\mathbf{Z}$ and the matrix obtained by eliminating $\mathbf{z}_j$ from $\mathbf{Z}$, respectively. The matrix $\mathbf{P}_{(j)}$ denotes the orthogonal projector defined by $\mathbf{Z}_{(j)}$, and $\mathbf{Q}_{(j)}$ denotes its orthogonal complement. Then, since $\text{Sp}(\mathbf{Z}_{(j)}) \subset \text{Sp}(\mathbf{Z})$,

$$\mathbf{P}_Z\mathbf{P}_{(j)} = \mathbf{P}_{(j)}, \tag{4.286}$$

and

$$\mathbf{P}_Z\mathbf{Q}_{(j)}\mathbf{z}_j = (\mathbf{P}_Z - \mathbf{P}_{(j)})\mathbf{z}_j = \mathbf{Q}_{(j)}\mathbf{z}_j \tag{4.287}$$

for $j = 1, \cdots, m$. Note that $\mathbf{P}_Z\mathbf{z}_j = \mathbf{z}_j$. These results imply that

$$\mathbf{Z}_I = \mathbf{P}_Z\mathbf{Z}_I, \tag{4.288}$$

and

$$\mathbf{Z}_A = \mathbf{P}_Z\mathbf{Z}_A. \tag{4.289}$$

We now show

$$\mathbf{Z}'\mathbf{Z}_A = \mathbf{D} = \text{diag}(d_1, d_2, \cdots, d_m), \tag{4.290}$$

where $\mathbf{D}$ is diagonal with $d_j = 1 - r^2_{j(j)}$ as the jth diagonal element, and $r^2_{j(j)}$ is the squared multiple correlation coefficient in predicting $\mathbf{z}_j$ from $\mathbf{Z}_{(j)}$. Note that

$$\mathbf{z}_j'\mathbf{z}_{Aj} = \mathbf{z}_j'\mathbf{Q}_{(j)}\mathbf{z}_j = 1 - \mathbf{z}_j'\mathbf{P}_{(j)}\mathbf{z}_j = 1 - r^2_{j(j)} = d_j,$$

and that for $i \neq j$,

$$\mathbf{z}_j'\mathbf{z}_{Ai} = \mathbf{z}_j\mathbf{Q}_{(i)}\mathbf{z}_i = \mathbf{z}_j'\mathbf{z}_i - \mathbf{z}_j'\mathbf{P}_{(i)}\mathbf{z}_i = 0.$$

Then, from (4.288), (4.289), and (4.290), we obtain

$$\mathbf{Z}_A = \mathbf{P}_Z\mathbf{Z}_A = \mathbf{Z}\mathbf{R}^{-1}\mathbf{D}, \tag{4.291}$$

and

$$\mathbf{Z}_I = \mathbf{P}_Z\mathbf{Z}_I = \mathbf{P}_Z(\mathbf{Z} - \mathbf{Z}_A) = \mathbf{Z} - \mathbf{Z}\mathbf{R}^{-1}\mathbf{D}, \tag{4.292}$$

which are identical to (4.279) and (4.278). To show (4.280), we note that $\mathbf{z}_j'\mathbf{z}_{Aj} = \mathbf{z}_{Aj}'\mathbf{z}_{Aj}$, which implies

$$\mathbf{D} = \text{diag}(\mathbf{Z}'\mathbf{Z}_A) = \text{diag}(\mathbf{Z}_A'\mathbf{Z}_A) = \text{diag}(\mathbf{D}\mathbf{R}^{-1}\mathbf{D}) = \mathbf{D}\text{diag}(\mathbf{R}^{-1})\mathbf{D}, \tag{4.293}$$

or $\mathbf{D} = (\text{diag}(\mathbf{R}^{-1}))^{-1}$. Yanai and Mukherjee (1987) extend the above results to a singular $\mathbf{R}$. See also Tucker et al. (1972) and Kaiser (1976).

Chapter 5

Related Topics of Interest

Several nontrivial decisions must be made in practical uses of PCA. In this chapter we discuss some of them, including dimensionality selection, reliability assessment, determination of the value of the regularization parameter, and treatment of missing data. Later in this chapter we discuss other topics of interest such as robust estimation, data transformation, joint plots of row and column representations, and probabilistic PCA.

5.1 Dimensionality Selection

Determining how many components should be retained in PCA is one of the most important decisions to make in practical applications of this technique, particularly when solutions are rotated (Section 4.1). Whereas the first $r-1$ components in the r-dimensional solution in unrotated PCA remain the same as the $r-1$ components in the $(r-1)$-dimensional solution, this is no longer true in the rotated case. (That is, the first $r-1$ components in the rotated r-dimensional solution are typically different from the $r-1$ components in the rotated $(r-1)$-dimensional solution.) Despite its paramount importance, there are currently no definitive methods for selecting dimensionality.

There are two heuristic tests in common use: One retains as many components as will jointly explain 95% of the total SS (variance) in the data set. The other keeps all components associated with singular values greater than unity. The latter rule is motivated by the fact that each observed variable has unit variance (after standardization), and a component extracted should be able to explain at least as much variance as a single observed variable. These heuristic criteria are useful sometimes, but they tend to suggest keeping too many components to be interpretable.

There are a class of techniques called scree tests (Cattell 1966), which are theoretically more palatable. They were initially motivated by an empirical observation that the eigenvalues (squared singular values) corresponding to scree components (reflecting random noise) tend to form a linearly decreasing sequence when ordered from the largest to smallest. If they do not follow a linear trend, then the largest component is retained, and subsequent eigenvalues are further inspected for linearity. Hong et al. (2006) compared several criteria for linearity, and found that the CNG (Cattell-Nelson-Gorsuch) criterion (Gorsuch and Nelson 1981) worked best among other criteria. In this criterion, five consecutive eigenvalues are examined at a time.

A line is fitted to the largest three eigenvalues, and another to the smallest three eigenvalues. The significance of the difference in slope is tested using the bootstrap method. (See the next section for the bootstrap method.) If the difference is significant, the largest component is considered significantly nonrandom, and the next five largest eigenvalues are examined in a similar manner, until no significant difference in slope is found. The bootstrap scree test works best when there are a few significant components and many scree components. At least five scree components are necessary for this procedure to work properly.

Horn (1965) proposed an alternative procedure called parallel analysis. In this method a number of data sets (of the same size as the original data set) are generated independently to follow the standard normal distribution. These data sets are analyzed by PCA to find benchmark distributions of singular values under the null condition. In Horn's original proposal, median singular values are calculated across the generated data sets and compared with the singular values obtained from the original data set. If the latter are found larger than their respective medians, they are judged significantly "nonnull." Buja and Eyuboglu (1992) assert that this procedure works remarkably well even for nonnormal data, but suggest using more conservative critical values than the median singular values. For example, the 95 or 99 percentile point can be used to reduce the Type I error rate, which is 50% if the medians are used as the critical values. Buja and Eyuboglu (1992) tabled these critical values for certain "representative" numbers of rows and columns in data matrices.

Parallel analysis, however, can only be theoretically justified for testing the largest singular value, whereas permutation tests may be used to test the significance of all the singular values successively from the largest to smallest. In the permutation tests, elements of column vectors in the data matrix $\mathbf{Z}$ are randomly permuted many times except for one column. PCA is applied to each permuted data set to find the null distribution of the largest singular value. The p-value is given by the relative frequency with which the largest singular values from the permuted data sets exceed the largest value from the original data set. If the p-value is smaller than a prescribed α level (.05 or .01), then the largest component is considered significant at the α level. The effect of the largest component is then eliminated from the original data (this is called deflation; see below), and the permutation test is applied to the deflated data set in a similar manner to test the second largest component. This process is repeated until a nonsignificant singular value is found or the maximum possible number of components is reached. See Legendre and Legendre (1998) and Takane and Hwang (2002) for more general discussions on permutation tests in similar contexts.

A cautionary remark is in order for the use of permutation tests. These tests are based on the assumption of exchangeability of observations (cases) in a data set. The most representative cases of exchangeability include those in which observations are independently and identically distributed (*iid*), and those in which observations are normally distributed with homogeneous variances and covariances (Good 2005, p. 24). There is some evidence, however, showing that permutation tests are not robust when the exchangeability assumption is violated. For example, it cannot be used for heterogeneously correlated observations (Boik 1987; Romano 1990).

Permutation tests similar to the above can also be used in methods other than

(C)PCA, provided that permuted data are generated appropriately. In RA (Section 4.2) and CANO (Section 4.3) there are two sets of variables. In these methods all columns of either one of the two sets must be permuted simultaneously. In GCANO and MCA (Sections 4.9 and 4.10) there are more than two sets of variables. In these methods all columns of all but one data set must be permuted simultaneously within the sets, but independently across the sets.

The deflation procedure to test subordinate components is also different in different methods. In PCA we obtain the first round deflated data matrix by $\mathbf{Z}^{(1)} = \mathbf{Z} - d_1\mathbf{u}_1\mathbf{v}_1' = \mathbf{Q}_{u_1}\mathbf{Z} = \mathbf{Z}\mathbf{Q}_{v_1}$, where $\mathbf{Z}$ is the original data matrix, and $\mathbf{u}_1$ and $\mathbf{v}_1$ are the left and right singular vectors of $\mathbf{Z}$ corresponding to the largest singular value d_1. Subsequent deflations are carried out in a similar manner. In CANO we need to deflate only one of the two sets of variables, $\mathbf{G}$ and $\mathbf{H}$. That is, $\mathbf{G}^{(1)} = \mathbf{Q}_{u_1}\mathbf{G}$, or $\mathbf{H}^{(1)} = \mathbf{Q}_{v_1}\mathbf{H}$, where $\mathbf{u}_1$ and $\mathbf{v}_1$ are the left and right singular vectors of $\mathbf{P}_G\mathbf{P}_H$ corresponding to the largest singular value. In GCANO we deflate the data matrix $\mathbf{G} = [\mathbf{G}_1, \cdots, \mathbf{G}_K]$ by $\mathbf{G}^{(1)} = \mathbf{Q}_{u_1}\mathbf{G}$, where $\mathbf{u}_1$ is the left singular vector from $\text{GSVD}(\mathbf{G}\mathbf{C}^{-1})_{I,C}$ corresponding to the largest (generalized) singular value. Here $\mathbf{C}$ is the block diagonal matrix with $\mathbf{G}_k'\mathbf{G}_k$ as the kth diagonal block. We then perform $\text{GSVD}(\mathbf{G}^{(1)}\mathbf{C}^{-1})_{I,C}$. That is, no deflation is necessary for $\mathbf{C}$. Finally in RA, we deflate the set of original criterion variables $\mathbf{Z}$ by $\mathbf{Z}^{(1)} = \mathbf{Z} - \mathbf{Z}\mathbf{v}_1\mathbf{v}_1'$, where $\mathbf{v}_1$ is the right singular vector of $\mathbf{P}_G\mathbf{Z}$ corresponding to the largest singular value. We then perform $\text{SVD}(\mathbf{P}_G\mathbf{Z}^{(1)})$. No deflation is necessary for the matrix of predictor variables $\mathbf{G}$. The above deflation techniques can easily be extended to the regularized LS estimation method.

It may be noted that there also are dimensionality selection methods based on cross validation (e.g., Eastment and Kruzanowski 1982). (See Section 5.3 for a more general account of cross validation methods.) Under certain conditions, these methods are known (Stone 1976) to be equivalent to the dimensionality selection by AIC (Akaike 1974). See also Yanai (1980) and Jolliffe (2002, Section 6.1) for other approaches to dimensionality selection problems in PCA.

5.2 Reliability Assessment

Data analysis is not meaningful unless the results replicate. Hence it is important to be able to assess the degree of their reliability. Several techniques are available for this purpose. The bootstrap method (Efron 1979; Efron and Tibshirani 1993) is one such procedure that we have already seen in Figures 4.2 and 4.3. This method is easy to apply and can be used in a wide variety of contexts. Its popularity partly stems from the fact that it involves no rigid distributional assumptions such as normality. Note, however, that standard implementations of the bootstrap method require independence among observations (cases).

The bootstrap method works as follows:

(i) From an original data set we repeatedly sample data sets of the same size as the original data set. The sampling is done with replacement, meaning that cases (sampling units) can be sampled repeatedly (more than once) in a data set. The sampled data sets are called bootstrap samples.

(ii) The sampled data sets are separately analyzed by the method of interest (e.g., PCA). Different bootstrap samples lead to different estimates of model parameters, but this yields the important information about their distributions. Caution must be exercised, however, in deriving the distributions. In some methods of analysis, model parameters can be reflected (sign-reversed) with no essential consequences in the model. In PCA, for example, a pair of corresponding left and right singular vectors can be jointly reflected without any change in model predictions (i.e., $d\mathbf{uv}' = d(-\mathbf{u})(-\mathbf{v}')$). Generally, it is not possible to control the direction of the solution obtained in a particular situation. This means that the bootstrap estimates must be jointly "aligned" in the direction consistent with the estimates from the original data set. The problem becomes more aggravated when two or more solution vectors can be rotated without essential consequences, as in a multi-component PCA solution. In such cases, bootstrap estimates may be rotated, so that they are in best agreement with the original estimates. The method described in Section 4.19 (Procrustes rotation) may be used for this purpose.

There are two distinct ways of utilizing the distributional information obtained by the bootstrap method. One is a parametric approach, and the other is a nonparametric approach:

Parametric approach. If the sample size of the original data set is sufficiently large, the estimates of parameters may be assumed to follow normal distributions, and we can summarize these distributions by calculating their means and standard deviations. The differences between the means and the estimates from the original data set indicate biases of the estimates, and the standard deviations indicate the stability of the estimates. (The biases are usually negligible, if the original sample is sufficiently large.) We may also calculate the ratio of the estimates to their standard errors. This quantity may be assumed to follow asymptotically the standard normal distribution, and may be used to test the significance of individual parameters. Roughly speaking, if this ratio exceeds ± 2, the corresponding estimate is considered significantly different from zero. (We may instead use the critical value of t with $n-1$ df, where n is the sample size. This will make the test slightly more conservative.)

We may also construct confidence intervals for individual parameters using the same distributional information as above. In some cases, however, we may be interested in the joint behavior of two parameter estimates at a time. For example, in a two-dimensional PCA solution, the stability of a point (vector) defined by two parameter estimates may be investigated by constructing two-dimensional confidence regions (ellipses). Let θ denote a two-component vector of parameters, and let $\hat{\theta}$ denote its estimate. Let $\mathbf{S}_\theta$ represent the estimate of the covariance matrix of $\hat{\theta}$. Then the $(1-\alpha)100\%$ asymptotic confidence region for θ can be expressed as

$$(\hat{\theta} - \theta)'\mathbf{S}_{\hat{\theta}}^{-1}(\hat{\theta} - \theta) = \chi^2_{2,1-\alpha}, \tag{5.1}$$

where $\chi^2_{2,1-\alpha}$ is the critical value of chi-square with 2 df, corresponding to the $(1-\alpha)100$ percentile. (This critical value may be replaced by $2F_{(2,n-2),1-\alpha}$, where

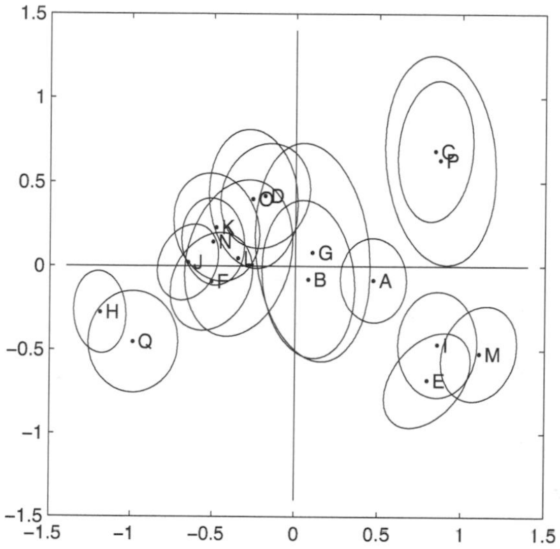

Figure 5.1 *The 95% confidence regions for unconstrained PCA of Mezzich's data.*

$F_{(2,n-2),1-\alpha}$ is the critical value of F with 2 and $n-2$ df, corresponding to the $(1-\alpha)100$ percentile.)

Figure 5.1 shows the 95% confidence regions for the points estimated from the Mezzich data analyzed in Section 1.1. The confidence regions indicate that the corresponding true (population) points are assured to be inside the ellipses with probability .95. Thus, the smaller the confidence regions, the more reliably the points are estimated. For example, A, H, J, and N are among the most reliably estimated points, while G, F, and L are among the least reliably estimated. Psychiatrists were in better agreement in their rating on the items corresponding to the former, while they disagreed considerably on the items corresponding to the latter.

Nonparametric approach. The significance test may also be carried out nonparametrically if the assumption of asymptotic normality is in suspect. We may simply count the number of times bootstrap estimates "cross" over zero (i.e., we count the number of times bootstrap estimates are negative if the original estimate is positive, and the number of times bootstrap estimates are positive if the original estimate is negative). If the relative frequency of crossovers (the p-value) is less than the prescribed α level, we conclude that the coefficient is significantly positive (or negative). This test is one-tailed, but it can be made two-tailed by dividing α by 2. Confidence intervals and regions can also be constructed nonparametrically. In the case of (one-dimensional) confidence intervals, we may simply set the upper and lower boundaries of the intervals equal to the bootstrap estimates at the highest and lowest $\alpha/2$ percentiles. Two-dimensional confidence regions are much more difficult to construct nonparametrically. One approach is to plot the bootstrap estimates of points in the two-dimensional solution space, and construct a convex hull of the smallest size around the points in such a way that it contains $(1-\alpha)100\%$ of the bootstrap estimates of the points. This procedure is rather complicated and most often requires special computer software (e.g., Linting et al. 2007b).

Significance tests and confidence intervals can also be constructed for any contrasts of interest between parameters, either parametrically or nonparametrically. We simply derive an empirical distribution of the contrast by the bootstrap method. The rest of the procedure remains essentially the same as in individual parameters.

The bootstrap method is time consuming, particularly when the sample size is large. (This aspect of the bootstrap method is shared by the permutation tests discussed in the previous section, and the jackknife method to be described in the next section.) There are alternative approaches to reliability assessment, which are less computation bound. One approach is called sensitivity analysis, and is based on the perturbation theory. Sensitivity analysis investigates how a small change (or changes) in a data set affects estimates of model parameters. It has been usefully applied to regression analysis (Belsley et al. 1980; Chatterjee and Hadi 1988; Cook and Weisberg 1982), and other multivariate analysis techniques that involve eigendecompositions (Critchley 1985; Tanaka 1984, 1988; Tanaka and Tarumi 1986; Tarumi 1986). It would be interesting if similar analysis could be developed for CPCA. A perturbation theory for CPCA must accommodate perturbations in row and column information matrices and metric matrices as well as in the main data matrices. This means that we need a perturbation theory for projectors (Golub and Pereyra 1973), one for the Cholesky factors of metric matrices (Tanaka and Tarumi 1989), and one for SVD (Stewart and Sun 1990), all put together in one comprehensive procedure.

Another line of work worth mentioning concerns asymptotic distributions of eigenvalues (squared singular values) and vectors (right singular vectors) of covariance or correlation matrices. These distributions have been derived under both the normality and nonnormality assumptions. See Konishi (1978, 1979), Kollo and Neudecker (1994, 1997), and Ogasawara (2000, 2002, 2004, 2006). These results may also be extended to CPCA.

5.3 Determining the Value of δ

We have discussed a number of multivariate techniques in which a ridge type of regularization is incorporated to improve their predictive performance. In the regularized estimation methods, we need to choose an "optimal" value of the regularization parameter in such a way that the model's predictive power is maximized. Predictions, however, involve future observations not available at the time of analysis. Even if they are available, they should be reserved for evaluating the accuracy of predictions, and should not be used to make predictions.

We may split the data at hand into two groups, one (called the calibration sample) to be used for estimating model parameters, and the other (called the test sample) for evaluating the magnitude of prediction errors. This is called the split-half validation method. The idea can be easily generalized to splitting the data into J (≥ 2) subsets.

The method is then called the J-fold cross validation method (Hastie et al. 2001). In the J-fold cross validation method, one of the J subsets is set aside as the test sample, while the remaining $J-1$ subsets are used to estimate model parameters. The estimates from the latter are then applied to the test sample to estimate the size of prediction errors. This is repeated J times with the test sample changed systematically, and the prediction errors accumulated over the J test samples. This entire process is repeated with different values of the regularization parameter δ, and the value of δ which gives the smallest prediction errors is chosen as the "optimal" value. In particular, when $J = n$ (the sample size in the original data), this method is called the jackknife or leaving-one-out (LOO) method.

The cross validation methods described above are computationally heavy, particularly for the jackknife method ($J = n$) when n is large. Fortunately in this case, the magnitude of prediction errors can be expressed in closed form without going through repeated computations. This is called the cross validation (CV) index. For the OLS estimation of regression coefficients, this index is given by

$$CV = \text{tr}[(\mathbf{Z}-\hat{\mathbf{Z}})'(\mathbf{I}-\text{diag}(\mathbf{G}(\mathbf{G}'\mathbf{G})^{-1}\mathbf{G}'))^{-2}(\mathbf{Z}-\hat{\mathbf{Z}})/n], \tag{5.2}$$

where $\mathbf{Z}$ is the matrix of observed criterion variables (for multivariate criterion variables; in the univariate case, $\mathbf{Z}$ is replaced by a vector of observations on a criterion variable $\mathbf{z}$), $\hat{\mathbf{Z}}$ is the OLS prediction of $\mathbf{Z}$, $\mathbf{G}$ is the matrix of predictor variables, and n is the sample size. Note that $\text{diag}(\mathbf{G}(\mathbf{G}'\mathbf{G})^{-1}\mathbf{G}')$ is the diagonal matrix obtained from the orthogonal projector $\mathbf{P}_G$. This index can be readily extended to the regularized LS estimation by simply replacing $\mathbf{P}_G$ by the corresponding ridge operator $\mathbf{R}_G(\delta)$. The index is then a function of δ. We may minimize the index with respect to δ to find its optimal value. This may not be easy if we use a standard numerical optimization method. However, it is usually sufficient to try a few values of δ and choose the best one among them, since the optimal value of δ does not need to be determined very precisely. There is usually a wide range of values of δ for which the regularization method works almost equally well.

Note 5.1 We derive the CV index given above for a single criterion variable $\mathbf{z}$. The generalization to the multivariate case is straightforward. Let

$$\mathbf{z} = \mathbf{G}\mathbf{b} + \mathbf{e} \tag{5.3}$$

be the regression model for $\mathbf{z}$, where $\mathbf{G}$ is the matrix of predictor variables (assumed columnwise nonsingular), $\mathbf{b}$ is the vector of regression coefficients, and $\mathbf{e}$ is the vector of disturbance terms. Let $\mathbf{G}_{(i)}$ denote the matrix of predictor variables with the ith row vector $\mathbf{g}_i'$ removed from $\mathbf{G}$, and let $\mathbf{z}_{(i)}$ denote the vector of observations on the criterion variable with the ith observation z_i removed from $\mathbf{z}$. Then the ith jackknife estimate of $\mathbf{b}$ (this is simply the OLSE of $\mathbf{b}$ estimated from $\mathbf{G}_{(i)}$ and $\mathbf{z}_{(i)}$) is given by

$$\begin{aligned}
\hat{\mathbf{b}}_{(i)} &= (\mathbf{G}_{(i)}'\mathbf{G}_{(i)})^{-1}\mathbf{G}_{(i)}'\mathbf{z}_{(i)} \\
&= [(\mathbf{G}'\mathbf{G})^{-1} + (\mathbf{G}'\mathbf{G})^{-1}\mathbf{g}_i(1-p_i)^{-1}\mathbf{g}_i'(\mathbf{G}'\mathbf{G})^{-1}](\mathbf{G}'\mathbf{z}-\mathbf{g}_i z_i), \\
&= (1+\frac{1}{c_i}(\mathbf{G}'\mathbf{G})^{-1}\mathbf{g}_i\mathbf{g}_i')\hat{b}_{OLSE} - (1+\frac{p_i}{c_i})(\mathbf{G}'\mathbf{G})^{-1}\mathbf{g}_i z_i,
\end{aligned} \tag{5.4}$$

where

$$p_i = \mathbf{g}_i'(\mathbf{G}'\mathbf{G})^{-1}\mathbf{g}_i \tag{5.5}$$

is the ith diagonal element of $\mathbf{P}_G = \mathbf{G}(\mathbf{G}'\mathbf{G})^{-1}\mathbf{G}'$, and $c_i = 1 - p_i$. In the above derivation, we first used

$$\mathbf{G}_{(i)}'\mathbf{G}_{(i)} = \mathbf{G}'\mathbf{G} - \mathbf{g}_i\mathbf{g}_i', \tag{5.6}$$

$$\mathbf{G}_{(i)}'\mathbf{z}_{(i)} = \mathbf{G}'\mathbf{z} - \mathbf{g}_i z_i, \tag{5.7}$$

and then identity (2.27). The jackknife prediction of the ith observation is now given by

$$\begin{aligned}\hat{z}_{i(i)} = \mathbf{g}_i'\hat{\mathbf{b}}_{(i)} &= (1+\frac{p_i}{c_i})\hat{z}_i - p_i(1+\frac{p_i}{c_i})z_i \\ &= \frac{1}{c_i}(\hat{z}_i - p_i z_i),\end{aligned} \tag{5.8}$$

where $\hat{z}_i = \mathbf{g}_i'\hat{\mathbf{b}}_{OLSE}$ is the OLSE prediction of z_i based on the entire data set. The prediction error for the ith observation by the jackknife method is now expressed as

$$e_i^* = z_i - \hat{z}_{i(i)} = z_i - \frac{1}{c_i}(\hat{z}_i - p_i z_i) = \frac{1}{c_i}(z_i - \hat{z}_i). \tag{5.9}$$

The jackknife mean square error, defined by the sum of squares (SS) of e_i^* divided by the total number of observations n can be expressed as

$$\mathrm{CV} = \sum_{i=1}^{n} \frac{1}{(1-p_i)^2}(z_i - \hat{z}_i)^2/n, \tag{5.10}$$

which can be written, using matrix notation, as

$$\mathrm{CV} = (\mathbf{z} - \hat{\mathbf{z}})'(\mathbf{I} - \mathrm{diag}(\mathbf{P}_G))^{-2}(\mathbf{z} - \hat{\mathbf{z}})/n. \tag{5.11}$$

The relationship between (5.11) and (5.2) is obvious. In a completely analogous manner, we can derive the CV index for the RLS estimation. We simply use the RLS estimate of $\mathbf{z}$ for $\hat{\mathbf{z}}$ and the ridge operator $\mathbf{R}_G(\delta)$ for $\mathbf{P}_G$ in (5.11). Unfortunately, no formula analogous to (5.11) is available for PCA.

The p_i defined in (5.5) is often called the leverage of observation i (Belseley et al. 1980; Chatterjee and Hadi 1988; Cook and Weisberg 1982). It is used as an indicator of how influential the ith observation is in the OLS estimation of regression coefficients. The p_i takes a value between 0 and 1 inclusive with a mean value of m/n, where m is the number of predictor variables. (This assumes that $m \leq n$.) As a rule of thumb, an observation whose p_i exceeds twice its mean is considered a high leveraged observation (case). Note, however, that the leverage as defined in (5.5) indicates an influential observation based on the information on predictor variables alone. The notion of influential observations should also necessarily encompass the criterion variable(s) as well as the relationships between the criterion and predictor variables. See the references given above for further details on this point.

If we assume that all diagonal elements of $\mathbf{P}_G$ are equal, the formula (5.3) is simplified considerably. The resultant index is called the generalized cross validation (GCV) index (Craven and Wahba 1979), and is written as

$$\mathrm{GCV} = \frac{1}{(1-p)^2}\mathrm{SSE}/n, \tag{5.12}$$

where p is the mean of the diagonal elements of $\mathbf{P}_G$, and SSE $= \text{tr}[(\mathbf{Z}-\hat{\mathbf{Z}})'(\mathbf{Z}-\hat{\mathbf{Z}})]$. Again, it is straightforward to extend this expression to the case of regularized LS estimation. All we need is to replace p by the mean of the diagonal elements of $\mathbf{R}_G(\delta)$. The GCV index is easier to manipulate than CV. However, the assumption that all diagonal elements of $\mathbf{P}_G$ or $\mathbf{R}_G(\delta)$ are equal is unrealistic in many situations. Some argue that GCV does not work as well as CV (e.g., Breiman and Friedman 1997).

5.4 Missing Data

Missing data occur in virtually all data analysis situations. This is a "nasty" problem for which no definitive solutions exist that can be applied to all situations. There are several heuristic strategies that have traditionally been used:

(i) Remove all cases having at least one missing entry from subsequent analyses.
(ii) Substitute means for missing entries.
(iii) Predict missing data from other variables by regression analysis.

These operations make the data complete, and so once they have been applied, an analysis can proceed as if there were no missing data. They are easy to apply, but are somewhat ad hoc and difficult to justify, theoretically.

More recently, iterative methods are gaining popularity (e.g., van de Velden and Takane 2012). These procedures iteratively estimate "optimal" values of missing data based on the models being fitted. More specifically, model parameters are estimated using tentative estimates of missing data (as if there were no missing data), and the "best" predictions of missing data are derived using the estimated parameters. Those predictions are then used as the values of missing data in the next round of parameter estimation. This process is iterated until convergence is reached. This is a kind of ALS (alternating LS) algorithm that has already been discussed several times in this book (see, in particular, Section 4.18). An ALS algorithm is monotonically convergent, easy to use, and widely applicable. However, it can be time consuming, and in some cases, can lead to degenerate solutions (i.e., solutions that are dominated by estimated missing data).

Below we discuss a noniterative method specifically designed to deal with missing data in PCA. It is a multidimensional extension (Takane and Oshima-Takane 2003) of the method originally developed for test equating (the TE method; Shibayama 1995). In university entrance examinations (administered in Japan), not all applicants take the same examinations. The applicants are often allowed to choose which examinations to take (e.g., English vs French as a foreign language) within certain limits. This gives rise to a data set with systematic missing observations. Nonetheless, the test administrators should be able to come up with a set of scores that (at least partially) rank-order all applicants. The method we discuss provides closed-form solutions, which is a definite advantage, although it is only "applicable" to standardized data.

Let $\mathbf{D}_{z_i}$ denote the diagonal matrix with the ith row of the data matrix $\mathbf{Z}$ ($n \times m$)

as its diagonal elements. Define

$$\mathbf{Y}_i = \mathbf{D}_{z_i}\mathbf{V} + \mathbf{V}_0, \tag{5.13}$$

where $\mathbf{V}$ $(m \times r)$ and $\mathbf{V}_0$ $(m \times r)$ are the weight matrices that satisfy the following constraints:

$$\mathbf{1}_m'\mathbf{V}_0 = \mathbf{0}_r', \tag{5.14}$$

and

$$\mathbf{V}'\mathbf{S}\mathbf{V} = \mathbf{I}_r, \tag{5.15}$$

where r is the number of components, and $\mathbf{S}$ is the diagonal matrix of variations of the m observed variables calculated from nonmissing portions of the data. (The matrix $\mathbf{S}$ is n times the diagonal portions of the sample covariance matrix.) The conditions in (5.14) and (5.15) are necessary for identifiability of $\mathbf{V}$ and $\mathbf{V}_0$. Let $\mathbf{D}_{w_i}$ denote the diagonal matrix indicating missing data. If the jth diagonal element of $\mathbf{D}_{z_i}$ is observed, the corresponding element in $\mathbf{D}_{w_i}$ takes the value of 1, and if the jth diagonal element of $\mathbf{D}_{z_i}$ is unobserved, the corresponding element in $\mathbf{D}_{w_i}$ takes the value of 0. Let $\mathbf{u}_i$ denote a vector of component scores for observation (case) i, and consider minimizing

$$g_1(\mathbf{V}, \mathbf{V}_0, \mathbf{u}_i) = \sum_{i=1}^{n} \mathrm{SS}(\mathbf{Y}_i - \mathbf{1}_m\mathbf{u}_i')_{D_{w_i}, I} \tag{5.16}$$

with respect to $\mathbf{V}$, $\mathbf{V}_0$, and $\mathbf{u}_i$ subject to the constraints (5.14) and (5.15). We first minimize g_1 with respect to $\mathbf{u}_i$ for fixed $\mathbf{V}$ and $\mathbf{V}_0$. This leads to

$$\hat{\mathbf{u}}_i' = \mathbf{1}_m'\mathbf{D}_{w_i}\mathbf{Y}_i/m_i, \tag{5.17}$$

where $m_i = \mathbf{1}_m'\mathbf{D}_{w_i}\mathbf{1}_m$. Let $g_2(\mathbf{V}, \mathbf{V}_0)$ denote the minimum of $g_1(\mathbf{V}, \mathbf{V}_0, \mathbf{u}_i)$ with respect to $\mathbf{u}_i$ for given $\mathbf{V}$ and $\mathbf{V}_0$. Then it can be written as

$$g_2(\mathbf{V}, \mathbf{V}_0) = \sum_{i=1}^{n} \mathrm{tr}(\mathbf{Y}_i'\mathbf{C}_i\mathbf{Y}_i), \tag{5.18}$$

where

$$\mathbf{C}_i = \mathbf{Q}_{1_m/D_{w_i}}'\mathbf{D}_{w_i}\mathbf{Q}_{1_m/D_{w_i}} = \mathbf{Q}_{1_m/D_{w_i}}'\mathbf{D}_{w_i} = \mathbf{D}_{w_i}\mathbf{Q}_{1_m/D_{w_i}}. \tag{5.19}$$

Here, $\mathbf{Q}_{1_m/D_{w_i}} = \mathbf{I}_m - \mathbf{1}_m\mathbf{1}_m'\mathbf{D}_{w_i}/m_i$. Substituting this for $\mathbf{C}_i$ in (5.18) and letting

$$\begin{aligned} \mathbf{A}_1 &= \sum_i \mathbf{D}_{z_i}\mathbf{C}_i\mathbf{D}_{z_i} \\ \mathbf{A}_2 &= \sum_i \mathbf{D}_{z_i}\mathbf{C}_i \\ \mathbf{A}_3 &= \sum_i \mathbf{C}_i, \end{aligned} \tag{5.20}$$

$g_2(\mathbf{V}, \mathbf{V}_0)$ can be further rewritten as

$$g_2(\mathbf{V}, \mathbf{V}_0) = \mathrm{tr}(\mathbf{V}'\mathbf{A}_1\mathbf{V}) + 2\mathrm{tr}(\mathbf{V}'\mathbf{A}_2\mathbf{V}_0) + \mathrm{tr}(\mathbf{V}_0\mathbf{A}_3\mathbf{V}_0). \tag{5.21}$$

Minimizing $g_2(\mathbf{V}, \mathbf{V}_0)$ with respect to $\mathbf{V}_0$ for fixed $\mathbf{V}$ leads to

$$\hat{\mathbf{V}}_0 = -\mathbf{A}_3^+\mathbf{A}_2'\mathbf{V}, \tag{5.22}$$

where $\mathbf{A}_3^+$ is the Moore-Penrose inverse of $\mathbf{A}_3$. Note that this $\mathbf{V}_0$ satisfies (5.14). Let $g_3(\mathbf{V})$ denote the minimum of $g_2(\mathbf{V}, \mathbf{V}_0)$ with respect to $\mathbf{V}$ for fixed $\mathbf{V}_0$. Then it can be written as

$$g_3(\mathbf{V}) = \text{tr}(\mathbf{V}'(\mathbf{A}_1 - \mathbf{A}_2\mathbf{A}_3^+\mathbf{A}_2')\mathbf{V}) = \text{tr}(\mathbf{V}'\mathbf{A}\mathbf{V}), \tag{5.23}$$

where $\mathbf{A} = \mathbf{A}_1 - \mathbf{A}_2\mathbf{A}_3^+\mathbf{A}_2'$. Minimizing $g_3(\mathbf{V})$ with respect to $\mathbf{V}$ subject to (5.15) leads to

$$\mathbf{AV} = \mathbf{SV}\Delta, \tag{5.24}$$

which is a generalized eigenequation for $\mathbf{A}$ with respect to $\mathbf{S}$. Since g_3 is to be minimized, the eigenvectors corresponding to the r smallest eigenvalues are retained. The generalized eigenequation above can be turned into an ordinary eigenequation by

$$\mathbf{S}^{-1/2}\mathbf{A}\mathbf{S}^{-1/2}\tilde{\mathbf{V}} = \tilde{\mathbf{V}}\Delta, \tag{5.25}$$

where $\tilde{\mathbf{V}} = \mathbf{S}^{1/2}\mathbf{V}$, or $\mathbf{V} = \mathbf{S}^{-1/2}\tilde{\mathbf{V}}$.

It is instructive to see what happens if there are no missing data (i.e., $\mathbf{D}_{w_i} = \mathbf{I}_n$ for all i) in the above procedure. The matrix $\mathbf{A}$ becomes

$$\mathbf{A} = \text{diag}(\mathbf{Z}'\mathbf{Q}_{1_n}\mathbf{Z}) - \frac{1}{n}\mathbf{Z}'\mathbf{Q}_{1_n}\mathbf{Z}. \tag{5.26}$$

Let $\mathbf{R}$ denote the correlation matrix derived from $\mathbf{Z}$. Then

$$\mathbf{S}^{-1/2}\mathbf{A}\mathbf{S}^{-1/2} = \mathbf{I}_m - \mathbf{R}/n. \tag{5.27}$$

Hence, the eigenvectors of $\mathbf{S}^{-1/2}\mathbf{A}\mathbf{S}^{-1/2}$ and those of $\mathbf{R}$ coincide. The eigenvalues are different, although they are related in a simple way. Let Δ and $\tilde{\Delta}$ denote the diagonal matrices of the eigenvalues of $\mathbf{S}^{-1/2}\mathbf{A}\mathbf{S}^{-1/2}$ and $\mathbf{R}$, respectively. Then,

$$\Delta = \mathbf{I} - \tilde{\Delta}/n. \tag{5.28}$$

That is, larger eigenvalues of $\mathbf{R}$ correspond with smaller eigenvalues of $\mathbf{S}^{-1/2}\mathbf{A}\mathbf{S}^{-1/2}$.

Takane and Oshima-Takane (2003) have shown that the above method is essentially equivalent to the missing data passive (MDP) approach (Meulman 1982) in homogeneity analysis, which is a variant of MCA (see Section 4.10). There is a subtle difference between the two, however. Whereas the TE method deals with raw data (although it works like PCA of standardized data as seen in the previous paragraph), the MDP method deals with columnwise centered data. Because of the initial centering of the data, the MDP method assumes that $\mathbf{V}_0 = \mathbf{O}$. Takane and Oshima-Takane (2003), however, have shown that having nonzero $\mathbf{V}_0$ is still necessary in the latter to make it completely equivalent to the TE method. Allowing nonzero $\mathbf{V}_0$ is important unless there is evidence that the means calculated from nonmissing portions of the data are close to the means that would have resulted if there were no missing data. This is not likely to happen in test equating situations.

It is relatively straightforward to incorporate column constraints in the above method. Let $\mathbf{P}_H$ denote the orthogonal projector defined by the column constraint matrix $\mathbf{H}$. Then, we may simply obtain SVD($\mathbf{P}_H\mathbf{A}\mathbf{P}_H$). On the other hand, it is almost impossible to incorporate row-side constraints. This is because the method described above assumes that the matrix of row means $\mathbf{U}$ is separable (i.e., $\mathbf{u}_i'$ can be estimated separately for each i).

5.5 Robust Estimations

Regression analysis (projection) and PCA (SVD) are the main building blocks in CPCA. In a majority of situations, these two analysis tools are applied in conjunction with ordinary least squares estimation (OLSE), which is intrinsically nonrobust against outliers. Outliers are observations that are far from other observations. They seem to operate under different principles from the rest of observations, and tend to have undue influence on solutions. This is quite disturbing if our interest lies in the "other" (nonoutlying) observations, although in some cases, outliers may be our focus of interest. Perhaps the easiest way to identify outliers is to fit the model (either the regression model or PCA with a prescribed number of components), and take residuals from the fitted model. Observations associated with excessively large residuals are likely to be outliers. Once the outliers are detected, there are two ways to proceed. One possibility is to discard the outliers from subsequent analyses. Alternatively, we may use the weighted least squares (WLS) estimation, where smaller weights are given to observations with larger residuals, while larger weights are given to observations with smaller residuals, thereby equalizing the influence of observations. The weights do not have to be fixed. They may be adjusted, as the influence of observations is more accurately estimated. Resultant techniques are a kind of robust estimation methods called iteratively reweighted LS (IRLS) methods. Here, the word "robust" means "less susceptible to outliers." This topic is in fact not entirely new. In Section 4.18, we discussed algorithms for WLRA (the weighted low rank approximation), where the weights were fixed. If, however, the weights were allowed to be adjusted, we would have obtained a robust low rank approximation method. In the following two subsections, we briefly discuss robust regression and robust SVD.

5.5.1 Robust regression

Let

$$\mathbf{z} = \mathbf{G}\mathbf{b} + \mathbf{e} \tag{5.29}$$

denote a linear regression model. As before, $\mathbf{z}$ is the vector of observations on the criterion variable, $\mathbf{G}$ the matrix of observations on the predictor variables, $\mathbf{b}$ the vector of regression coefficients, and $\mathbf{e}$ the vector of disturbance terms. The WLSE of $\mathbf{b}$ is obtained by

$$\hat{\mathbf{b}}_{WLSE} = (\mathbf{G}'\mathbf{W}\mathbf{G})^{-1}\mathbf{G}'\mathbf{W}\mathbf{z}, \tag{5.30}$$

where $\mathbf{W}$ is a *nnd* matrix of weights (the row-side metric matrix). A special case of this, in which $\mathbf{W}$ is set equal to the inverse of the covariance matrix among the elements of $\mathbf{z}$, is called the generalized LS (GLS) estimation, which yields the BLUE of $\mathbf{b}$. In robust regression analysis, $\mathbf{W}$ is usually restricted to be a diagonal matrix whose diagonal elements are adjusted according to the influence of the elements of $\mathbf{z}$. Typically, the OLSE of $\mathbf{b}$ is obtained first, and residuals are calculated, based on the estimate. Weights are then adjusted to be inversely proportional to the residuals. In the second iterative cycle, the WLSE of $\mathbf{b}$ is obtained under the new weights, residuals are recalculated, new weights are determined, and so on. This iterative process is repeated until convergence.

There are a number of proposals for weight adjustments. We briefly mention a few of them below. Tukey's biweights, Hampel et al.'s (1986) methods based on influence functions, Huber's M estimators, and least trimmed squares (LTS; Rousseeuw and Leroy 1987) are among the most popular ones. We refer to Verboon (1994) for a quick introduction to IRLS, and McKeen (2004) for an updated survey on robust analysis of linear models. Coleman et al. (1980) implemented seven different weighting functions in the form of FORTRAN subroutines to be used in IRLS.

5.5.2 Robust PCA/SVD

There are at least two representative approaches to robust PCA/SVD: One approach estimates covariance matrices robustly (e.g., Devlin et al. 1981) for subsequent PCA/SVD. This includes simple methods such as 1) dividing the sum of squares (SS) and products matrix by $n+1$ rather than n or $n-1$, thereby shrinking both variances and covariances, 2) adding some positive constants to diagonals of a sample covariance matrix, and 3) weighting observations (cases) according to their importance or reliability in calculating a covariance matrix. The other approach obtains SVD in a robust way. This may involve iterative calculations of the weights applied to observations (cases) in the weighted LS approximation of a data matrix by the bilinear model (i.e., WLRA in Section 4.18). The weights could be such that the WLS criterion in effect reduces to minimizing the sum of absolute residuals, as shown below.

In several places, we discussed an ALS (alternating LS) algorithm for fitting the bilinear model $\mathbf{ba}'$ to the data matrix $\mathbf{Z}$, where $\mathbf{b}$ ($||\mathbf{b}|| = 1$) is the vector of component scores, and $\mathbf{a}$ the vector of component loadings. Let

$$\phi(\mathbf{b}, \mathbf{a}) = \mathrm{SS}(\mathbf{Z} - \mathbf{ba}'). \tag{5.31}$$

The ALS algorithm minimizes the above criterion iteratively by alternately updating $\mathbf{b}$ for fixed $\mathbf{a}$, and $\mathbf{a}$ for fixed $\mathbf{b}$. To recapitulate, the algorithm proceeds as:

Step 1. Initialize $\mathbf{a}$.
Step 2. Update $\mathbf{b}$ by first obtaining

$$\mathbf{b}^* = \mathbf{Za}, \tag{5.32}$$

then normalizing it by $\mathbf{b} = \mathbf{b}^*/||\mathbf{b}^*||$.
Step 3. Update $\mathbf{a}$ by

$$\mathbf{a} = \mathbf{Z}'\mathbf{b}. \tag{5.33}$$

Steps 2 and 3 are repeatedly applied until convergence. This algorithm was specifically referred to as the Daugavet algorithm in Section 4.5. As in the case of regression analysis in the previous section, the LS criterion used in (5.31) can be extended to the weighted LS criterion, and the weights in WLRA may also be adjusted to obtain robust SVD. In principle, any one of the adjustment schemes discussed in the previous section may be used. In what follows, we discuss a procedure which adjusts the weights in a specific way.

Hawkins et al. (2001) considered minimizing the sum of absolute residuals in fitting $\mathbf{ba}'$ to $\mathbf{Z}$. The sum of absolute residuals is sometimes called the L_1 norm residuals, as opposed to the L_2 norm residuals used in the LS criterion. The criterion based on the L_1 norm can be expressed, using the elementwise notation, as

$$\psi(\mathbf{b},\mathbf{a}) = \sum_{i,j} |z_{ij} - b_i a_j|, \tag{5.34}$$

where z_{ij} is the ijth element of $\mathbf{Z}$, b_i is the ith element of $\mathbf{b}$, and a_j is the jth element of $\mathbf{a}$. This criterion is appealing in that it does not weigh large residuals as much as the LS criterion, which is bound to be sensitive to outliers due to the squaring of residuals. The L_1 norm criterion above can be rewritten, in the form of a WLS criterion, as

$$\psi(\mathbf{b},\mathbf{a}) = \sum_{i,j} \frac{1}{|z_{ij} - b_i a_j|}(z_{ij} - b_i a_j)^2, \tag{5.35}$$

where the weight $w_{ij} = 1/|z_{ij} - b_i a_j|$ is a function of b_i and a_j. This rewriting, however, has no practical benefit except that it shows that the minimization of the L_1 norm is a special case of WLS estimation. Since the weights are adjustable, it is also a special case of IRLS.

Hawkins et al. (2001) has developed an alternating least L_1 norm algorithm to minimize (5.34). We first update a_j for fixed b_i ($i = 1, \cdots, n$) by minimizing

$$\psi(a_j) = \sum_i |z_{ij} - b_i a_j| = \sum_i |b_i||z_{ij}/b_i - a_j|. \tag{5.36}$$

Such an a_j can be obtained by the weighted median of z_{ij}/b_i. The weighted median is the value of z_{ij}/b_i which lies right at $\sum_i |b_i|/2$. We arrange z_{ij}/b_i from the smallest to the largest, and take the cumulative sums of $|b_i|$ rearranged in order by z_{ij}/b_i. We then identify z_{ij}/b_i whose corresponding cumulative sum is closest to $\sum_i |b_i|/2$ but not exceeding it. In other words, we treat $|b_i|$ as if it were the observed frequency of z_{ij}/b_i. We update other a_j's in a similar fashion. We then update b_i by minimizing

$$\psi(b_i) = \sum_j |z_{ij} - b_i a_j| = \sum_j |a_j||z_{ij}/a_j - b_i|. \tag{5.37}$$

The b_i that minimizes (5.37) can be found, in a manner similar to the above, by the weighted median of z_{ij}/a_j. Once all the b_i's are updated, we normalize them to satisfy $||\mathbf{b}|| = 1$. The alternating updating process above is repeated until convergence. (Hawkins et al. update $\mathbf{a}$ first and then $\mathbf{b}$. The order is essentially arbitrary.) If more than one component are wanted, the effect of the first component is eliminated (deflation), and the same procedure is applied to the deflated data matrix. In contrast to the L_2 case, successive components are not mutually orthogonal. Also, they are not necessarily in the order of "importance" in terms of variance accounted for by the components.

There are several variants to the above method. Choulakian (2001) used an enumeration method to minimize the L_1 norm in PCA. Liu et al. (2003) employed LTS (least trimmed squares) for robust PCA, while hinting at other possibilities. Heiser (1987) and Nishisato (1984, 1987, 1988) used the least L_1 norm in CA (correspondence analysis; Section 4.6) and MCA (multiple CA; Section 4.10), while Sachs and Kong (1994) used Tukey's biweights for robust MCA. Ammann (1993) proposed a robust SVD by an iteratively reweighted transposed QR algorithm. (See Note 5.2 below.)

Note 5.2 The transposed QR algorithm for ordinary SVD is interesting in its own right. Let

$$\mathbf{Z} = \mathbf{Q}_0\mathbf{R}_0' \tag{5.38}$$

denote the QR decomposition (Section 2.1.12) of the data matrix $\mathbf{Z}$. We then obtain the QR decomposition of $\mathbf{R}_0$, denoted as

$$\mathbf{R}_0 = \mathbf{Q}_1\mathbf{R}_1'. \tag{5.39}$$

This leads to

$$\mathbf{Z} = \mathbf{Q}_0\mathbf{R}_1\mathbf{Q}_1'. \tag{5.40}$$

We next find the QR decomposition of $\mathbf{R}_1$, denoted as

$$\mathbf{R}_1 = \mathbf{Q}_2\mathbf{R}_2', \tag{5.41}$$

which leads to

$$\mathbf{Z} = \mathbf{Q}_0\mathbf{Q}_2\mathbf{R}_2'\mathbf{Q}_1'. \tag{5.42}$$

We then find the QR decomposition of $\mathbf{R}_2$, denoted as

$$\mathbf{R}_2 = \mathbf{Q}_3\mathbf{R}_3', \tag{5.43}$$

which leads to

$$\mathbf{Z} = \mathbf{Q}_0\mathbf{Q}_2\mathbf{R}_3\mathbf{Q}_3'\mathbf{Q}_1'. \tag{5.44}$$

This iterated process is known to converge to the SVD of $\mathbf{Z}$ (Ammann 1993), i.e., $\mathbf{R}_i$ ($i = 1, 2, \cdots$) converges to the diagonal matrix of singular values, $\mathbf{Q}_0\mathbf{Q}_2\mathbf{Q}_4\cdots$ to the matrix of left singular vectors, and $\mathbf{Q}_1\mathbf{Q}_3\mathbf{Q}_5\cdots$ to the matrix of right singular vectors. Ammann (1993) suggests using a weighted QR decomposition in the above iterative process to obtain robust SVD.

5.6 Data Transformations

So far we have treated the data as given, except that we have discussed some options for preprocessing of the data in Section 4.1. However, it is often possible to find more parsimonious representations of the data by allowing elaborate data transformations. The data, for example, may be measured on an ordinal scale. In such cases the data may be monotonically transformed, so that the transformed data better conform to required properties of fitted models. This is called optimal scaling or optimal transformations of data (Young 1981; see also Takane 2005).

Optimal scaling involves parameterizations of transformations and their optimal estimation based on the data. We discuss a particular kind of optimal scaling called the LS (least squares) monotonic transformation in this section. This kind of transformation is appropriate when the data are measured on an ordinal scale, meaning that only ordinal relations among data elements (i.e., which data elements are larger or smaller than others) are meaningful. Consider, as an example, the situation in which subjective largeness of rectangles is measured by the frequency with which the rectangles are judged large over a group of subjects. The frequency does not represent an exact measurement of the subjective largeness of the rectangles, but merely its ordinal measure. In such cases we may transform the data monotonically, so that the rank-order among the data elements is preserved, but the transformed data are in better agreement with (more linearly related to) the predictions of a fitted model that captures the basic mechanism of how subjective largeness judgments are formed. A concrete example of this is presented in Section 6.4.

Optimal monotonic transformations can be easily incorporated in traditional multivariate data analysis techniques based on the principle of ALS (alternating LS) algorithms. In these algorithms, conventional (linear) multivariate data analysis methods (e.g., PCA) are first applied to the observed data matrix to find initial (tentative) predictions of the data. The predictions of the data are then transformed to satisfy monotonicity with the observed data, which serve as the transformed data in the next cycle of iterations. This alternating process is repeated until convergence. The second phase of the alternating process is called the LS monotonic transformation. An important point to keep in mind is that in the ALS algorithms, both model estimation and data transformations are performed so as to minimize a single common LS criterion. This assures monotonic convergence of the algorithms.

Suppose that we wish to apply the optimal scaling idea to PCA. Let $\mathbf{Z}$ denote an observed data matrix, and let $\mathbf{Z}^*$ denote an optimally (monotonically) transformed data matrix such that $\text{diag}(\mathbf{Z}^{*\prime}\mathbf{Z}^*/n) = \mathbf{I}$. Let $\mathbf{Z}_0$ represent a matrix of predictions under a PCA model with a prescribed number of components r. Under this condition, $\mathbf{Z}_0$ can be parameterized as $\mathbf{Z}_0 = \mathbf{BA}'$, where $\mathbf{B}$ is an $n \times r$ columnwise nonsingular matrix of component scores such that $\mathbf{B}'\mathbf{B}/n = \mathbf{I}$, and $\mathbf{A}$ is an $m \times r$ columnwise nonsingular matrix of component loadings. Consider minimizing

$$\phi(\mathbf{B}, \mathbf{A}) = \text{SS}(\mathbf{Z}^* - \mathbf{BA}') \tag{5.45}$$

with respect to $\mathbf{B}$, $\mathbf{A}$, and $\mathbf{Z}^*$ subject to the normalization restriction on $\mathbf{B}$, and the monotonicity restriction on the columns of $\mathbf{Z}^*$. This criterion can be minimized by the following ALS algorithm:

Step 1. Set $\mathbf{Z}^* = \mathbf{Z}$, where $\mathbf{Z}$ is the observed data matrix.
Step 2. Find $\mathbf{B}$ and $\mathbf{A}$ that minimize (5.45) by SVD($\mathbf{Z}^*$), and calculate the matrix of model predictions $\mathbf{Z}_0 = \mathbf{BA}'$.
Step 3. Apply the LS monotonic transformation (to be explained in details below) to each column of $\mathbf{Z}_0$. The transformed data are then normalized to obtain new $\mathbf{Z}^*$ to be

used in the next iteration.
Step 4. Stop, if converged. Otherwise, go back to Step 2.

The above algorithm is implemented in the form of a computer program (PRINCIPALS; Young et al. 1978). A variant of the program with more attractive input/output features exists in SPSS under the name CATPCA (PCA of categorical data; Linting 2007a). Note that in Step 2, it is not absolutely necessary to minimize $\phi(\mathbf{B},\mathbf{A})$ in (5.45) completely with respect $\mathbf{B}$ and $\mathbf{A}$ for given $\mathbf{Z}^*$ by the SVD of $\mathbf{Z}^*$ to preserve the monotonic convergence of the ALS algorithm. It suffices to merely decrease the value of $\phi(\mathbf{B},\mathbf{A})$. Consequently, Step 2 may be replaced by the following less computationally intensive substeps:

Step 2.1. Update $\mathbf{B}$ by first calculating $\mathbf{B}^* = \mathbf{Z}^*\mathbf{A}(\mathbf{A}'\mathbf{A})^{-1}$, and then orthonormalizing $\mathbf{B}^*$ to obtain $\mathbf{B}$ by the Schmidt orthogonalization procedure (Section 2.1.11).
Step 2.2 Update $\mathbf{A}$ by $\mathbf{A} = \mathbf{Z}^{*\prime}\mathbf{B}$.

Step 2.1 obtains the LS update of $\mathbf{B}$ for given $\mathbf{A}$ and $\mathbf{Z}^*$ subject to the normalization restriction of $\mathbf{B}$, while Step 2.2 obtains the LS update of $\mathbf{A}$ for fixed $\mathbf{B}$ and $\mathbf{Z}^*$.

It remains to be seen exactly how Step 3 is carried out. This procedure will be discussed for a column of the data matrix, since the transformation is applied for each column separately. Let $\mathbf{z}$ be a column vector in $\mathbf{Z}$, and $\mathbf{z}_0$ the corresponding column vector in $\mathbf{Z}_0$ in a particular iterative cycle of the algorithm. We explain how the optimal $\mathbf{z}^*$ is obtained based on these two vectors. The elements of $\mathbf{z}^*$ are (weakly) monotonically related to the elements in $\mathbf{z}$. Here the weak monotonicity means that if $z_j > z_k$ (where z_j and z_k are the jth and kth elements of $\mathbf{z}$), it must hold that $z_{0j} \geq z_{0k}$ (where z_j^* and z_k^* are the corresponding elements of $\mathbf{z}^*$).

For explanation, we first assume that no two elements in $\mathbf{z}$ have the same magnitude, i.e., there are no ties in $\mathbf{z}$. Before the iterative cycle starts, we sort the elements of $\mathbf{z}$ in ascending order of magnitude. In each iterative cycle, we order the elements of $\mathbf{z}_0$ in accordance with the ordered elements of $\mathbf{z}$. For simplicity, we denote these reordered vectors as $\mathbf{z}$ and $\mathbf{z}_0$. In what follows, we illustrate the transformation using concrete numerical examples.

Suppose that $\mathbf{z} = (1,2,3,4,5,6,7)'$, and $\mathbf{z}_0 = (1,3,4,5,6,7,9)'$. In this case, the elements of $\mathbf{z}_0$ are already perfectly monotonic with those in $\mathbf{z}$, so there is no need to change $\mathbf{z}_0$ to obtain $\mathbf{z}^*$. There is no loss in fit in this case. Suppose next that $\mathbf{z}_0 = (1,4,2,6,5,4,7)$, while $\mathbf{z}$ remains the same. In this case, the elements of $\mathbf{z}_0$ are not perfectly monotonic with those of $\mathbf{z}$, so some modifications are necessary in $\mathbf{z}_0$ to find $\mathbf{z}^*$, the elements of which are supposed to be weakly monotonic with those of $\mathbf{z}$. To find such a $\mathbf{z}^*$, we identify where a violation of monotonicity occurs. We begin with the first two elements in $\mathbf{z}_0$ corresponding to the two smallest elements of $\mathbf{z}$. We observe that the first two elements in $\mathbf{z}_0$ satisfy the required monotonicity, so we tentatively set $z_1^* = 1$ and $z_2^* = 4$. We next compare the third element of $\mathbf{z}_0$ with z_2^*, and find that the former is smaller than the latter. This means that we cannot set z_3^* equal to the third element of $\mathbf{z}_0$ because that will incur a violation of monotonicity. Instead, we take the average of these two elements, (4 + 2)/2 = 3, and give this value to both the second and third elements of $\mathbf{z}^*$. Taking the average between the two represents the minimal change necessary to "restore" the (weak) monotonicity. Before going further, however, we need to check that this averaging operation did not create any new violation of monotonicity among the elements of $\mathbf{z}^*$ obtained so far. This is done by comparing $z_2^* = 3$ with $z_1^* = 1$, which satisfies the monotonicity. So we tentatively accept $(1,3,3)'$ as the first three elements of $\mathbf{z}^*$. We may now proceed further by comparing the fourth element of $\mathbf{z}_0$ with z_3^*, which satisfies the monotonicity, so we may set $z_4^* = 6$. We next compare this element with the fifth element of $\mathbf{z}_0$, and find a violation of monotonicity. So we take the average between them, (6 + 5)/2 = 5.5, and assign this value to both the fourth and fifth elements of $\mathbf{z}^*$. We again find that this operation does not create any new violation among the elements of $\mathbf{z}^*$ so far obtained. So we may proceed to the next step, which compares the sixth element of $\mathbf{z}_0$ with $\mathbf{z}_5^*$. We find a violation of monotonicity, so we take the average of 4 and 5.5, but we find this average is smaller than $\mathbf{z}_4^*$, creating another violation of monotonicity. So we end up with taking the average of the three numbers (5.5 + 5.5 + 4)/3 = (6 + 5 + 4)/3 = 5, and set $\mathbf{z}_4^*$, $\mathbf{z}_5^*$, and $\mathbf{z}_6^*$ equal to this value. We confirm that this operation creates no new violation of monotonicity. Finally, we compare the seventh element of $\mathbf{z}_0$ with $\mathbf{z}_6^*$, and find the monotonicity is satisfied. We set $z_7^* = 7$. We now have $\mathbf{z}^* = (1,3,3,5,5,5,7)'$ whose elements are weakly monotonic with those of $\mathbf{z}$. The sum of squared discrepancies between $\mathbf{z}_0$ and $\mathbf{z}^*$ indicates the (squared) loss in fit due to the transformation. This turns out to be $\text{SS}(\mathbf{z}^* - \mathbf{z}_0) = 0+1+1+1+0+1+0 = 4$ in the present case. The algorithm described above is called Kruskal's (1964b) LS monotonic transformation (MATLAB codes available at http://takane.brinkster.net/Yoshio/). The $\mathbf{z}^*$ is normalized before going to the next stage.

So far we have assumed that there are no ties in the observed data. What happens if there are ties? Kruskal (1964b) suggests two alternative approaches: the primary approach and the secondary approach. In the former, the elements of $\mathbf{z}^*$ corresponding to tied observations in $\mathbf{z}$ can be in any order. Algorithmically, this amounts to reordering the elements of $\mathbf{z}_0$ corresponding to the tied observations according to their size. In the secondary approach, the elements of $\mathbf{z}^*$ corresponding to tied observations are forced to be equal. Algorithmically, this amounts to taking the average of the elements of $\mathbf{z}_0$ corresponding to tied observations prior to the monotonic transformation. Note, however, that when assessing the loss in fit incurred by the transformation, the discrepancy has to be measured between $\mathbf{z}^*$ and the original $\mathbf{z}_0$ (before the averaging operations for tied observations are carried out). Suppose, as an example, that we have $\mathbf{z} = (1,2,2,2,3,4,4)'$, while $\mathbf{z}_0$ remains the same as in the previous paragraph. The primary approach redefines $\mathbf{z}_0$ as $(1,2,4,6,5,4,7)'$ by reordering its elements corresponding to tied elements by their size, which is then subjected to the LS monotonic transformation algorithm described above to obtain $\mathbf{z}^* = (1,2,4,5,5,5,7)'$. The loss incurred by the transformation is calculated by $0+0+0+1+0+1+0 = 2$. The secondary approach, on the other hand, redefines $\mathbf{z}_0$ as $(1,4,4,4,5,5.5,5.5)'$ by taking the averages of its elements corresponding to tied elements in $\mathbf{z}$. This redefined $\mathbf{z}_0$ is then subjected to the LS monotonic transformation algorithm. In this particular example, the redefined $\mathbf{z}_0$ already satisfies the weak monotonicity, so $\mathbf{z}^*$ may be simply set equal to the redefined $\mathbf{z}_0$. The loss incurred

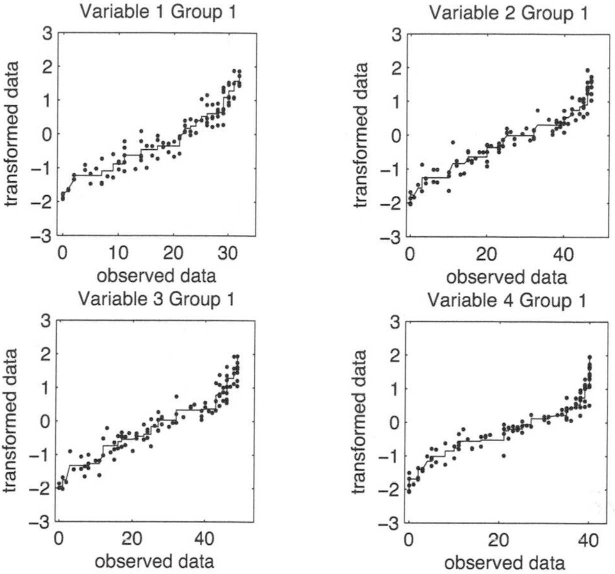

Figure 5.2 *Examples of the LS monotonic transformation for Kempler's data in Section 6.4. (Reproduced from Figure 2 of Hwang and Takane (2010) with permission from the Behaviormetric Society of Japan.)*

by the transformation, however, has to be calculated between $\mathbf{z}^*$ and the original $\mathbf{z}_0$, which would be $0+0+4+4+0+1.5^2+1.5^2 = 12.5$ in this case.

Figure 5.2 shows examples of the LS monotonic transformation. In this figure, dots indicate the observed frequency data (on the x-axis) plotted against the corresponding model predictions (on the y-axis). Connected line segments show the LS monotonic transformation (the elements of $\mathbf{z}^*$ plotted against the corresponding elements of $\mathbf{z}$ connected by the line segments). It is observed that the connected line segments traced from left to right never go downwards, implying that they collectively represent a monotonically nondecreasing function. It may also be observed that the transformed data are stretched out at both ends, indicating the ceiling and floor effects in the original data. See Section 6.4 for more detailed descriptions of the data and the fitted model.

The idea of optimal scaling originated from nonmetric multidimensional scaling (MDS; Section 4.5) by the pioneering work of Shepard (1962) and Kruskal (1964a, b), who demonstrated that a multidimensional stimulus configuration could be constructed from proximity data measured on an ordinal scale by transforming the data monotonically. The idea of optimal scaling quickly spread to other multivariate techniques such as PCA (Kruskal and Shepard 1974; Young et al. 1978), culminating in the ALSOS system (Young 1981) and the Gifi system (Gifi 1990) of nonmetric multivariate analysis.

The LS monotonic transformation is optimal in the LS sense under the weak monotonicity restriction. However, it may at times be very "steppy" in contrast to our intuition that it must be smoother. Monotonic spline transformations, being smoother than the LS monotonic transformations, may provide a good alternative to the latter. Spline transformations are piecewise polynomial functions of certain degrees connected at prescribed points called knots. The connections are made in such a way that the functions are differentiable at the knots a prescribed number of times, guaranteeing the desired degrees of smoothness. See Ramsay (1988), Winsberg and Ramsay (1983), and de Leeuw et al. (1981) for technical details of the spline transformations.

Ramsay (1982) and Ramsay and Silverman (2005) proposed PCA of functional data, in which the data are regarded as continuous functions over time. In this approach, z_{ij}, the ijth element of the data matrix $\mathbf{Z}$, represents a sample of $z_i(t)$ for case i and at time $t = t_j$, where t_j is the time point at which the measurement is taken. Spline functions and other basis functions (e.g., sine and cosine, wavelet, etc.) are used as building blocks to approximate the continuous functions underlying the observed discrete data. Besse and Ramsay (1986) derived a metric matrix $\mathbf{L}$ that has the effect of analyzing the supposed continuous function underlying the data, while in effect analyzing the latter. They call their approach a functional PCA, and their approach in general a functional data analysis (FDA). (See Ramsay and Silverman (2005) for a comprehensive account of FDA, and visit http://www.psych.mcgill.ca/faculty/ramsay/ramsay.html for software to perform FDA.) Note that the information about the basis functions is captured in the column side information matrix $\mathbf{H}$ in the CPCA framework.

The data transformations used in MVA are mostly univariate. That is, each variable is transformed separately. Nonlinear transformations performed by neural network models, on the other hand, are multivariate in the sense that simultaneous transformations of more than one variable are realized. This allows us to deal with complex interactions among variables (Takane and Oshima-Takane 2003).

5.7 Biplot

We have already seen several examples of graphical displays of PCA results. Many of them plotted component loadings and scores separately. It is possible, however, to display both in a single graph, as in Figure 4.3. Gabriel (1971) called the joint plot of loadings and scores a biplot if it preserves inner products. He demonstrated that there were infinitely many possible biplots depending on how we "scale" $\mathbf{A}$ and $\mathbf{B}$ to keep $\mathbf{Z}_0 = \mathbf{BA}'$ intact. Let $\mathbf{T}$ denote an arbitrary nonsingular square matrix. Then, as noted earlier, $\mathbf{Z}_0 = \mathbf{BTT}^{-1}\mathbf{B}' = \mathbf{B}^*\mathbf{A}^{*'}$, where $\mathbf{B}^* = \mathbf{BT}$ and $\mathbf{A}^* = \mathbf{AT}^{-1}$. The matrices $\mathbf{B}^*$ and $\mathbf{A}^*$ are as good to produce $\mathbf{Z}_0$ as $\mathbf{B}$ and $\mathbf{A}$. Thus, the joint plot of any pair of $\mathbf{B}^*$ and $\mathbf{A}^*$ with an arbitrary nonsingular transformation matrix $\mathbf{T}$ may be called a biplot. Limiting the scope of permissible transformations to pure "scaling", $\mathbf{T}$ can still be an arbitrary *pd* diagonal matrix with considerable freedom in its choice. In particular, the effects of singular values can be distributed in any way between $\mathbf{B}^*$ and $\mathbf{A}^*$. Because of this arbitrariness, inner products are the only interpretable properties of a biplot, no matter what scaling conventions are used. (One should strictly refrain from interpreting distances between row and column points.)

In Section 1.3 we explained the scaling conventions for $\mathbf{B}$ and $\mathbf{A}$ used in this book. To recapitulate, $\mathbf{B} = \sqrt{n}\mathbf{U}_r$ and $\mathbf{A} = \mathbf{V}_r\mathbf{D}_r/\sqrt{n}$, where $\mathbf{D}_r$ is the diagonal matrix of the r most dominant singular values of $\mathbf{A}$, and $\mathbf{U}_r$ and $\mathbf{V}_r$ are the matrices of the left and right singular vectors corresponding to the r largest singular values. We must always keep in mind that other choices are possible, and try to identify what scaling convention is used in a particular application. It is important to know what is interpretable and what is not. Recall that under the scaling convention adopted in this book, $\mathbf{B}'\mathbf{B}/n = \mathbf{I}$ (components are orthonormal), and $\mathbf{B}'\mathbf{Z}_0/n = \mathbf{A}'$ (the elements of $\mathbf{A}$ represent the covariances or correlations between the components and the observed variables), Our scaling convention also preserves inner products between the columns of $\mathbf{Z}_0$, as seen from $\mathbf{Z}_0'\mathbf{Z}_0/n = \mathbf{A}\mathbf{A}'$. Note, however, that inner products between cases (rows of $\mathbf{Z}_0$) are not preserved since $m\mathbf{Z}_0\mathbf{Z}_0' = \mathbf{B}\mathbf{D}_r^2\mathbf{B}'$.

5.8 Probabilistic PCA

So far, the estimation problem in PCA has been dealt with exclusively from a LS perspective. There is, however, an approach in which component scores and disturbance terms are regarded as random variables. Let

$$\mathbf{z} = \mathbf{A}\mathbf{b} + \mathbf{e} \tag{5.46}$$

be such a model, where $\mathbf{z}$ is a random vector of observations on m observed variables, $\mathbf{A}$ an $m \times r$ matrix of component loadings (fixed), where r is the number of components, $\mathbf{b}$ is an r-element vector of component scores (random), and $\mathbf{e}$ an m-element vector of random disturbance terms. Assume that $\mathbf{b} \sim \mathcal{N}(\mathbf{0}, \mathbf{I}_r)$, $\mathbf{e} \sim \mathcal{N}(\mathbf{0}, \sigma^2\mathbf{I}_m)$, and $\mathbf{b}$ and $\mathbf{e}$ are independent of each other. Then, $\mathbf{z} \sim \mathcal{N}(\mathbf{0}, \mathbf{A}\mathbf{A}' + \sigma^2\mathbf{I}_m)$.

Let $\mathbf{Z} = \mathbf{U}\mathbf{D}\mathbf{V}'$ denote the SVD of the data matrix $\mathbf{Z}$. Let $\mathbf{U}_r$, $\mathbf{V}_r$, and $\mathbf{D}_r$ represent the portions of $\mathbf{U}$, $\mathbf{V}$, and $\mathbf{D}$ pertaining to the r largest singular values, and let $\mathbf{D}_{m-r}$ represent the portions of $\mathbf{D}$ corresponding to the $m-r$ smallest singular values. Then under the distributional assumptions made above, the maximum likelihood (ML) estimates of parameters are given by (Tipping and Bishop 1999):

$$\hat{\sigma}^2_{MLE} = \mathrm{tr}(\mathbf{D}^2_{m-r})/(m-r),$$

and

$$\hat{\mathbf{A}}_{MLE} = \mathbf{V}_r(\mathbf{D}_r^2 - \hat{\sigma}^2_{MLE}\mathbf{I}_r)^{1/2}.$$

Component scores, being random, cannot be estimated in the usual sense. Nonetheless they can be “predicted” from the data for each case (subject), which can be collectively expressed as

$$\hat{\mathbf{B}}_{MLE} = \sqrt{n}\mathbf{U}_r\mathbf{D}_r(\mathbf{D}_r^2 - \hat{\sigma}^2_{MLE}\mathbf{I}_r)^{-1/2}.$$

When $\mathbf{e} \sim \mathcal{N}(\mathbf{0}, \Psi)$, where Ψ is an arbitrary *pd* diagonal matrix, the formulation above leads to ML common factor analysis. The matrix Ψ is called the matrix of unique variances. Allowing a nonconstant diagonal matrix for Ψ, however, makes the parameter estimation problem much more difficult and time consuming. Complicated iterative algorithms have to be used to estimate a nonconstant diagonal Ψ. The probabilistic approach to PCA is not very popular among PCA advocates because of its restrictive distributional assumption. It is also not very popular among factor analysis advocates because of its restrictive assumption of constant unique variances across observed variables.

Chapter 6

Different Constraints on Different Dimensions (DCDD)

As is observed in Figure 1.4b (Section 1.3), factorial structures of rows and columns of a data matrix do not necessarily emerge as rectangular configurations in CPCA. This is because the row design matrices $\mathbf{G}_j$ $(j = 1, \cdots, J)$ representing the factorial structures are put together into a single row block matrix $\mathbf{G} = [\mathbf{G}_1, \cdots, \mathbf{G}_J]$ in CPCA. The column design matrices $\mathbf{H}_j$ $(j = 1, \cdots, J)$ are similar. Consequently, the dimensions extracted from CPCA do not correspond one-to-one with any particular $\mathbf{G}_j$. To obtain such dimensions, we have to use DCDD (different constraints on different dimensions; Takane et al. 1995), the topic of this chapter.

Figure 6.1 shows the results of a DCDD analysis of Greenacre's (1993) car purchase data previously analyzed by CPCA. The first dimension (the x-axis) was purposefully made to represent car sizes and ages of purchasers, while the second dimension (the y-axis) was made to represent luxuriousness of cars and purchaser's income level. Specifically, we used the left and right halves of $\mathbf{G}$ defined in Section 1.3 as $\mathbf{G}_1$ representing the age groups, and $\mathbf{G}_2$ representing the income levels. The matrices $\mathbf{H}_1$ representing the car sizes, and $\mathbf{H}_2$ representing the luxuriousness were similarly obtained by splitting $\mathbf{H}$ into two parts. Note, however, that the dummy variable vector representing the passenger utility car was included in both $\mathbf{H}_1$ as indicating passenger cars and $\mathbf{H}_2$ as indicating utility cars. As is seen in Figure 6.1b, the age and income groups form a perfect rectangle. Note that DCDD does not impose order restrictions on age (or income), so that the last two age groups show reversals in tendency to prefer larger cars as one gets older. The types of cars do not form a complete factorial structure, but an incomplete one (i.e., cars with the same level within each factor have the same coordinate value within each dimension). The derived two-dimensional solution accounts for 74.3% of the total SS, which is 12% (percentage points) lower than for the corresponding two-dimensional CPCA solution.

6.1 Model and Algorithm

Let $\mathbf{Z}$ be an $n \times m$ data matrix. In DCDD we approximate $\mathbf{Z}$ by

$$\mathbf{Z}_0 = \sum_{j=1}^{J} \mathbf{G}_j \mathbf{M}_j \mathbf{H}_j', \tag{6.1}$$

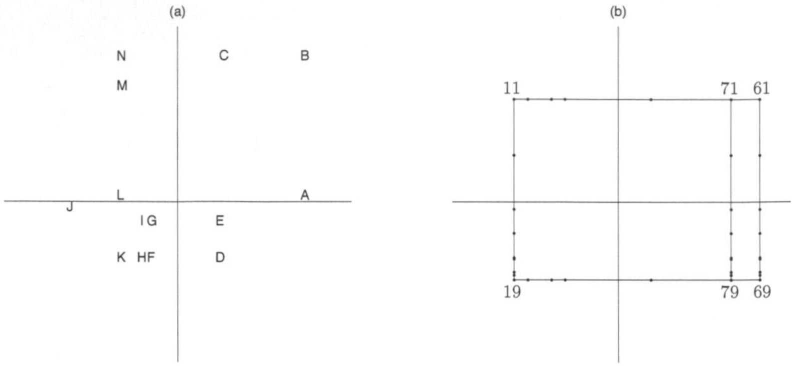

Figure 6.1 *Two-dimensional configurations of size classes and luxuriousness of cars (a), and of age and income groups of purchasers (b), obtained by DCDD.*

where $\mathbf{G}_j$ and $\mathbf{H}_j$ are, respectively, $n \times p_j$ $(p_j \leq n)$ and $m \times q_j$ $(q_j \leq m)$ row and column design (information) matrices, and $\mathbf{M}_j$ is the matrix of regression coefficients such that

$$\text{rank}(\mathbf{M}_j) = r_j \leq \min(p_j, q_j) \quad (j = 1, \cdots, J). \tag{6.2}$$

This model generalizes the model proposed by Velu (1991) to multiple sets of variables. It is similar to ExGCM discussed in Section 4.14, except that in ExGCM we usually do not impose the rank restriction on $\mathbf{M}_j$. The value of r_j is chosen to be 1 in most applications, although it could be larger than 1. When r_j is equal to its upper bound (i.e., $r_j = \min(p_j, q_j)$), there is no rank constraint on $\mathbf{M}_j$. When there are no obvious row or column constraints, we set $\mathbf{G}_j = \mathbf{I}_n$ or $\mathbf{H}_j = \mathbf{I}_m$. We may estimate $\mathbf{M}_j$ $(j = 1, \cdots, J)$ in such a way as to minimize

$$f(\mathbf{M}_1, \cdots, \mathbf{M}_J) = \text{SS}(\mathbf{Z} - \mathbf{Z}_0)_{K,L} \tag{6.3}$$

under the rank restriction (6.2). Here $\mathbf{K}$ and $\mathbf{L}$ are, as before, metric matrices. Let $\mathbf{K} = \mathbf{R}_K \mathbf{R}_K'$ and $\mathbf{L} = \mathbf{R}_L \mathbf{R}_L'$ represent square root decompositions of $\mathbf{K}$ and $\mathbf{L}$. Then (6.3) can be rewritten as

$$f(\mathbf{M}_1, \cdots, \mathbf{M}_J) = \text{SS}(\mathbf{R}_K'(\mathbf{Z} - \mathbf{Z}_0)\mathbf{R}_L) = \text{SS}(\mathbf{Z}^* - \sum_j^J \mathbf{G}_j^* \mathbf{M}_j \mathbf{H}_j^{*\prime}), \tag{6.4}$$

where $\mathbf{Z}^* = \mathbf{R}_K' \mathbf{Z} \mathbf{R}_L$, $\mathbf{G}_j^* = \mathbf{R}_K' \mathbf{G}_j$, and $\mathbf{H}_j^* = \mathbf{R}_L' \mathbf{H}_j$.

The criterion above cannot be minimized in closed form except for a few special cases. One special case is when $\mathbf{G}_j = \mathbf{G}$ and $\mathbf{H}_j = \mathbf{H}$ for all j $(j = 1, \cdots, J)$. In this case DCDD reduces to CPCA. A second special case occurs when all $\mathbf{G}_j$'s are mutually orthogonal, and/or all $\mathbf{H}_j$'s are mutually orthogonal (see below for the solutions in this case). The following three cases (third, fourth, and fifth cases) involve no rank restrictions on the $\mathbf{M}_j$'s: The third special case assumes both $\mathbf{K}$ and $\mathbf{L}$ are known, while the fourth assumes that $\mathbf{K}$ is known, but $\mathbf{L}$ is only partially known, that is, it is a constant diagonal matrix. The fifth case is equivalent to the Banken model (Section 4.14), in which $\mathbf{K} = \mathbf{I}_n$, $\mathbf{L}$ is an unknown *pd* matrix, and the $\mathbf{G}_j$'s satisfy the nesting structure given by (4.160).

In the second case above, the minimization of f in (6.4) reduces to that of

$$g(\mathbf{M}_1, \cdots, \mathbf{M}_J) = \sum_j^J \text{SS}(\mathbf{Z}^* - \mathbf{G}_j^* \mathbf{M}_j \mathbf{H}_j^{*'}), \tag{6.5}$$

in which each term in the summation can be minimized independently from the other terms. Each minimization problem is a separate CPCA problem, which can be solved by first obtaining the rank-free LS estimate of $\mathbf{M}_j$ by

$$\hat{\mathbf{M}}_j = (\mathbf{G}_j^{*'} \mathbf{G}_j^*)^{-1} \mathbf{G}_j^{*'} \mathbf{Z}^* \mathbf{H}_j^* (\mathbf{H}_j^{*'} \mathbf{H}_j^*)^{-1}, \tag{6.6}$$

and then applying a GSVD to $\hat{\mathbf{M}}_j$, i.e., $\text{GSVD}(\hat{\mathbf{M}}_j)_{G_j^{*'} G_j^*, H_j^{*'} H_j^*}$ to obtain the reduced rank estimate. In the third case (i.e, both $\mathbf{K}$ and $\mathbf{L}$ are known), we define

$$\mathbf{A} = [\mathbf{H}_1^* \otimes \mathbf{G}_1^*, \cdots, \mathbf{H}_J^* \otimes \mathbf{G}_J^*]. \tag{6.7}$$

Then, (6.4) can be rewritten as

$$f(\mathbf{m}) = \text{SS}(\mathbf{z}^* - \mathbf{A}\mathbf{m}), \tag{6.8}$$

where $\mathbf{m}$ is a supervector of $\mathbf{m}_j = \text{vec}(\mathbf{M}_j)$, and $\mathbf{z}^* = \text{vec}(\mathbf{Z}^*)$. The OLSE of $\mathbf{m}$ can be obtained by

$$\hat{\mathbf{m}} = (\mathbf{A}'\mathbf{A})^{-1} \mathbf{A}' \mathbf{z}^*. \tag{6.9}$$

The fourth case reduces to the third by setting $\mathbf{L} = \mathbf{I}$. For estimation of the Banken model (the fifth case), see Section 4.14.

In all other cases, we need an iterative minimization procedure like the one described below. We first define

$$\mathbf{Z}_{(k)}^* = \mathbf{Z}^* - \sum_{j \neq k} \mathbf{G}_j^* \mathbf{M}_j \mathbf{H}_j^{*'} \tag{6.10}$$

for a specific k. Then, (6.4) can be rewritten as

$$f(\mathbf{M}_k) = \text{SS}(\mathbf{Z}_{(k)}^* - \mathbf{G}_k^* \mathbf{M}_k \mathbf{H}_k^{*'}). \tag{6.11}$$

The $\mathbf{M}_k$ can be updated by minimizing the above criterion, assuming temporarily that all other $\mathbf{M}_j$'s ($j \neq k$) are known. For a specific k, this is a CPCA problem. We first initialize all $\mathbf{M}_k$'s (for example, by first obtaining the rank-free estimate of $\mathbf{M}_k$ ($k = 1, \cdots, J$), and then applying GSVD). For $k = 1, \cdots, J$, we successively update $\mathbf{M}_k$ with the other $\mathbf{M}_j$'s fixed. We repeat the process until convergence.

Note that in the algorithm described above, (6.11) is completely minimized with respect to $\mathbf{M}_k$ conditionally on all other $\mathbf{M}_j$'s ($j \neq k$). However, this is not an absolute necessity. It is sufficient to construct a sequence of updates that will consistently decrease the minimization criterion so as to obtain a monotonically convergent algorithm. This suggests that we can construct an algorithm that does not involve GSVD in each step. Assume, for simplicity, that $r_k = 1$. Then, $\mathbf{M}_k$ can be reparameterized as

$$\mathbf{M}_k = \mathbf{u}_k^* d_k \mathbf{v}_k^{*'}, \tag{6.12}$$

so that $\mathbf{G}_k^* \mathbf{M}_k \mathbf{H}_k^{*'} = \mathbf{G}_k^* \mathbf{u}_k^* d_k \mathbf{v}_k^{*'} \mathbf{H}_k^{*'} = \mathbf{u}_k d_k \mathbf{v}_k'$, where $\mathbf{u}_k = \mathbf{G}_k^* \mathbf{u}_k^*$ and $\mathbf{v}_k = \mathbf{H}_k^* \mathbf{v}_k^*$ are of unit length (i.e., $||\mathbf{u}_k|| = 1$ and $||\mathbf{v}_k|| = 1$). Then we obtain the following algorithm:

Step 1. Set $s = 0$ (where s is the iteration number).
Step 2. Set $s = s + 1$.
Step 3. For $k = 1, \cdots, J$, repeat Steps 3.1 through 3.5.
  Step 3.1. Calculate $\mathbf{Z}_{(k)}^*$ according to (6.10).
  Step 3.2. (Only when $s = 1$), find an initial estimate of $\mathbf{v}_k$. (This can be a set of random numbers normalized to satisfy $||\mathbf{v}_k|| = 1$; it can also be the right singular vector of $\mathbf{P}_{G_k^*} \mathbf{Z}_{(k)}^* \mathbf{P}_{H_k^*}$ corresponding to the largest singular value.)
  Step 3.3. Update $\mathbf{u}_k$ by $\mathbf{u}_k = \hat{\mathbf{u}}_k / ||\hat{\mathbf{u}}_k||$, where $\hat{\mathbf{u}}_k = \mathbf{P}_{G_k^*} \mathbf{Z}_{(k)}^* \mathbf{v}_k$.
  Step 3.4. Update $\mathbf{v}_k$ by $\mathbf{v}_k = \hat{\mathbf{v}}_k / ||\hat{\mathbf{v}}_k||$, where $\hat{\mathbf{v}}_k = \mathbf{P}_{H_k^*} \mathbf{Z}_{(k)}^{*'} \mathbf{u}_k$.
  Step 3.5. Update d_k by $d_k = \mathbf{u}_k' \mathbf{Z}_{(k)}^* \mathbf{v}_k$.
Step 4. Check convergence. If converged, stop. Otherwise, go back to Step 2.

The algorithm presented above is an ALS (alternating LS) algorithm, in which the entire set of parameters are partitioned into several subsets, and each subset of parameters is sequentially updated in such a way as to minimize a single common LS criterion with all other subsets of parameters being fixed. Other subsets of parameters are updated in similar ways, and the entire process is repeated until convergence. The value of the minimization criterion monotonically decreases, and hence is convergent. (A monotonically decreasing bounded sequence converges to some accumulation point.) There is no assurance, however, that the convergence point is the one we wish to reach.

When only $\mathbf{H}_j$'s exist (i.e., $\mathbf{G}_j = \mathbf{I}_n$ for all j), and the number of rows in $\mathbf{Z}$ is much larger than the number of columns, the computation time may be substantially cut down by the following procedure: Let $\mathbf{Z}^* = \mathbf{F}\mathbf{R}'$ denote the compact QR decomposition of $\mathbf{Z}^*$. Then (6.4) can be rewritten as

$$\begin{aligned} f(\mathbf{M}_j) &= \mathrm{SS}(\mathbf{Z}^* - \sum_j \mathbf{M}_j \mathbf{H}_j^{*'}) \\ &= \mathrm{SS}(\mathbf{Z}^* - \sum_j \mathbf{M}_j \mathbf{H}_j^{*'})_{P_F} + \mathrm{SS}(\mathbf{Z}^* - \sum_j \mathbf{M}_j \mathbf{H}_j^{*'})_{Q_F} \qquad (6.13) \\ &= \mathrm{SS}(\mathbf{R}' - \sum_j \mathbf{M}_j^* \mathbf{H}_j^{*'}), \qquad (6.14) \end{aligned}$$

where $\mathbf{M}_j^* = \mathbf{F}'\mathbf{M}_j$, $\mathbf{P}_F = \mathbf{F}\mathbf{F}'$, and $\mathbf{Q}_F = \mathbf{I} - \mathbf{P}_F$. The second term in (6.13) vanishes because $\mathrm{Sp}(\mathbf{F}) \supset \mathrm{Sp}(\mathbf{Z}^*)$ and $\mathrm{Sp}(\mathbf{F}) \supset \sum \mathbf{M}_j \mathbf{H}_j^{*'}$. We then minimize (6.14) with respect to $\mathbf{M}_j^*$. In most cases, the number of rows in $\mathbf{R}'$ is much smaller than that of $\mathbf{Z}^*$, and this minimization problem can be solved more efficiently than the original one. Once the optimal $\mathbf{M}_j^*$ is found, $\mathbf{M}_j$ can be recovered by

$$\mathbf{M}_j = \mathbf{F}\mathbf{M}_j^*. \tag{6.15}$$

See Takane et al. (1995) for how to deal with the case in which $r_j > 1$ for some j.

Takane et al. (1995) discuss the necessary and sufficient (*ns*) condition for uniqueness of parameters in the DCDD model. Assume for the moment that $r_j = 1$ for all j. Then, this condition is stated as all $\mathbf{G}_j$'s ($j = 1, \cdots, J$) being disjoint from all $\mathbf{u}_k$'s for $k \neq j$, or all $\mathbf{H}_j$ ($j = 1, \cdots, J$) being disjoint from all $\mathbf{v}_k$'s for $k \neq j$, assuming that all $\mathbf{u}_j$'s and all $\mathbf{v}_j$'s are linearly independent. See Kiers and Takane (1993) for a proof, where this condition was first presented in a similar (but different) context. This condition is useful, but it is stated in terms of unknown parameters. Consequently, it can only be used after the DCDD model is fitted. Kiers and Takane (1993) also developed a sufficient condition for identifiability that can be used before DCDD is run. This sufficient condition stipulates that all $\mathbf{G}_j$'s are disjoint, or that all $\mathbf{H}_j$'s are disjoint. Note, however, that this is a sufficient condition, so that there is a chance that parameters may still be identifiable even if it fails. When $r_j > 1$ for some j, the *ns* condition above is still valid with $\mathbf{u}_j$ and $\mathbf{v}_j$ replaced by $\mathbf{U}_j$ and $\mathbf{V}_j$. However, $\mathbf{U}_j$ and $\mathbf{V}_j$ must be properly orthogonalized to make them unique within each j.

6.2 Additional Constraints

The criterion f in (6.4) can be minimized under additional restrictions on $\mathbf{M}_j$'s. This feature is convenient when inter-dimensional constraints are to be incorporated. For example, we may have a hypothesis that an interval on dimension 1 is equal to an interval on dimension 2, and want to investigate its empirical validity. For such inter-dimensional constraints to be meaningful, however, the measurement unit must be comparable across dimensions. The restrictions that $||\mathbf{u}_j|| = 1$ and $||\mathbf{v}_j|| = 1$ are rather arbitrary, and it is unlikely that their elements are comparable across different dimensions. However, $\bar{\mathbf{v}}_j = d_j\mathbf{v}_j$ is a weight vector applied to $\mathbf{u}_j$ to find the best rank 1 approximation of $\mathbf{P}_{G_j^*}\mathbf{Z}_{(k)}^*\mathbf{P}_{H_j^*}$. Hence, the elements of $\bar{\mathbf{v}}_j$ may be regarded as comparable across dimensions.

Let $\bar{\mathbf{v}}$ and $\bar{\mathbf{v}}^*$ denote the supervectors of $\bar{\mathbf{v}}_j$ and $\bar{\mathbf{v}}_j^*$ (i.e., $\bar{\mathbf{v}} = (\bar{\mathbf{v}}_1', \cdots, \bar{\mathbf{v}}_J')'$ and $\bar{\mathbf{v}}^* = (\bar{\mathbf{v}}_1^{*'}, \cdots, \bar{\mathbf{v}}_J^{*'})'$). Then, inter-dimensional constraints can generally be expressed as

$$\mathbf{C}'\bar{\mathbf{v}} = \mathbf{C}'\mathbf{D}_H\bar{\mathbf{v}}^* = \mathbf{C}^{*'}\bar{\mathbf{v}}^* = \mathbf{0}, \tag{6.16}$$

where $\mathbf{C}'$ is a constraint matrix, $\mathbf{D}_H$ is the block diagonal matrix of $\mathbf{H}_j^*$ as the jth diagonal block, and $\mathbf{C}^{*'} = \mathbf{C}'\mathbf{D}_H$. Define

$$\mathbf{A} = [\mathbf{H}_1^* \otimes \mathbf{u}_1, \cdots, \mathbf{H}_J^* \otimes \mathbf{u}_J]. \tag{6.17}$$

Then, (6.4) can be rewritten as

$$f = \text{SS}(\text{vec}(\mathbf{Z}^*) - \mathbf{A}\bar{\mathbf{v}}^*). \tag{6.18}$$

We then find $\bar{\mathbf{v}}^*$ that minimizes (6.18) under (6.16). One simple way to solve this constrained LS problem is to find a $\mathbf{T}$ such that $\mathrm{Sp}(\mathbf{T}) = \mathrm{Ker}(\mathbf{C}^{*'})$. (This matrix can be obtained by a square root factor of $\mathbf{Q}_{C^*} = \mathbf{I} - \mathbf{C}^*(\mathbf{C}^{*'}\mathbf{C}^*)^{-1}\mathbf{C}^{*'}$.) Then $\bar{\mathbf{v}}^*$ can be reexpressed as

$$\bar{\mathbf{v}}^* = \mathbf{Ta} \tag{6.19}$$

for some $\mathbf{a}$. By substituting (6.19) for $\bar{\mathbf{v}}^*$ in (6.18) and minimizing (6.18) with respect to $\mathbf{a}$, we obtain

$$\hat{\mathbf{a}} = (\mathbf{T}'\mathbf{A}'\mathbf{AT})^{-1}\mathbf{T}'\mathbf{A}'\mathrm{vec}(\mathbf{Z}^*), \tag{6.20}$$

and

$$\bar{\mathbf{v}}^* = \mathbf{T}\hat{\mathbf{a}} = \mathbf{T}(\mathbf{T}'\mathbf{A}'\mathbf{AT})^{-1}\mathbf{T}'\mathbf{A}'\mathrm{vec}(\mathbf{Z}^*), \tag{6.21}$$

from which it follows that $\bar{\mathbf{v}} = \mathbf{D}_H\bar{\mathbf{v}}^*$. The vector $\mathbf{u}_j$, on the other hand, is updated sequentially for fixed $\bar{\mathbf{v}}$ as before.

6.3 Example 1

An analysis of Greenacre's car purchase data by DCDD was presented at the beginning of this chapter. We offer another example in this section. The stimuli used in this example may be characterized by a number of hypothetical structures, and we wish to see which ones fare well in the light of empirical data.

Delbeke (1978) asked 82 Belgian university students to rank-order 16 different family compositions in terms of their preference. The 16 family compositions were created by factorially combining four levels (0, 1, 2, and 3) each in the number of boys and the number of girls that the subjects would like to have as their children. Table 6.1 shows the construction of the 16 stimuli. The preference rank-order data are given in Table 6.2. By construction, the 16 stimuli have a factorial structure in terms of the number of boys (B) and the number of girls (G), which we call the B $\times$ G factorial design. The stimuli can also be characterized by the total number of children (T = B + G) and the sex bias (S = B - G), which we call the T $\times$ S factorial design. (This is necessarily an incomplete factorial design due to some structural constraints. For example, it is impossible to have T = 2 and S = 1.) These two factorial structures are represented as contrast matrices given in Tables 6.3 and 6.4. Note that the contrasts representing these structures are not unique. The last two columns of these tables indicate contrast vectors representing the hypothesis that the adjacent intervals between levels of a factor are equal under the B $\times$ G and T $\times$ S factorial structures. The equal interval hypothesis implies that the difference between 1 boy and 2 boys, for example, is the same as the difference between 2 boys and 3 boys.

The data in Table 6.2 were first transformed by subtracting all observed values from 17 to make larger values indicate larger preferences. The transformed data were then analyzed by constrained CA, a special case of CPCA in which diagonal matrices of row and column totals of the data matrix were used as metric matrices (see Section 4.7), and by DCDD under the same metric matrices. In DCDD, the B $\times$ G and T $\times$ S factorial structures shown in Tables 6.3 and 6.4 were used as column constraints $\mathbf{H}_j$ $(j = 1, 2)$. In CPCA, $\mathbf{H}_1$ and $\mathbf{H}_2$ were combined into a single $\mathbf{H}$. In no instances were row constraints imposed.

Table 6.1 *The construction of stimuli used in Delbeke (1978). Main entries in the table give stimulus numbers used in Table 6.2.*

		The # of boys			
		0	1	2	3
	0	1	2	3	4
The #	1	5	6	7	8
of girls	2	9	10	11	12
	3	13	14	15	16

Figure 6.2 displays stimulus configurations obtained under a variety of hypotheses. Ellipses surrounding stimulus points indicate 95% confidence regions obtained by the bootstrap method (Section 5.2). These regions indicate that the population stimulus points are within the ellipses with probability .95.

Panel (a) displays the two-dimensional stimulus configuration obtained by the unconstrained CA. The x-axis roughly corresponds with the total number of children (T) and the y-axis with the sex bias (S). However, the estimated point locations are rather unstable, as indicated by the huge confidence regions. This is primarily caused by stimulus 1 (the combination of 0 boys and 0 girls), which is very unpopular with almost everyone, and its location is rather unstable. Indeed, as shown in Panel (b), much more reliable results were obtained when the analysis was redone with this stimulus removed from the data. Panel (c) shows the stimulus configuration obtained by CPCA under the T $\times$ S factorial structure. This is strikingly similar to the one presented in Panel (a). As mentioned earlier, the T $\times$ S factorial hypothesis is not a complete factorial structure, and does not provide strong constraints in CPCA. In contrast, DCDD yields a much more stable configuration for this structure, as shown in Panel (d). Stimulus 1 still has quite a large fluctuation along the horizontal direction because there are no other stimuli with T = 0. Panel (e) gives the stimulus configuration obtained by CPCA under the B $\times$ G factorial hypothesis. This solution rotates the B $\times$ G dimensions by 45°. As a result, the x-axis roughly corresponds with the T dimension, while the y-axis with the S dimension. This is similar to Panel (d), although the shape of confidence regions is generally different. Panel (f) displays the stimulus configuration obtained by DCDD under the B $\times$ G hypothesis.

The three configurations shown in the bottom row of Figure 6.2 were all derived under the equal interval hypotheses. CPCA yields an identical configuration regardless of the structure that is incorporated, whether the B $\times$ G or T $\times$ S hypothesis (Panel (g)). This is because the constraint matrix $[\mathbf{h}_1, \mathbf{h}_2]$ spans the same subspace under the two hypotheses. In contrast, DCDD treats $\mathbf{h}_1$ and $\mathbf{h}_2$ as separate constraints, so the two hypotheses yield different configurations. Panels (h) and (i) show the stimulus configurations derived by DCDD under the T $\times$ S and the B $\times$ G hypotheses, respectively. More stable configurations tend to be obtained under more stringent constraints. Note, however, that they also tend to produce larger biases.

Table 6.5 presents the %SS accounted for by the solutions given in Figure 6.2. Unconstrained CA accounts for approximately 2/3 of the total SS in the data set. This

Table 6.2 *Delbeke's family composition preference data.*

	Stimuli															
Sub.	1	2	3	4	5	6	7	8	9	10	11	12	13	14	15	16
1	16	6	8	4	7	1	2	12	5	3	10	13	9	11	14	15
2	16	7	4	5	15	1	3	8	10	9	2	6	14	13	11	12
3	16	14	9	7	15	6	4	2	10	5	1	11	8	3	12	13
4	16	15	13	12	14	8	3	5	11	2	1	7	10	4	6	9
5	16	12	10	13	14	8	2	7	11	6	5	1	15	9	3	4
6	16	7	5	6	15	11	1	2	13	10	4	3	14	9	12	8
7	16	12	11	6	15	5	2	3	14	4	1	8	13	7	9	10
8	16	8	6	5	14	9	4	2	15	11	3	1	13	12	7	10
9	14	15	10	12	16	4	2	5	11	3	1	7	13	6	8	9
10	15	16	13	9	14	10	7	8	11	5	4	2	12	6	1	3
11	16	13	4	3	15	1	2	9	11	8	5	7	12	14	10	6
12	16	14	6	3	15	5	2	4	12	8	1	7	13	9	11	10
13	16	14	9	8	15	4	1	11	10	2	5	3	13	12	6	7
14	16	14	10	11	15	9	5	6	13	8	4	2	12	7	3	1
15	15	14	12	16	13	11	9	10	8	7	5	6	2	3	1	4
16	16	6	2	7	8	1	4	10	3	5	9	11	13	14	12	15
17	16	15	13	10	14	9	8	7	12	5	3	6	11	4	1	2
18	16	8	5	2	15	3	1	10	9	4	6	7	14	11	12	13
19	16	14	12	13	15	10	7	6	11	8	1	3	9	5	4	2
20	15	12	10	14	11	6	2	5	9	1	3	8	13	4	7	16
21	16	15	10	11	13	9	8	5	12	7	1	3	14	6	2	4
22	16	14	11	10	15	9	5	6	13	7	1	3	12	8	4	2
23	16	11	6	12	13	4	1	5	14	2	3	7	15	10	9	8
24	16	15	13	12	14	7	8	10	4	1	5	9	2	3	6	11
25	16	14	10	9	15	8	6	5	12	7	1	2	13	11	4	3
26	16	12	11	10	15	5	6	8	14	7	4	2	13	9	3	1
27	14	15	11	9	16	8	1	5	13	6	2	3	12	7	10	4
28	16	13	12	14	10	1	4	15	5	3	2	8	11	6	7	9
29	16	14	8	7	15	10	6	4	13	9	5	1	12	11	3	2
30	1	2	5	7	4	3	6	10	14	12	8	9	16	15	13	11
31	14	15	10	11	16	6	4	7	12	8	1	3	13	9	5	2
32	16	14	11	9	15	10	2	4	13	3	1	6	12	5	7	8
33	16	4	3	8	6	1	2	10	7	5	9	12	13	11	14	15
34	15	7	6	9	4	1	3	12	5	2	10	14	8	11	13	16
35	16	7	6	5	15	8	3	1	14	9	2	4	13	10	11	12
36	16	2	4	5	11	3	1	6	13	7	8	9	15	14	12	10
37	16	7	8	9	6	3	2	12	4	1	10	14	5	11	13	15
38	16	4	9	10	7	3	2	5	11	6	1	13	12	8	14	15
39	16	12	11	10	15	7	5	8	14	6	1	3	13	9	4	2

Table 6.2 continued

40	16	2	3	12	4	1	6	10	5	7	8	9	13	11	14	15
41	8	2	3	10	5	1	4	11	6	7	9	12	13	14	15	16
42	16	9	4	7	12	1	2	8	5	3	6	10	11	13	14	15
43	16	15	13	12	14	8	5	7	10	4	1	3	11	6	2	9
44	16	14	6	5	15	9	7	1	13	10	2	3	12	11	8	4
45	16	10	8	9	12	5	4	2	14	11	1	3	15	13	7	6
46	16	13	1	2	15	3	4	6	11	5	7	8	14	12	9	10
47	1	2	5	9	3	4	7	12	6	8	11	14	10	13	16	15
48	10	3	1	5	4	2	7	11	6	8	12	14	9	13	15	16
49	16	13	4	1	15	6	2	3	14	7	5	8	12	10	9	11
50	16	2	7	9	3	1	5	11	4	6	8	13	12	10	14	15
51	16	15	10	13	14	5	7	8	12	6	2	4	11	9	3	1
52	16	15	12	11	14	10	3	4	13	2	1	6	9	8	5	7
53	16	12	11	9	13	10	4	5	14	6	1	2	15	8	3	7
54	16	12	11	9	15	7	5	6	14	10	1	2	13	8	3	4
55	16	15	13	11	14	4	2	6	12	1	3	8	10	5	7	9
56	16	14	13	11	15	6	2	8	12	4	1	3	10	7	5	9
57	16	15	13	10	14	11	6	8	12	5	3	4	9	7	2	1
58	16	15	13	9	14	11	8	5	12	7	3	2	10	4	1	6
59	16	14	8	7	15	12	6	2	13	10	3	1	11	9	4	5
60	12	16	14	10	15	11	8	6	13	7	4	2	9	5	1	3
61	16	15	13	12	14	9	8	6	11	7	2	4	10	5	3	1
62	16	14	10	13	15	9	7	2	11	8	1	4	12	3	5	6
63	16	14	11	13	15	8	6	5	12	7	4	1	10	9	3	2
64	16	12	5	2	15	11	6	1	14	10	7	3	13	9	8	4
65	16	10	11	13	12	9	7	5	14	8	4	2	15	6	3	1
66	16	10	8	3	15	9	5	2	14	11	6	1	13	12	7	4
67	16	10	12	14	11	9	7	3	13	8	1	2	15	6	4	5
68	15	5	3	6	10	2	1	9	8	7	4	11	12	13	14	16
69	16	15	13	10	14	6	8	9	12	5	1	7	11	3	2	4
70	16	12	8	9	15	10	6	4	14	7	5	2	13	11	3	1
71	6	3	5	9	2	1	8	13	4	7	11	15	10	12	14	16
72	16	12	11	10	15	9	5	7	14	6	4	1	13	8	2	3
73	14	13	8	11	12	2	4	5	15	3	1	9	16	7	6	10
74	16	10	11	14	12	9	1	3	13	5	2	4	15	7	6	8
75	16	14	13	15	12	6	10	8	11	5	1	4	9	7	3	2
76	16	12	10	11	14	4	1	3	13	5	2	6	15	7	8	9
77	16	12	4	3	15	9	5	2	14	10	6	1	13	11	8	7
78	16	10	11	13	15	4	2	6	14	3	1	5	12	7	8	9
79	1	2	3	4	9	5	6	7	14	11	10	8	15	16	13	12
80	16	14	11	10	15	9	7	5	13	8	4	2	12	6	3	1
81	16	15	13	11	14	9	8	6	12	7	4	3	10	5	2	1
82	16	12	8	4	15	11	7	3	14	10	6	2	13	9	5	1

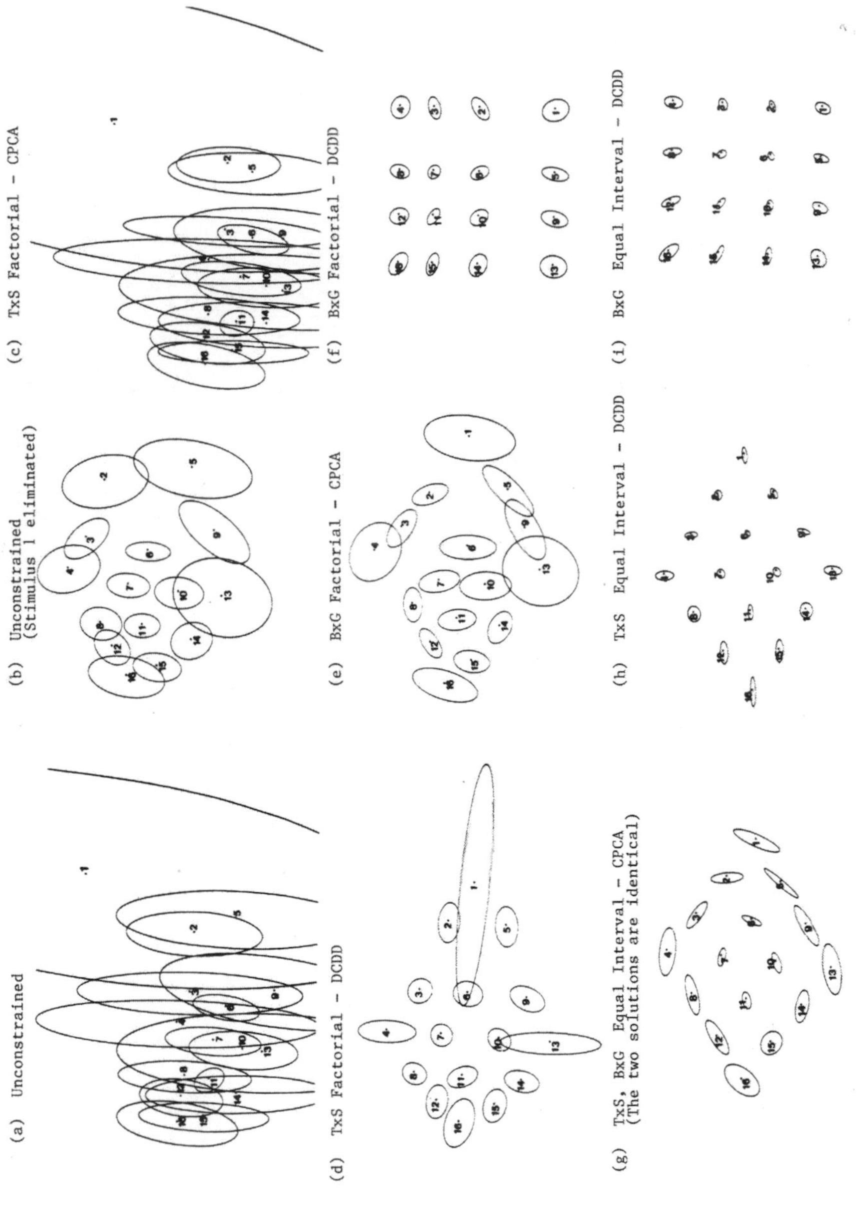

Figure 6.2 *Solutions derived from Delbeke's data under a variety of hypotheses. (Adapted from Figure 2 of Takane et al. (1995) with permission from the Psychometric Society.)*

Table 6.3 *Design matrices (B × G) for Delbeke's data**.

	Factorial						Equal Interval	
	Total ($\mathbf{H}_1$)			Bias ($\mathbf{H}_2$)			Total ($\mathbf{H}_1$)	Bias ($\mathbf{H}_2$)
St.	$\mathbf{h}_{11}$	$\mathbf{h}_{12}$	$\mathbf{h}_{13}$	$\mathbf{h}_{21}$	$\mathbf{h}_{22}$	$\mathbf{h}_{23}$	$\mathbf{h}_1$	$\mathbf{h}_2$
1	1	0	1	1	0	1	−3	−3
2	−1	0	1	1	0	1	−1	−3
3	0	1	−1	1	0	1	1	−3
4	0	−1	−1	1	0	1	3	−3
5	1	0	1	−1	0	1	−3	−1
6	−1	0	1	−1	0	1	−1	−1
7	0	1	−1	−1	0	1	1	−1
8	0	−1	−1	−1	0	1	3	−1
9	1	0	1	0	1	−1	−3	1
10	−1	0	1	0	1	−1	−1	1
11	0	1	−1	0	1	−1	1	1
12	0	−1	−1	0	1	−1	3	1
13	1	0	1	0	−1	−1	−3	3
14	−1	0	1	0	−1	−1	−1	3
15	0	1	−1	0	−1	−1	1	3
16	0	−1	−1	0	−1	−1	3	3

*Adapted from Table 3 of Takane et al. (1995) with permission from the Psychometric Society.

percentage is decreased by only 4% under the most stringent hypotheses ((g), (h), and (i)). As already mentioned, the two CPCA solutions yield an identical configuration (Panel (g)). The two DCDD solutions, on the other hand, are distinct from each other, and are also distinct from the CPCA configuration. Yet all four solutions have an identical fit. This is because no rank restrictions are in effect in equal interval cases, and no row constraints are imposed in any of the analyses conducted.

An inter-dimensional equal interval hypothesis can be incorporated as described in the previous section. In addition to $\mathbf{h}_1$ and $\mathbf{h}_2$ representing intra-dimensional equal interval hypothesis, we supply

$$\mathbf{C}' = (1, -1). \tag{6.22}$$

Using this $\mathbf{C}'$ in combination with the B × G equal interval hypothesis (the difference in the number of boys has the same effect as the difference in the number of girls) did not change the %SS very much (only a reduction of .07%). However, using the same $\mathbf{C}'$ in combination with the T × S equal interval hypothesis considerably (6.4%) reduced the %SS explained. It may be observed from Panel (h) that a one-unit change in T is roughly equivalent to a two-unit change in S.

There was some inevitable reduction in %SS as a result of imposing hypothetical

Table 6.4 *Design matrices (T × S) for Delbeke's data**.

	Factorial												Equal Interval	
	Total ($\mathbf{H}_1$)						Bias ($\mathbf{H}_2$)						Total ($\mathbf{H}_1$)	Bias ($\mathbf{H}_2$)
St.	$\mathbf{h}_{11}$	$\mathbf{h}_{12}$	$\mathbf{h}_{13}$	$\mathbf{h}_{14}$	$\mathbf{h}_{15}$	$\mathbf{h}_{16}$	$\mathbf{h}_{21}$	$\mathbf{h}_{22}$	$\mathbf{h}_{23}$	$\mathbf{h}_{24}$	$\mathbf{h}_{25}$	$\mathbf{h}_{26}$	$\mathbf{h}_1$	$\mathbf{h}_2$
1	1	0	0	2	1	−1	0	0	0	0	0	3	−3	0
2	0	1	0	−1	1	−1	0	0	1	0	−1	−1	−2	−1
3	0	0	1	0	−1	−1	0	1	0	−1	1	−1	−1	−2
4	0	0	0	0	0	3	1	0	0	2	1	−1	0	−3
5	0	1	0	−1	1	−1	0	0	−1	0	−1	−1	−2	1
6	0	0	1	0	−1	−1	0	0	0	0	0	3	−1	0
7	0	0	0	0	0	3	0	0	1	0	−1	−1	0	−1
8	0	0	−1	0	−1	−1	0	1	0	−1	1	−1	1	−2
9	0	0	1	0	−1	−1	0	−1	0	−1	1	−1	−1	2
10	0	0	0	0	0	3	0	0	−1	0	−1	−1	0	1
11	0	0	−1	0	−1	−1	0	0	0	0	0	3	1	0
12	0	−1	0	−1	1	−1	0	0	1	0	−1	−1	2	−1
13	0	0	0	0	0	3	−1	0	0	2	1	−1	0	3
14	0	0	−1	0	−1	−1	0	−1	0	−1	1	−1	1	2
15	0	−1	0	−1	1	−1	0	0	−1	0	−1	−1	2	1
16	−1	0	0	2	1	−1	0	0	0	0	0	3	3	0

*Adapted from Table 4 of Takane et al. (1995) with permission from the Psychometric Society.

structures on the stimulus configuration. The decrease, however, was almost always less than 5%. The only exception was the inter-dimensional equal interval hypothesis on the T × S factorial structure. This means that most of the structural constraints considered were fairly consistent with the observed data.

6.4 Example 2

The second example concerns an application of DCDD to fit the weighted additive model proposed by Takane et al. (1980; see also Takane 1984). This model captures individual differences in the perception of stimuli in a manner similar to the individual differences (ID) MDS discussed in Section 4.5. Recall that in ID-MDS, the stimulus configuration is assumed common across individuals, while the dimensions defining the common configuration are differentially weighted to generate different dissimilarity judgments. In the weighted additive model, on the other hand, stimuli are constructed by factorial combinations of two or more attributes, and it is assumed that the scores assigned to levels of the attributes remain the same across individuals, but that the attributes are differentially weighted by the individuals to produce different judgments. The procedure developed by Takane et al. (1980) to fit the weighted additive model is called WADDALS. It has been pointed out that WADDALS is a special case of DCDD, as will be shown below.

Table 6.5 *Summary results for Delbeke's data**.

				bootstrap on %SS Explained	
	Analysis	%SS Explained	#para.	mean	sd
(a)	Unconstrained	67.45	29	69.24	2.65
	T×S: Factorial				
(c)	CPCA	64.69	21	68.40	2.68
(d)	DCDD	65.60	12	66.26	2.67
	T×S: Equal Interval				
(g)	CPCA	63.41*	2	64.05	2.82
(h)	DCDD	63.41*	2	64.05	2.82
	T×S: Equal Interval Across Dimensions				
	DCDD (T:S=1:1)	56.99	1	58.40	2.15
	DCDD (T:S=2:1)	63.41	1	63.08	3.18
	B×G: Factorial				
(e)	CPCA	64.81	11	65.48	2.65
(f)	DCDD	64.76	6	65.26	2.92
	B×G: Equal Interval				
(g)	CPCA	63.41*	2	64.05	2.82
(i)	DCDD	63.41*	2	64.05	2.82
	B×G: Equal Interval Across Dimensions				
	DCDD	63.34	1	63.51	3.00

*Adapted from Table 5 of Takane et al. (1995) with permission from the Psychometric Society.

Let us begin with an empirical motivation behind the model. Many developmental psychologists believe that subjective size of rectangles changes systematically as a function of children's age. Young children are more sensitive to the height of rectangles than to their width when they make subjective area judgments. This tendency gradually decreases as the children get older until they reach the point where they take into account both dimensions equally. This is one possible interpretation of Piaget's conservation of mass (or the lack thereof) in young children from a developmental psychology perspective.

Kempler (1971) generated a set of 100 rectangles by factorial combinations of 10 height levels (varied from 10 inches to 14.5 inches in .5 inch gradations) and 10 width levels varied similarly. He asked four groups of children (1st graders, 3rd graders, 5th graders, and 7th graders) to judge whether the rectangles looked large or small. This particularly simple format of response was dictated by the age of subjects, some of whom were quite young. Takane et al. (1980) counted the frequencies with which the rectangles were judged to be large for each age group, and applied WADDALS to the observed frequency data as the criterion variable. Let $\mathbf{z}^{(k)}$ denote the column vector of the observed frequencies, and let $\mathbf{G}_j$ ($j = 1,2$) denote the matrices of dummy variables indicating levels of the height and width of rectangles. The WADDALS model may then be written as

$$\mathbf{z}^{(k)} = \sum_{j=1}^{J} \mathbf{G}_j \mathbf{u}_j^* w_{kj} + \mathbf{e}^{(k)} \quad (k = 1, \cdots, K), \tag{6.23}$$

where $\mathbf{u}_j^*$ is the vector of scores assigned to the levels of factor j (i.e., the vector of weights applied to $\mathbf{G}_j$), w_{kj} is the weight applied to attribute j in group k, and $\mathbf{e}^{(k)}$ is the vector of disturbance terms. Note that $\mathbf{u}_j^*$ has no subscript k indicating that it is common across all k.

Figure 6.3 shows the plot of the estimated weights from Kempler's data as a function of age group. As expected, the weight for height is initially (at a younger age) much larger than that for width, but decreases gradually as the children get older until it becomes approximately the same as the weight for width at grade 7. In this analysis, LS monotonic transformations (see Section 5.6) were applied to $\mathbf{z}^{(k)}$ to monotonically transform the data, since the observed frequency data are likely to be only an ordinal measure of the subjective area of rectangles. Note that this analysis uses cross sectional data aggregated over subjects within age groups. Consequently, it is unclear from the analysis if the change in weights is gradual within individual subjects. It may be that the change is sudden within subjects, but that the proportion of subjects who consider only height decreases gradually with age. Further analysis, which sheds some light on this issue, can be found in Takane (1984).

It only remains to show that WADDALS is a special case of DCDD. Let $\mathbf{Z}$ denote the matrix of $\mathbf{z}^{(k)}$ as the kth column vector. Then $\mathbf{Z}$ can be written as

$$\mathbf{Z} = \mathbf{G}\mathbf{D}_{u^*}\mathbf{W}' + \mathbf{E}, \tag{6.24}$$

where

$$\mathbf{W}' = \begin{bmatrix} w_{11} & \cdots & w_{1K} \\ \vdots & \ddots & \vdots \\ w_{J1} & \cdots & w_{JK} \end{bmatrix}, \tag{6.25}$$

$$\mathbf{D}_{u*} = \begin{bmatrix} \mathbf{u}_1^* & \cdots & \mathbf{0} \\ \vdots & \ddots & \vdots \\ \mathbf{0} & \cdots & \mathbf{u}_J^* \end{bmatrix}, \tag{6.26}$$

$$\mathbf{G} = [\mathbf{G}_1, \cdots, \mathbf{G}_J], \tag{6.27}$$

and $\mathbf{E}$ is the matrix of disturbance terms. Model (6.24) indicates that it is a special case of DCDD with $\mathbf{W} = \mathbf{V} = [\mathbf{v}_1, \cdots, \mathbf{v}_J]$.

DCDD can also handle symmetric data, which may arise in MDS (Section 4.5). Some care is necessary in this case, since $\mathbf{u}_j$ and $\mathbf{v}_j$ should logically be the same. One way to deal with the problem is to keep them as separate parameters in the anticipation that they converge to identical quantities. This is indeed the case if $\mathbf{Z}_{(k)}$ is *nnd* for all k and throughout the iterations. Unfortunately, this condition is not guaranteed. Another strategy is to treat them as one. This requires combining Steps 3.3

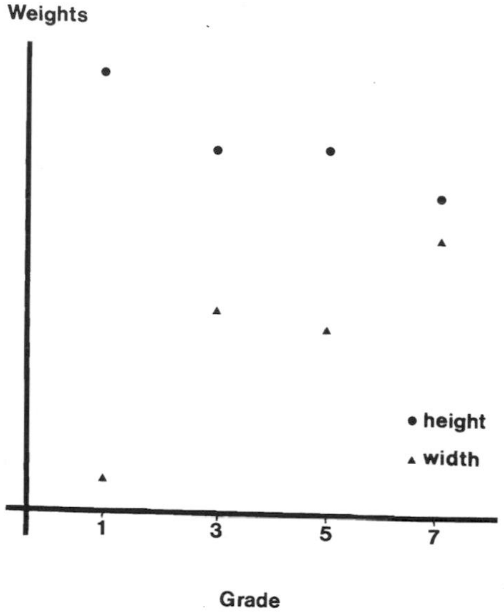

Figure 6.3 *Weights attached to the height and width of rectangles derived from Kempler's data. (Reproduced from Figure 9 of Takane et al. (1980) with permission from the Psychometric Society.)*

and 3.4 into one. This combined step, however, violates the basic principle of an ALS algorithm, because the update of $\mathbf{u}_j = \mathbf{v}_j$ is also a function of the previous update. The monotonic convergence is still assured, if $\mathbf{Z}_{(k)}$ is *nnd* throughout the iterations, which is not necessarily the case, as has been noted above. A special majorization technique (Kiers 1990; Kiers and ten Berge 1992) may be used to ensure monotonic convergence in such cases. More details as well as an example of DCDD with a symmetric data matrix can be found in Takane et al. (1995).

6.5 Residual Analysis

Residual analysis in CPCA means analysis (mostly by PCA) of the fourth term of Eq. (3.23) in the External Analysis. In some cases (e.g., GCM), the second and third terms may also be put in the residual term. (Of course, separate analyses of these terms are also possible.) In DCDD, we may analyze

$$\tilde{\mathbf{Z}} = \mathbf{Z}^* - \sum_{j=1}^{J} \mathbf{P}_{G_j^*} \mathbf{Z}^* \mathbf{P}_{H_j^*}, \tag{6.28}$$

and/or

$$\tilde{\mathbf{Z}} = \mathbf{Z}^* - \sum_{j=1}^{J} d_j \mathbf{u}_j \mathbf{v}_j'. \tag{6.29}$$

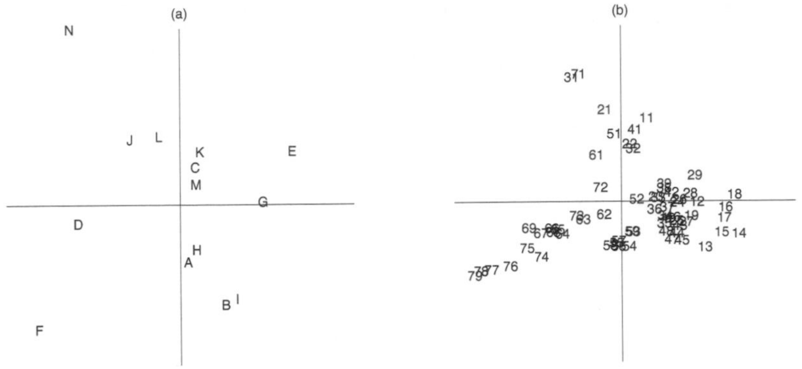

Figure 6.4 *Two-dimensional configurations of size classes of cars (a) and purchaser profiles (b) derived from residuals from DCDD.*

Residual analysis is always worth conducting, although the chance of a spectacular discovery may not be high. Since everything that is known about the data has likely been incorporated into the analysis of the structural parts, there is often little left to understand in the residuals. This can make the results of residual analysis difficult to interpret. Nonetheless there is always a chance of finding some "hidden treasure," and it is recommended that some kind of residual analysis be carried out as a matter of routine practice.

Figure 6.4 displays two-dimensional configurations of size classes of cars (a) and purchaser profiles (b) derived by PCA of residuals ((6.29) is used) from DCDD of Greenacre's (1993) data. This figure indicates that the low income seniors (74 through 79) buy regular compact cars (F) more frequently than expected from the two-dimensional DCDD configurations given in Figure 6.1. Similarly, very high income seniors (71) and a young generation of purchasers (31) tend to buy luxurious imported (N) cars more often than expected from Figure 6.1. The two-dimensional solution in Figure 6.4 accounts for 17.2% of the total variation in the data set.

6.6 Graphical Display of Oblique Components

Components extracted by DCDD are usually not orthogonal. It is often misleading to plot component "loadings" (see below for why quotation marks are put on the word loadings) as if the components are orthogonal, when in fact they are highly correlated. For example, the rectangular grid in dotted line segments in Figure 6.5 shows the component "loadings" plotted as if the components were orthogonal. While in (C)PCA, the components are always **K**-orthogonal, nonorthogonal components are the rule rather than the exception in DCDD. For example, in Figure 6.2(f) that fits the B $\times$ G factorial structure by DCDD, the correlation between the two components is as high as .6.

When the columns of **U** are not mutually orthogonal, the matrix of weights applied to **U** to approximate the data matrix and the matrix of covariances between **U**

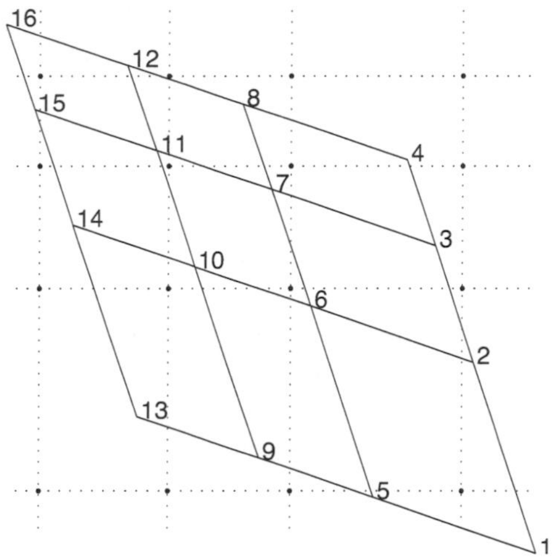

Figure 6.5 *A graphical display in oblique axes.*

and the data matrix should be distinguished. The former is called the pattern matrix, and the latter the structure matrix. When **U** is columnwise orthogonal, these two matrices coincide, and are called the matrix of (component) loadings. Although this was not made explicit, all configurations derived from DCDD in Figure 6.2 were in fact plots of the pattern matrices. (Thus, strictly speaking, it is incorrect to refer to these matrices as the matrices of component loadings as was done in the previous paragraph.)

There is an interesting relationship between the pattern and structure matrices. Let $\tilde{\mathbf{V}}$ denote a pattern matrix, and define

$$\mathbf{Z}_0 = \mathbf{U}\tilde{\mathbf{V}}'. \tag{6.30}$$

Then

$$\mathbf{Z}_0 = \mathbf{U}(\mathbf{U}'\mathbf{U})^{-1}\mathbf{U}'\mathbf{U}\tilde{\mathbf{V}}' = \mathbf{U}^*\tilde{\mathbf{V}}^{*'}, \tag{6.31}$$

where $\mathbf{U}^* = \mathbf{U}(\mathbf{U}'\mathbf{U})^{-1}$, and $\tilde{\mathbf{V}}^{*'} = \mathbf{U}'\mathbf{U}\tilde{\mathbf{V}}'$. By regressing $\mathbf{Z}_0$ onto $\mathbf{U}$, we obtain the matrix of regression coefficients

$$(\mathbf{U}'\mathbf{U})^{-1}\mathbf{U}'\mathbf{Z}_0 = \tilde{\mathbf{V}}'. \tag{6.32}$$

Covariances between $\mathbf{U}$ and $\mathbf{Z}_0$, on the other hand, are given by

$$\mathbf{U}'\mathbf{Z}_0 = \mathbf{U}'\mathbf{U}\tilde{\mathbf{V}}' = \tilde{\mathbf{V}}^{*'}. \tag{6.33}$$

By comparing (6.30) and (6.31), we notice that they are essentially of the same form. This means that $\tilde{\mathbf{V}}^*$ can be regarded as the weights applied to $\mathbf{U}^*$ to obtain $\mathbf{Z}_0$. The corresponding structure matrix, on the other hand, is obtained by

$$\mathbf{U}^{*'}\mathbf{Z}_0 = \mathbf{U}^{*'}\mathbf{U}^*\tilde{\mathbf{V}}^{*'} = (\mathbf{U}'\mathbf{U})^{-1}\mathbf{U}'\mathbf{U}\tilde{\mathbf{V}}' = \tilde{\mathbf{V}}'. \tag{6.34}$$

This means that when one of $\tilde{\mathbf{V}}$ and $\tilde{\mathbf{V}}^*$ is taken as the pattern matrix, the other becomes the corresponding structure matrix. For $\mathbf{U}$, $\tilde{\mathbf{V}}$ is the pattern matrix and $\tilde{\mathbf{V}}^*$ is the structure matrix, while for $\mathbf{U}^*$, this relation is reversed, i.e., $\tilde{\mathbf{V}}^*$ is the pattern matrix and $\tilde{\mathbf{V}}$ is the structure matrix. There is no reason to favor one or the other between $\mathbf{U}$ and $\mathbf{U}^*$.

This relationship is analogous to that between primary and reference factors in oblique factor solutions. The pattern matrix in primary factors corresponds to the structure matrix in reference factors, and the pattern matrix in reference factors corresponds to the structure matrix in primary factors (Browne and Kristof 1969). Which one of $\mathbf{U}$ and $\mathbf{U}^*$ we call primary or reference is essentially arbitrary. More fundamentally, the same relationship holds between a set of linearly independent basis vectors and its dual basis. If $\mathbf{U}$ represents the original basis, its dual basis is given by $\mathbf{U}^*$, and if $\mathbf{U}^*$ represents the original basis, $\mathbf{U}$ represents its dual. Thus, which we call original (primary) or dual is completely arbitrary.

A related problem also arises in regression analysis (and canonical correlation analysis), where predictor variables are generally correlated. There has been a long standing controversy between two "camps": One advocates that the pattern vector (the vector of regression weights) should be interpreted because it shows how the best prediction vector is constructed from the predictor variables. The other argues that the pattern vector does not accurately reflect the relationship between the best prediction vector and the original predictor variables because of the collinearity among the latter, and advocates that the structure vector should be interpreted. To illustrate, let $\hat{\mathbf{z}}$ denote the estimated best prediction vector and let $\mathbf{G}$ represent the matrix of predictor variables in regression analysis. Then

$$\hat{\mathbf{z}} = \mathbf{G}\hat{\mathbf{b}} \quad ((\mathbf{G}'\mathbf{G})^{-1}\mathbf{G}'\hat{\mathbf{z}} = \hat{\mathbf{b}} \text{ is a pattern vector}) \tag{6.35}$$

$$= \mathbf{G}(\mathbf{G}'\mathbf{G})^{-1}\mathbf{G}'\mathbf{G}\hat{\mathbf{b}} \quad ((\mathbf{G}'\hat{\mathbf{z}} = \mathbf{G}'\mathbf{G}\hat{\mathbf{b}} \text{ is a structure vector}). \tag{6.36}$$

The problem is whether $\hat{\mathbf{b}}$ should be interpreted, or $\mathbf{G}'\mathbf{G}\hat{\mathbf{b}}$ should be interpreted. This controversy has not yet been resolved in a satisfactory way. The point here is that $\mathbf{G}$ and $\mathbf{G}(\mathbf{G}'\mathbf{G})^{-1}$ are dual to each other, and that the relationship between $\hat{\mathbf{b}}$ and $\mathbf{G}'\mathbf{G}\hat{\mathbf{b}}$ is reversed if we take $\mathbf{G}(\mathbf{G}'\mathbf{G})^{-1}$ as the predictor variables.

As a possible resolution of the problem, we suggest the following: Let

$$\mathbf{Z}_0 = \mathbf{U}(\mathbf{U}'\mathbf{U})^{-1/2}(\mathbf{U}'\mathbf{U})^{1/2}\tilde{\mathbf{V}}' = \bar{\mathbf{U}}\bar{\mathbf{V}}^{*\prime}, \tag{6.37}$$

where $\bar{\mathbf{U}} = \mathbf{U}(\mathbf{U}'\mathbf{U})^{-1/2}$ and $\bar{\mathbf{V}}^* = \tilde{\mathbf{V}}(\mathbf{U}'\mathbf{U})^{1/2}$, and we plot $\bar{\mathbf{V}}^*$. Note that $\bar{\mathbf{U}}$ is orthogonal columnwise, so that $\bar{\mathbf{V}}^*$ is both a pattern and structure matrix for $\bar{\mathbf{U}}$. (That is, $\bar{\mathbf{V}}^*$ is a matrix of component loadings in the true sense of the word.) This is analogous to interpreting $\bar{\mathbf{b}}$ in regression analysis in

$$\hat{\mathbf{z}} = \mathbf{G}(\mathbf{G}'\mathbf{G})^{-1/2}(\mathbf{G}'\mathbf{G})^{1/2}\hat{\mathbf{b}} = \bar{\mathbf{G}}\bar{\mathbf{b}}. \tag{6.38}$$

The elements of $\hat{\mathbf{b}}$ are generally not directly comparable. However, the elements of $\hat{\mathbf{z}} = \mathbf{G}\hat{\mathbf{b}}$ are comparable, and so are the elements of $\hat{\mathbf{b}}$ under the metric matrix $\mathbf{G}'\mathbf{G}$. The latter is equivalent to comparing the elements of $\bar{\mathbf{b}} = (\mathbf{G}'\mathbf{G})^{1/2}\hat{\mathbf{b}}$ directly. The matrix $\bar{\mathbf{G}}$ is the best (in the LS sense) orthogonal approximation to $\mathbf{G}$ (Section 2.3.4).

The parallelogram in Figure 6.5 shows a plot of $\tilde{\mathbf{V}}^*$ transformed from $\tilde{\mathbf{V}}$ in Example 1 using (6.37). The original plot of $\tilde{\mathbf{V}}$ is reproduced as the rectangular grid in the figure. This configuration ignores the correlation between the columns of $\mathbf{U}$. The two configurations give quite different impressions. The parallelogram configuration better resembles the configurations given in Figure 6.2(d) and (e), if rotated counter clockwise about 45°.

6.7 Extended Redundancy Analysis (ERA)

In the previous section, we discussed a graphical display of a stimulus (variable) configuration characterized by correlated components. In this and the next sections, we present methods that attempt to model the relationships among the correlated components.

We discuss extended redundancy analysis (ERA) in this section. In Section 4.2, we described RA as a method to predict a set of criterion variables from a reduced subspace of predictor variables. Suppose that there are more than one set of predictor variables, and that we predict a set of criterion variables from their reduced subspaces. This is equivalent to DCDD, and the simplest form of ERA. We have already seen an example of this case in Section 6.4. Recall that there were two sets of dummy variables indicating the levels of height and width of rectangles. Components representing the subjective height and width of the rectangles were extracted from them. The components are then used to predict the subjective area of rectangles in four groups of subjects. The model given in (6.24) may be written in a more standard notation of ERA as

$$\mathbf{Z}_3 = \mathbf{Z}_1\mathbf{w}_1\mathbf{c}'_{13} + \mathbf{Z}_2\mathbf{w}_2\mathbf{c}'_{23} + \mathbf{E}_3, \tag{6.39}$$

where $\mathbf{Z}_j$ $(j = 1, \cdots, 3)$ are three sets of observed variables, $\mathbf{w}_1$ and $\mathbf{w}_2$ are the weights applied to the sets of predictor variables $\mathbf{Z}_1$ and $\mathbf{Z}_2$ to derive predictor components, $\mathbf{c}'_{13}$ and $\mathbf{c}'_{23}$ are the weights (called component loadings) applied to the predictor components to derive the best prediction of the criterion variables $\mathbf{Z}_3$, and $\mathbf{E}_3$ is the matrix of disturbance terms. Here $\mathbf{Z}_3$ corresponds with $\mathbf{Z}$, $\mathbf{Z}_1$ with $\mathbf{G}_1$, and $\mathbf{Z}_2$ with $\mathbf{G}_2$ in (6.24), $\mathbf{w}_1$ and $\mathbf{w}_2$ with $\mathbf{u}_1^*$ and $\mathbf{u}_2^*$, $\mathbf{c}'_{13}$ and $\mathbf{c}'_{23}$ with the rows of $\mathbf{W}'$, and $\mathbf{E}_3$ with $\mathbf{E}$ in (6.24).

ERA, however, is more general than DCDD. It allows some sets of variables to be both predictors and criterion variables (Hwang and Takane 2002b; Takane and Hwang 2005). For example, $\mathbf{Z}_2$ may be predicted from a linear combination of $\mathbf{Z}_1$, while $\mathbf{Z}_3$ may in turn be predicted from linear combinations of both $\mathbf{Z}_1$ and $\mathbf{Z}_2$. The model in this case can be written as:

$$\begin{aligned} \mathbf{Z}_2 &= \mathbf{Z}_1\mathbf{w}_1\mathbf{c}'_{12} + \mathbf{E}_2, \\ \mathbf{Z}_3 &= \mathbf{Z}_1\mathbf{w}_1\mathbf{c}'_{13} + \mathbf{Z}_2\mathbf{w}_2\mathbf{c}'_{23} + \mathbf{E}_3. \end{aligned} \tag{6.40}$$

Notice that $\mathbf{Z}_2$ serves as both predictor and criterion variables.

The ERA model may be generally written as

$$\mathbf{Z}^{(1)} = \mathbf{Z}^{(2)}\mathbf{W}\mathbf{C}' + \mathbf{E}, \tag{6.41}$$

where $\mathbf{Z}^{(1)}$ is a row block matrix of criterion variables, $\mathbf{Z}^{(2)}$ is a row block matrix of all predictor variables. The variables serving as both criterion and predictor variables are included in both $\mathbf{Z}^{(1)}$ and $\mathbf{Z}^{(2)}$. The matrix $\mathbf{W}$ is a matrix of weights, and $\mathbf{C}$ is a matrix of loadings. Both of these matrices are patterned in the sense that there are many zero elements at prescribed locations. In the first example of the ERA model (6.39), we set $\mathbf{Z}^{(1)} = \mathbf{Z}_3$, $\mathbf{Z}^{(2)} = [\mathbf{Z}_1, \mathbf{Z}_2]$, and $\mathbf{W} = \text{bdiag}(\mathbf{w}_1, \mathbf{w}_2) = \begin{bmatrix} \mathbf{w}_1 & \mathbf{0} \\ \mathbf{0} & \mathbf{w}_2 \end{bmatrix}$, $\mathbf{C} = [\mathbf{c}_{13}, \mathbf{c}_{23}]$, and $\mathbf{E} = \mathbf{E}_3$. In the second example (6.40), we set $\mathbf{Z}^{(1)} = [\mathbf{Z}_2, \mathbf{Z}_3]$, $\mathbf{Z}^{(2)} = [\mathbf{Z}_1, \mathbf{Z}_2]$ and $\mathbf{W} = \text{bdiag}(\mathbf{w}_1, \mathbf{w}_2)$ as before, and $\mathbf{C} = \begin{bmatrix} \mathbf{c}_{12} & \mathbf{0} \\ \mathbf{c}_{13} & \mathbf{c}_{23} \end{bmatrix}$, and $\mathbf{E} = [\mathbf{E}_2, \mathbf{E}_3]$. It is customary to require $\mathbf{w}_j'\mathbf{Z}_j'\mathbf{Z}_j\mathbf{w}_j = n$ $(j = 1, 2)$ in order to remove the scale indeterminacies between $\mathbf{W}$ and $\mathbf{C}'$.

Model parameters (i.e., elements in $\mathbf{W}$ and $\mathbf{C}$) can be estimated using an ALS algorithm similar to that for ERA. We first rewrite (6.41) as

$$\begin{aligned} \text{vec}(\mathbf{Z}^{(1)}) &= (\mathbf{I} \otimes \mathbf{Z}^{(2)}\mathbf{W})\text{vec}(\mathbf{C}') + \text{vec}(\mathbf{E}) &\quad (6.42) \\ &= (\mathbf{C} \otimes \mathbf{Z}^{(2)})\text{vec}(\mathbf{W}) + \text{vec}(\mathbf{E}). &\quad (6.43) \end{aligned}$$

Each of the above equations constitutes a constrained regression problem. The first one involves estimating $\mathbf{C}$ for fixed $\mathbf{W}$, and the second involves estimating $\mathbf{W}$ for fixed $\mathbf{C}'$. In (6.42), let $\mathbf{c}$ denote the portion of $\text{vec}(\mathbf{C}')$ corresponding to its nonzero elements, and let $\mathbf{B}^*$ denote a matrix consisting of only the columns of $\mathbf{I} \otimes \mathbf{Z}^{(2)}\mathbf{W}$ corresponding to the nonzero elements of $\text{vec}(\mathbf{C}')$. Then, (6.42) can be rewritten as

$$\mathbf{z}^{(1)} \equiv \text{vec}(\mathbf{Z}^{(1)}) = \mathbf{B}^*\mathbf{c} + \mathbf{e}, \quad (6.44)$$

where $\mathbf{e} = \text{vec}(\mathbf{E})$. From this, the OLSE of $\mathbf{c}$ can be found by $\hat{\mathbf{c}} = (\mathbf{B}^{*\prime}\mathbf{B}^*)^{-1}\mathbf{B}^{*\prime}\mathbf{z}^{(1)}$. The OLSE of $\mathbf{C}'$ can be found by rearranging the elements of $\hat{\mathbf{c}}$ appropriately.

Now let $\mathbf{w}^*$ denote the portion of $\text{vec}(\mathbf{W})$ corresponding to its nonzero elements, and let $\mathbf{A}^*$ denote the matrix consisting of only the columns of $\mathbf{C} \otimes \mathbf{Z}^{(2)}$ corresponding to the nonzero elements of $\text{vec}(\mathbf{W})$. Note that $\mathbf{w}^*$ is a super vector of the $\mathbf{w}^* = [\mathbf{w}_1', \mathbf{w}_2']'$, assuming that there are only two predictor sets. Let $\mathbf{A}^*$ be partitioned as $\mathbf{w}^*$, that is, $\mathbf{A}^* = [\mathbf{A}_1, \mathbf{A}_2]$. Then (6.43) can be rewritten as

$$\mathbf{z}^{(1)} = \mathbf{A}^*\mathbf{w}^* + \mathbf{e} = [\mathbf{A}_1, \mathbf{A}_2]\begin{pmatrix} \mathbf{w}_1 \\ \mathbf{w}_2 \end{pmatrix} + \mathbf{e}. \quad (6.45)$$

For convenience of incorporating the normalization restrictions ($\mathbf{w}_j'\mathbf{Z}_j'\mathbf{Z}_j\mathbf{w}_j = n$ for $j = 1, 2$), we update $\mathbf{w}_1$ and $\mathbf{w}_2$ sequentially. We first define $\mathbf{y} = \mathbf{z}^{(1)} - \mathbf{A}_2\mathbf{w}_2$, so that (6.45) can be further rewritten as $\mathbf{y} = \mathbf{A}_1\mathbf{w}_1 + \mathbf{e}$. Then the (unnormalized) OLSE of $\mathbf{w}_1$ can be found by $\hat{\mathbf{w}}_1 = (\mathbf{A}_1'\mathbf{A}_1)^{-1}\mathbf{A}_1'\mathbf{y}$, which is then normalized to satisfy the normalization restriction that $\hat{\mathbf{w}}_1'\mathbf{Z}_1'\mathbf{Z}_1\hat{\mathbf{w}}_1 = n$. The $\mathbf{w}_2$ is updated similarly by interchanging the roles of $\mathbf{A}_1$ and $\mathbf{w}_1$ with those of $\mathbf{A}_2$ and $\mathbf{w}_2$. The updated $\mathbf{w}_1$ and $\mathbf{w}_2$ are then arranged in the form of $\mathbf{W}$. The explanation above assumed that there were only two predictor sets. A generalization to more than two predictor sets is rather obvious.

The alternate updates of $\mathbf{C}'$ and $\mathbf{W}$ are repeated until convergence. This algorithm is monotonically convergent.

It is possible to take more than one component from each predictor set. In such cases, $\mathbf{w}_j$ will be the matrix $\mathbf{W}_j$ with as many columns as there are components to be extracted, and $\mathbf{c}'_j$ will also be the matrix $\mathbf{C}'_j$ with as many rows as there are components. The matrices $\mathbf{W}$ and $\mathbf{C}'$ will be constructed accordingly. The former is a block diagonal matrix of $\mathbf{W}_j$ and the latter is a column block matrix of $\mathbf{C}'_j$. The algorithm works just as described above for a single component for each predictor set, except that the LS update of $\mathbf{W}_j$ must be orthonormalized so as to satisfy $\mathbf{W}'_j\mathbf{Z}'_j\mathbf{Z}_j\mathbf{W}_j = n\mathbf{I}$.

There are a couple of interesting special cases of ERA worth discussing. The first one is called principal covariate regression (de Jong and Kiers 1992; see also Merola and Abraham 2001). This model can be written as:

$$\begin{aligned} \mathbf{Z}_1 &= \mathbf{Z}_1\mathbf{W}_1\mathbf{C}'_{11} + \mathbf{E}_1. \\ \mathbf{Z}_2 &= \mathbf{Z}_1\mathbf{W}_1\mathbf{C}'_{12} + \mathbf{E}_2. \end{aligned} \tag{6.46}$$

Parameters in this model may be estimated by minimizing $\mathrm{SS}(\mathbf{E}_1) + \mathrm{SS}(\mathbf{E}_2)$. An iterative algorithm similar to the one described above can be used to minimize the criterion. However, there is a simple closed-form solution to this problem that involves an SVD of $\mathbf{A} = [\mathbf{Z}_1, \mathbf{P}_{Z_1}\mathbf{Z}_2]$. This may be easily seen by noting that (6.46) can be rewritten as

$$[\mathbf{Z}_1, \mathbf{Z}_2] = \mathbf{Z}_1\mathbf{W}_1[\mathbf{C}'_{11}, \mathbf{C}'_{12}], \tag{6.47}$$

which is a simple RA problem with $[\mathbf{Z}_1, \mathbf{Z}_2]$ as the criterion variables, and $\mathbf{Z}_1$ as the predictor variables. Merola and Abraham (2001) proposed to minimize a slightly more generalized criterion. Specifically, they proposed to minimize $\alpha\mathrm{SS}(\mathbf{E}_1) + (1-\alpha)\mathrm{SS}(\mathbf{E}_2)$, where α is the prescribed weight that regulates the importance of the two error terms. This also leads to a simple closed-form solution, in which $\mathbf{A}$ is redefined as $\mathbf{A} = [\sqrt{\alpha}\mathbf{Z}_1, \sqrt{1-\alpha}\mathbf{P}_{Z_1}\mathbf{Z}_2]$.

The second special case is called the MIMIC (multiple-indicator multiple-cause) model. Let $\mathbf{Z}_2$ denote the matrix of criterion variables, and let $\mathbf{Z}_1$ denote the matrix of predictor variables. In RA, components extracted from $\mathbf{Z}_1$ are defined as linear combinations $\mathbf{Z}_1\mathbf{W}$ of $\mathbf{Z}_1$, which are then used to predict $\mathbf{Z}_2$. That is, we look for $\mathbf{W}$ and $\mathbf{C}'$ such that $\mathbf{Z}_1\mathbf{W}\mathbf{C}'$ best approximates $\mathbf{Z}_2$. No errors are allowed in extracting the predictor components. In the MIMIC model, the components could be error-contaminated. Let $\mathbf{B}$ represent the matrix of predictor components. Then, $\mathbf{B} = \mathbf{Z}_1\mathbf{W} + \mathbf{E}_1$. We may impose the columnwise orthogonality constraint on $\mathbf{B}$, i.e., $\mathbf{B}'\mathbf{B} = \mathbf{I}$. This $\mathbf{B}$ is then used to predict $\mathbf{Z}_2$, which leads to $\mathbf{Z}_2 = \mathbf{B}\mathbf{C}' + \mathbf{E}_2$. We may estimate $\mathbf{B}$, $\mathbf{W}$ and $\mathbf{C}'$ in such a way that $\mathrm{SS}(\mathbf{E}_1) + \mathrm{SS}(\mathbf{E}_2)$ is minimized. This minimization problem has a closed-form solution, which amounts to finding SVD($\mathbf{A}$), where $\mathbf{A} = [\mathbf{Z}_1(\mathbf{Z}'_1\mathbf{Z}_1)^{-1/2}, \mathbf{Z}_2]$). Let $\mathbf{U}_r\mathbf{D}_r\mathbf{V}'_r$ denote the best rank r approximation to $[\mathbf{Z}_1(\mathbf{Z}'_1\mathbf{Z}_1)^{-1/2}, \mathbf{Z}_2]$. Then $\mathbf{B} = \mathbf{U}_r$, $\mathbf{W} = (\mathbf{Z}'_1\mathbf{Z}_1)^{-1}\mathbf{Z}'_1\mathbf{B}$, and $\mathbf{C}' = \mathbf{B}'\mathbf{Z}_2$. As in Merola and Abraham (2001) described above, it is straightforward to introduce the weighting parameter α into the minimization criterion.

6.8 Generalized Structured Component Analysis (GSCA)

One notable characteristic of ERA is that linear combinations are formed only for predictor sets. The criterion variables are always directly predicted. In GSCA (Hwang and Takane 2004), on the other hand, components are extracted from all variable sets, and the components extracted from criterion sets are predicted from the components extracted from the predictor sets. The predictive relationships among these components are expressed in the form of regression equations, called structural models.

Another notable characteristic of ERA is that components extracted from predictor sets are exclusively used for predicting criterion sets. It does not matter how well they can explain the predictor sets from which they are extracted. In GSCA, on the other hand, all extracted components must be able to explain the variable sets from which they are extracted. The relationships between the components and the variable sets from which they are extracted are called measurement models.

Let us begin with the simplest possible cases, in which we have only two sets of variables, $\mathbf{Z}_1$ and $\mathbf{Z}_2$. We extract a component from each set capturing representative variations in each set. We then predict one component from the other. Suppose that $\mathbf{Z}_1$ is the predictor set, and $\mathbf{Z}_2$ the criterion set. Then, the process described above may be expressed by the following three submodels:

$$\begin{aligned} \mathbf{Z}_1 &= \mathbf{Z}_1\mathbf{w}_1\mathbf{c}_{11}' + \mathbf{E}_1, \\ \mathbf{Z}_2 &= \mathbf{Z}_2\mathbf{w}_2\mathbf{c}_{22}' + \mathbf{E}_2, \\ \mathbf{Z}_2\mathbf{w}_2 &= \mathbf{Z}_1\mathbf{w}_1 c_{12} + \mathbf{e}_3. \end{aligned} \tag{6.48}$$

As before, we assume, for model identification, that $\mathbf{w}_1$ and $\mathbf{w}_2$ are normalized to satisfy $\mathbf{w}_j'\mathbf{Z}_j'\mathbf{Z}_j\mathbf{w}_j = n$ for $j = 1,2$. Let $\mathbf{E} = [\mathbf{E}_1, \mathbf{E}_2, \mathbf{e}_3]$. Parameters in the above model ($\mathbf{c}_{jj}$'s. $\mathbf{w}_j$'s, and c_{12}) may be estimated so as to minimize $\mathrm{SS}(\mathbf{E}) = \mathrm{tr}(\mathbf{E}_1'\mathbf{E}_1) + \mathrm{tr}(\mathbf{E}_2'\mathbf{E}_2) + \mathbf{e}_3'\mathbf{e}_3$. An ALS algorithm for minimizing this criterion can be constructed in a straightforward manner.

We first initialize $\mathbf{w}_i$ by taking the first principal component from $\mathbf{Z}_j$ $(j = 1,2)$. We then find $\mathbf{c}_j$'s and c_{12} for fixed $\mathbf{w}$'s. The formula for this can be derived by first rewriting the minimization criterion as

$$\mathrm{SS}(\mathbf{E}) = \mathrm{SS}\left(\begin{pmatrix} \mathbf{z}_1 \\ \mathbf{z}_2 \\ \mathbf{Z}_2\mathbf{w}_2 \end{pmatrix} - \mathbf{B}^* \begin{pmatrix} \mathbf{c}_{11} \\ \mathbf{c}_{22} \\ c_{12} \end{pmatrix}\right), \tag{6.49}$$

where $\mathbf{z}_j = \mathrm{vec}(\mathbf{Z}_j)$ $(j = 1,2)$, and

$$\mathbf{B}^* = \begin{bmatrix} \mathbf{I}\otimes\mathbf{Z}_1\mathbf{w}_1 & \mathbf{O} & \mathbf{0} \\ \mathbf{O} & \mathbf{I}\otimes\mathbf{Z}_2\mathbf{w}_2 & \mathbf{0} \\ \mathbf{O} & \mathbf{O} & \mathbf{Z}_1\mathbf{w}_1 \end{bmatrix}. \tag{6.50}$$

Note that the above LS problem reduces to three separate LS problems of smaller sizes since $\mathbf{B}^*$ is a block diagonal matrix. We then update $\mathbf{w}$'s sequentially, similarly to the previous section, although matrices involved are different: Let

$$\mathbf{y}^* = \begin{pmatrix} \mathbf{z}_1 \\ \mathbf{z}_2 \\ \mathbf{0} \end{pmatrix}, \tag{6.51}$$

and

$$\mathbf{A}^* = [\mathbf{A}_1, \mathbf{A}_2] = \begin{bmatrix} \mathbf{c}_{11} \otimes \mathbf{Z}_1 & \mathbf{O} \\ \mathbf{O} & \mathbf{c}_{22} \otimes \mathbf{Z}_2 \\ \mathbf{Z}_1 c_{12} & -\mathbf{Z}_2 \end{bmatrix}. \tag{6.52}$$

Then

$$\mathrm{SS}(\mathbf{E}) = \mathrm{SS}\left(\mathbf{y}^* - [\mathbf{A}_1, \mathbf{A}_2]\begin{pmatrix} \mathbf{w}_1 \\ \mathbf{w}_2 \end{pmatrix}\right). \tag{6.53}$$

Define $\mathbf{y} = \mathbf{y}^* - \mathbf{A}_2\mathbf{w}_2$. Then, the (unnormalized) OLSE of $\mathbf{w}_1$ is given by $\hat{\mathbf{w}}_1 = (\mathbf{A}_1'\mathbf{A}_1)^{-1}\mathbf{A}_1'\mathbf{y}$, which is then normalized to satisfy $\hat{w}_1'\mathbf{Z}_1'\mathbf{Z}_1\hat{w}_1 = n$. The $\mathbf{w}_2$ can be updated similarly. Again an extension to more than two predictor sets is straightforward. The alternate updating of $\mathbf{c}_j$'s, c_{12}, and $\mathbf{w}_j$ is repeated until convergence is reached.

The three submodels in (6.48) may be written as a single set of equations:

$$[\mathbf{Z}_1, \mathbf{Z}_2, \mathbf{0}] = [\mathbf{Z}_1, \mathbf{Z}_2]\begin{bmatrix} \mathbf{w}_1 & \mathbf{0} \\ \mathbf{0} & \mathbf{w}_2 \end{bmatrix}\begin{bmatrix} \mathbf{c}_{11}' & \mathbf{0}' & c_{12} \\ \mathbf{0}' & \mathbf{c}_{22}' & -1 \end{bmatrix} + \mathbf{E}, \tag{6.54}$$

that is,

$$\mathbf{Z}^{(1)} = \mathbf{Z}^{(2)}\mathbf{W}\mathbf{C}' + \mathbf{E}. \tag{6.55}$$

This is identical in form to (6.41), although the key matrices ($\mathbf{Z}^{(1)}$, $\mathbf{Z}^{(2)}$, $\mathbf{W}$, and $\mathbf{C}'$) are defined differently. An important point to emphasize is that (6.55) is so general that almost all empirically meaningful GSCA models can be written in this form by defining the constituent matrices appropriately.

Since Hwang and Takane's (2004) original proposal, GSCA has been extended in a number of directions, e.g., allowing nonlinear transformations (Hwang and Takane 2010), accommodating respondent heterogeneity (Hwang et al. 2007a), development of multilevel GSCA (Hwang et al. 2007b) for clustered data, and incorporating interaction effects among components (Hwang et al. 2010). More recently, it has been generalized to dynamic GSCA (Jung et al. 2012) to capture statistical dependencies among observations taken over time by incorporating autoregressive effects as part of the model. Dynamic GSCA also accommodates stimulus effects on various parts of the brain as well as on connectivity between them. Hwang and Takane (forthcoming) are working on a comprehensive textbook on GSCA.

Epilogue

This book highlighted various aspects of CPCA: its mathematical foundations, data requirements, computational tools, applications, and related techniques. In closing, I would like to offer a brief discussion on CPCA in comparison with analysis of covariance structures (ACOVS; Bock and Bergman 1966; Jöreskog 1970), a group of techniques frequently used in research in social and behavioral sciences.

1. CPCA makes no rigid distributional assumptions. ACOVS, on the other hand, typically assumes a multivariate normal distribution on the data, which is often unrealistic. Micceri (1989), for example, examined more than 400 psychological measures, and found that none of them passed the test of normality. He concluded that normal distributions were as improbable as unicorns, which only existed in our imagination. There have been some attempts to relax the normality assumption in ACOVS. One extends the normal distribution to an elliptical distribution. This distribution is, however, not very much different from the normal distribution. There has also been a development of asymptotically distribution-free estimation (ADF; Browne 1984). However, ADF requires an even larger sample size to reliably estimate model parameters.

2. CPCA makes no claim that fitted models are correct. In contrast, attractive properties of maximum likelihood (ML) or generalized LS (GLS) estimators in ACOVS are contingent on two major premises, a large sample and the correctness of fitted models. Models used in social and behavioral sciences are, however, often only crude approximations of reality, and asymptotic properties of the estimators are necessarily compromised because the fitted models tend to be invalidated in large samples. There has been some work investigating the robustness of ACOVS against violations of assumptions, and a general conclusion is that while the estimates of model parameters are fairly robust, the chi-square goodness of fit (GOF) statistic or the estimate of standard error is not.

3. The computation is quick and easy in CPCA. It consists of projection and SVD, for which efficient and reliable computational methods exist. ACOVS, on the other hand, typically requires elaborate iterative fitting algorithms, in which there always are possibilities of nonconvergence and convergence to suboptimal solutions. CPCA can often obtain results similar to those obtained by ACOVS in a fraction of computation time (Veilicer and Jackson 1999) needed by ACOVS.

4. No problems of improper solutions exists in CPCA. In contrast, ACOVS may obtain improper solutions, such as negative estimates of variances and correlations out of proper range. There is usually no built-in mechanism to prevent the estimates from wandering off the feasible regions of the parameter space.

5. There is no factorial indeterminacy problem in CPCA. Component scores can always be calculated uniquely. In contrast, factor scores cannot be calculated uniquely due to the well-known factorial indeterminacy in ACOVS (e.g., Schönemann and Steiger 1978).

6. CPCA is largely descriptive. While CPCA makes no rigid distributional assumptions, it has no criterion for model evaluation and statistical inferences. This argument is only half correct: As shown in Section 5.2, the stability of CPCA results can be assessed by a resampling method such as the bootstrap method (Efron 1979).

7. CPCA takes no account of measurement errors. This argument is also only half correct: Discarding components corresponding to small singular values has the effect of reducing measurement errors (Gleason and Staelin 1973). In ACOVS, the effects of measurement errors are eliminated, but so are the effects of specific factors, which can be even more harmful than ignoring measurement errors.

8. CPCA has no scale invariance. Maximum likelihood or GLS estimators in ACOVS, on the other hand, yield parameter estimates which are scale invariant. This argument is also only half correct: Scale invariance can be obtained in CPCA by a proper use of a column side metric matrix $\mathbf{L}$ (Rao 1964).

9. CPCA is less flexible. It is indeed true that CPCA is limited in the variety of models that can be specified. Note, however, that CPCA has been extended in various ways to allow more flexible model specifications without sacrificing its strengths too much. See Chapter 6 for such extensions.

In the mid-1980s (Takane 1987, 1994), I was interested in incorporating systematic individual differences in preference judgments in Thurstone's law of comparative judgments. In pair comparison studies, data are often gathered by multiple-judgment sampling, in which each of a group of subjects makes judgments for all possible pairs of stimuli, giving rise to repeated measurement data. In such cases, it is essential to separate within-subject and between-subject variabilities in the data. Thinking that ACOVS would provide an ideal framework, I used LISREL IV to fit an ACOVS pair comparison model (Section 4.11). The analysis was very successful in the sense that all expected structures in the data were clearly borne out. However, the chi-square GOF statistic indicated that the fitted model was awful (Takane 1994). It made me conclude that there was something crucial missing in my model, and yet I had little clue as to what was lacking. This experience prompted me to carry out a residual analysis (Section 6.5), which I thought could be more easily done by PCA than ACOVS. To get the residuals for PCA, however, I had to reformulate my original model from a component analysis perspective. This was easy enough, but I soon realized that it would provide an alternative (perhaps a more natural) approach to modeling individual differences in pair comparison data. And so CPCA was born (Takane and Shibayama 1991).

Appendix

In this appendix we briefly introduce computer software for CPCA, DCDD, and others. There are three major groups of software, which we discuss in turn:

1. As announced at the end of Chapter 1, MATLAB® programs that created all the examples in Chapter 1 are available from http://takane.brinkster.net/Yoshio. These programs are heavily annotated to make clear what operations are performed line by line. Example data sets are also provided. Note that MATLAB is a commercial software platform, and is required in order to run these programs. The information about this product is given in the preface of this book. For those of you without access to MATLAB, these programs have been translated into equivalent R programs, which can be downloaded from the same site. The R platform can be freeely downloaded from http://cran.r-project.org/.

2. General-purpose programs for CPCA and DCDD are available at the same site as above. These programs, also written in MATLAB, were used to create other examples in later chapters. Manuals and input/output examples can also be downloaded from the same site. These general-purpose programs are also being translated into equivalent R programs, and will be available for free download shortly. Equivalent programs in FORTRAN (source codes and executables) are also available, and were used to create the examples in Hunter and Takane (1998, 2002).

3. A CPCA program specifically designed to analyze fMRI data is available from http://www.nitrc.org/projects/fmricpca. This site is hosted by Professor Todd Woodward of the University of British Columnbia. The program comes with example data, an installation guide, a detailed manual, and tutorials. The program is written in MATLAB, but contains many user friendly features (e.g., a graphical user-interface).

4. A program called GeSCA can be accessed at http://www.sem-gesca.org/. This is a web-based program for generalized structured component analysis (GSCA; Section 6.8). This program provides a graphical user interface that allows users to specify their model as a path diagram and to view the estimates of model parameters. You need JAVA to run this program.

Bibliography

Abdi, H. and D. Valentin. 2007. The STATIS method. In *Encyclopedia of measurement and statistics*, ed. N. J. Salkind, 955-62. Thousand Oaks, CA: Sage. [121][1]

Adachi, K. 2009. Joint procrustes analysis for simultaneous nonsingular transformation of component score and loading matrices. *Psychometrika* 74:667–83. [105]

Akaike, H. 1974. A new look at the statistical model identification. *IEEE Transactions on Automatic Control* 19:716–23. [157]

Amemiya, T. 1985. *Advanced econometrics*. Cambridge, MA: Harvard University Press. [50, 85]

Ammann, L. P. 1993. Robust singular value decompositions: A new approach to projection pursuit. *Journal of the American Statistical Association* 88:505–14. [169]

Anderson, T. W. 1951. Estimating linear restrictions on regression coefficients for multivariate normal distribution. *Annals of Mathematical Statistics* 22:327–51. [97, 101]

Bechtel, G. G. 1976. *Multidimensional preference scaling*. The Hague: Mouton. [125]

Bechtel, G. G., L. R. Tucker, and W. Chang. 1971. A scalar product model for the multidimensional scaling of choice. *Psychometrika* 36:369–87. [126]

Belsley, D. A., E. Kuh, and R. E. Welsch. 1980. *Regression diagnostics*. New York: Wiley. [49, 160, 162]

Benzécri, J. P. 1973. *Analyse des Données*. Paris: Dunod. [109]

Bertrami, E. 1873. Sulle funzioni bilineari. *Giornale di matematische ud uso Degli Studenti Delle Universite* 11:98–106. [62]

Besse, P. and J. O. Ramsay. 1986. Principal components analysis of sampled functions. *Psychometrika* 51:285–311. [74, 75, 174]

Bock, R. D. 1960. Methods and applications of optimal scaling. Report No. 25, Psychometric Laboratory, The University of North Carolina. [109]

Bock, R. D. 1989. *Multilevel analysis of educational data*. San Diego, CA: Academic Press. [144]

Bock, R. D. and R. E. Bergman. 1966. Analysis of covariance structures. *Psychometrika* 31:507–34. [201]

Böckenholt, U. and I. Böckenholt. 1990. Canonical analysis of contingency tables with linear constraints. *Psychometrika* 55:633–9. [83, 109, 112]

Böckenholt, U. and Y. Takane. 1994. Linear constraints in correspondence analysis. In *Correspondence analysis in the social sciences: Recent developments and applications*, ed. M. J. Greenacre and J. Blasius, 112–27. New York: Academic Press. [109]

[1]Numbers in square brackets indicate page numbers on which the article is referenced.

Boik, R. J. 1987. The Fisher–Pitman permutation test: A nonrobust alternative to the normal theory F test when variances are heterogeneous. *British Journal of Mathematical and Statistical Psychology* 40:26–42. [156]

Boik, R. J. 2013. Model-based principal components of correlation matrices. *Journal of Multivariate Analysis* 116:310–331. [96]

Boik, R. J., K. Panishkan, and S. K. Hyde. 2010. Model-based principal components of covariance matrices. *British Journal of Mathematical and Statistical Psychology* 63:113–37. [96]

Bojanczyk, A. W., M. Ewerbring, F. T. Luk, and P. van Dooren. 1991. An accurate product SVD algorithm. In *SVD and signal processing, II*, ed. R. J. Vaccaro, 113–31. Amsterdam: Elsevier. [66]

Borg, I. and P. J. F. Groenen. 2005. *Modern multidimensional scaling*. New York: Springer. [104]

Breiman, L. and J. H. Friedman. 1997. Predicting multivariate responses in multiple linear regression. *Journal of the Royal Statistical Society, Series B* 57:3–54. [101, 163]

Brown, A. L. and A. Page. 1970. *Elements of functional analysis*. New York: Van Nostrand. [39]

Browne, M. W. 1967. On oblique procrustes rotation. *Psychometrika* 32:125-32. [151]

Browne, M. W. 1984. Asymptotically distribution-free methods for the analysis of covariance structures. *British Journal of Mathematical and Statistical Psychology* 32:62–83. [201]

Browne, M. W. 2001. An overview of analytic rotation in exploratory factor analysis. *Multivariate Behavioral Research* 36:111–50. [95]

Browne, M. W. and W. Kristof. 1969. On the oblique rotation of a factor matrix to a specified pattern. *Psychometrika* 34:237–48. [194]

Bryk, A. S. and S. W. Raudenbush. 1992. *Hierarchical linear models*. Newbury Park, CA: Sage. [144]

Buja, A. and N. Eyuboglu. 1992. Remarks on parallel analysis. *Multivariate Behavioral Research* 27:509–40. [156]

Busing, F. M. T. A., P. J. F. Groenen, and W. Heiser. 2005. Avoiding degeneracy in multidimensional unfolding by penalizing on the coefficient of variation. *Psychometrika* 70:71-98. [105]

Cailliez, F. and J. P. Pagés. 1976. *Introduction à l'analyse des donnée*. Paris: SMASH. [30, 66]

Carroll, J. D. 1968. A generalization of canonical correlation analysis to three or more sets of variables. *Proceedings of the 76th Annual Convention of the American Psychological Association*, 227–8. [121]

Carroll, J. D. 1972. Individual differences and multidimensional scaling. In *Multidimensional scaling, Vol I*, ed. R. N. Shepard, A. K. Romney, and S. B. Nerlove, 105–55. New York: Seminar Press. [87, 125]

Carroll, J. D., and J. J. Chang. 1970. Analysis of individual differences in multidimensional scaling via an N-way generalization of "Echart–Young" decomposition. *Psychometrika* 35:282–319. [105, 107, 129]

Carroll, J. D., P. E. Green, and A. Chaturvedi. 1997. *Mathematical tools for applied multivariate analysis (revised edition)*. New York: Academic Press. [21]

Carroll, J. D., S. Pruzansky, and J. B. Kruskal. (1980). A general approach to multidimensional analysis of many-way arrays with linear constraints on parameters. *Psychometrika* 45:3–24. [76, 83, 129]

Cattell, R. B. 1966. The scree test for the number of factors. *Multivariate Behavioral Research* 1:245–76. [155]

Chatterjee, S. and A. S. Hadi. 1988. *Sensitivity analysis in linear regression*. New York: Wiley. [160, 162]

Chipman, H. A., and H. Gu. 2005. Interpretable dimension reduction. *Journal of Applied Statistics* 32:969–87. [85]

Choulakian, V. 2001. Robust Q-mode principal component analysis in L_1. *Computational Statistics and Data Analysis* 37:135–50. [169]

Cliff, N. 1966. Orthogonal rotation to congruence. *Psychometrika* 31:33–42. [150, 151]

Clint, M. and A. Jennings. 1970. The evaluation of eigenvalues and eigenvectors of real symmetric matrices by simultaneous iteration. *Computer Journal* 13:145–58. [150]

Coleman, D., P. Holland, N. Kaden, V. Klema, and S. C. Peters. 1980. A system of subroutines for iteratively reweighted least squares computations. *ACM Transactions on Mathematical Software* 6:327–36. [167]

Conniffe, D. 1982. Covariance analysis and seemingly unrelated regression. *Journal of the American Statistical Association* 36:169–71. [139, 140]

Cook, R. D., and S. Weisberg. 1982. *Residuals and influence in regression*. London: Chapman and Hall. [49, 160, 162]

Coombs, C. H. 1964. *A theory of data*. New York: Wiley. [105]

Corsten, L. C. A. 1976. Matrix approximation, a key to approximation of multivariate methods. Invited paper presented at the 9th Biometric Conference, Boston. [83]

Corsten, L. C. A. and A. C. van Eijnsbergen. 1972. Multiplicative effects in two-way analysis of variance. *Statistica Neelandica* 26:61–8. [83]

Cramer, E. M. 1974. On Browne's solution for oblique Procrustes rotation. *Psychometrika* 39:159–63. [151]

Craven, P. and G. Wahba. 1979. Smoothing noisy data with spline functions: Estimating the correct degree of smoothing by the method of generalized cross-validation. *Numerische Mathematik* 31:377–403. [162]

Critchley, F. 1985. Influence in principal component analysis. *Biometrika* 72:627–36. [160]

d'Aspremont, A., L. El Ghaoui, M. I. Jordan, and G. R. G. Lanckriet. 2007. A direct formulation for sparse PCA using semidefinite programming. *SIAM Review* 49:434–48. [95]

Daugavet, V. A. 1968. Variant of the stepped exponential method of finding some of the first characteristic values of a symmetric matrix. *USSR Computational Mathematics and Mathematical Physics* 8:212–23. [129, 149]

Davies, P. T. and M. K. Tso. 1982. Procedures for reduced-rank regression. *Applied Statistics* 31:244–55. [161]

de Jong, S. and H. A. L. Kiers. 1992. Principal covariates regression, Part I: Theory. *Chemometrics and Intelligent Laboratory Systems* 14:155–64. [197]

Delbeke, L. 1978. Enkele analyses op voorkeuroorden voor gezinssamenstellingen. Cengtarum voor Mathematische Psychologie en Psychologische Methodologie, Katholike Universiteit Leuven. [182, 183]

de Leeuw, J. 1982. Generalized eigenvalues problems with positive semidefinite matrices. *Psychometrika* 47:87–93. [69]

de Leeuw, J. 1983. On the prehistory of correspondence analysis. *Statistica Neerlandica* 37:161–4. [93, 109]

de Leeuw, J. and S. Pruzansky. 1978. A new computational method to fit the weighted Euclidean distance model. *Psychometrika* 48:479–90. [106]

de Leeuw, J., J. van Rijckevorsel, and M. van der Wouden. 1981. Non-linear principal component analysis with B-splines. *Methods of Operations Research* 33:379–93. [174]

De Moor, B. L. and G. H. Golub. 1991. The restricted singular value decomposition: Properties and applications. *SIAM Journal on Matrix Analysis and Applications* 12:401–25. [66]

De Soete, G. and J. D. Carroll. 1983. A maximum likelihood estimation for fitting the wandering vector model. *Psychometrika* 48:553–66. [126]

Devlin, S. J., R. Gnanadesikan, and J. R. Kettenring. 1981. Robust estimation and outlier detection with correlation coefficients. *Biometrika* 62:531–45. [167]

Eastment, H. T. and W. J. Krzanowsky. 1982. Cross-validatory choice of the number of components from principal component analysis. *Technometrics* 24:73–7. [157]

Eckart, C. and G. Young. 1936. The application of one matrix by another of lower rank indexlow rank. *Psychometrika* 1:211–8. [62]

Efron, B. 1979. bootstrap methods: Another look at the Jackknife. *Annals of Statistics* 7:1–26. [128, 157, 202]

Efron, B. and R. J. Tibshirani. 1993. *An introduction to the bootstrap*. Boca Raton, FL: Chapman and Hall/CRC Press. [157]

Escoufier, Y. 1987. The duality diagram: A means for better practical applications. In *Development in numerical ecology*, ed. P. Legendre and L. Legendre, 139–56. Berlin: Springer. [75]

Ewerbring, L. M. and F. T. Luk. 1989. Canonical correlations and generalized SVD: Applications and new algorithms. *Journal of Computational and Applied Mathematics* 27:37–52. [66]

Faddeev, D. K. and V. N. Faddeeva. 1963. *Computational methods of linear algebra*. San Francisco: Freeman. [87]

Fernando, K. V. and S. J. Hammarling. 1988. A product induced singular value decomposition for two matrices and balanced realization. In *Linear algebra in signals, systems and control*, ed. B. N. Datta, 128–40. Philadelphia: SIAM. [69]

Fisher, R. A. 1936. The use of multiple measurements in taxonomic problems. *Annals of Human Genetics* 6:179–88. [102, 103]

Fisher, R. A. 1940. The precision of discriminant functions. *Annals of Human Genetics* 10:22–9. [14, 15, 109]

Flury, B. 1988. *Common principal components and related multivariate models*. New York: Wiley. [107]

Fortier, J. J. 1966. Simultaneous linear prediction. *Psychometrika* 31:369–81. [97]

Fowler, J. E. 2009. Compressive-projection principal component analysis. *IEEE Transactions on Image Processing* 8:2230–42. [95]

Friedman, J. H. 1987. Exploratory projection pursuit. *Journal of the American Statistical Association* 82:249–66. [95]

Friedman, J. H. and J. W. Tukey. 1974. A projection pursuit algorithm for exploratory data analysis. *IEEE Transactions on Computers* 23:881–90. [95]

Fujikoshi, Y. and K. Satoh. 1996. Estimation and model selection in an extended growth curve model. *Hiroshima Mathematical Journal* 26:635–47. [135]

Fukunaga, K. 1990. *Introduction to statistical pattern recognition (second edition).* London: Academic Press. [62]

Gabriel, K. R. 1971. The biplot graphic display of matrices with application to principal component analysis. *Biometrika* 58:453–67. [174]

Gabriel, K. R. 1978. Least squares approximation of matrices by additive and multiplicative models. *Journal of the Royal Statistical Society, Series B* 40:186–96. [80, 83]

Gabriel, K. R. and S. Zamir. 1979. Lower rank approximation of matrices by least squares with any choice of weights. *Technometrics* 21:489–98. [148]

Gifi, A. 1990. *Nonlinear multivariate analysis*. Chichester: Wiley. [121, 174]

Gleason, T. C. and R. Staelin. 1973. Improving the metric quality of questionnaire data. *Psychometrika* 38:393–410. [202]

Gold, E. M. 1973. Metric unfolding: Data requirement for unique solution and clarification of Schönemann's algorithm. *Psychometrika* 37:555–70. [105]

Goldstein, H. 1987. *Multilevel models in educational and social research*. London: Griffin. [144]

Gollob, H. F. 1968. A statistical model which combines features of factor analytic and analysis of variance techniques. *Psychometrika* 37:555–70. [83]

Golub, G. H. 1973. Some modified eigenvalue problems. *SIAM Journal: Review* 15:318–35. [83]

Golub, G. H. and W. Kahan. 1965. Calculating the singular values and pseudo-inverse of a matrix. *SIAM Journal on Numerical Analysis, Series B* 2:205–24. [64]

Golub, G. H. and V. Pereyra. 1973. The differentiation of pseudo-inverses and nonlinear least squares problems whose variable separate. *SIAM Journal on Numerical Analysis* 10:413–32. [160]

Golub, G. H. and C. Reinsch 1970. Singular value decomposition and least squares solutions. *Numerisch Mathematik* 14:134–51. [64]

Golub, G. H. and C F. van Loan. 1996. *Matrix computation (third edition).* Baltimore: The Johns Hopkins University Press. [35]

Golub, G. H., A. Hoffman, and G. W. Stewart. 1987. A generalization of the Eckart–Young–Mirsky matrix approximation theory. *Linear Algebra and Its Applications* 88/89:317–27. [62]

Good, P. I. 2005. *Permutation, parametric, and bootstrap tests of hypotheses (third edition).* New York: Springer. [156]

Goodman, L. A. and W. H. Kruskal. 1954. Measures of association for cross classifications. Part I. *Journal of the American Statistical Association* 49:732-64. [118]

Gorsuch, R. L. and J. Nelson. 1981. CNG scree test: An objective procedure for determining the number of factors. Paper presented at the Annual Meeting of the Society for Multivariate Experimental Psychology. [155]

Green, B. F. and J. C. Gower. 1979. A problem with congruence. Paper presented at the Annual Meeting of the Psychometric Society, Monterey, CA. [151]

Green, P. E. and J. D. Carroll. 1976. *Mathematical tools for applied multivariate analysis*. New York: Academic Press. [21]

Greenacre, M. J. 1984. *Theory and applications of correspondence analysis*. London: Academic Press. [12, 109, 123]

Greenacre, M. J. 1993. *Correspondence analysis in practice*. London: Academic Press. [9, 11, 19, 179, 192]

Grizzle, J. E. and D. M. Allen. 1969. Analysis of growth and dose response curves. *Biometrics* 25:357–81. [131]

Grob, J. and G. Trenkler. 1998. On the product of oblique projectors. *Linear and Multilinear Algebra* 30:1–13. [39]

Guttman, L. 1941. The quantification of a class of attributes: A theory and method of scale construction. In *The prediction of personal adjustment*, ed. P. Horst, 321–48. New York: Social Science Research Council. [109]

Guttman, L. 1944. General theory and methods for matric factoring. *Psychometrika* 9:1–16. [70, 140]

Guttman, L. 1952. Multiple group methods for common-factor analysis: Their basis, computation, and interpretation. *Psychometrika* 17:209–22. [140]

Guttman, L. 1953. Image theory for the structure of quantitative variables. *Psychometrika* 18:277–96. [75, 152]

Guttman, L. 1957. A necessary and sufficient formula for matric factoring. *Psychometrika* 22:79–81. [70, 140]

Haberman, S. J. 1979. *Analysis of qualitative data, Vol. 2*. New York: Academic Press. [112, 113]

Hampel, F. R., E. M. Ronchetti, P. J. Rousseeuw, and W. J. Stahel. 1986. *Robust statistics: T he approach based on influence functions*. New York: Wiley. [167]

Harville, D. A. 1997. *Matrix algebra from a statistician's perspective*. New York: Springer. [21, 26, 31]

Hastie, T., R. J. Tibshirani, and J. H. Friedman. 2001. *The elements of statistical learning*. New York: Springer. [161]

Hawkins, D. M., L. Liu, and S. S. Young. 2001. Robust singular value decomposition. Technical Report 122, National Institute of Statistical Sciences (http://www.niss.org/sites/default /files/pdfs/ technicalreports/tr122.pdf). [168]

Hayashi, C. 1952. On the prediction of phenomena from qualitative data and the quantification of qualitative data from the mathematico-statistical point of view. *Annals of the Institute of Statistical Mathematics* 3:69–98. [109]

Heiser, W. J. 1987. Correspondence analysis with least absolute residuals. *Computational Statistics and Data Analysis* 5:337–56. [169]

Heiser, W. J., and J. de Leeuw. 1981. Multidimensional mapping of preference data. *Mathematiqué et Sciences Humaines* 19:39–96. [126]

Hirschfeld, H. O. 1935. A connection between correlation and contingency. *Proceedings of the Cambridge Philosophical Society* 31:520–4. [109]

Hoerl, A. E. and R. W. Kennard. 1970. Ridge regression: Biased estimation for nonorthogonal problems. *Technometrics* 12:55–67. [56]

Hong, S., S. Mitchell, and R. A. Harshman. 2006. Bootstrap scree tests: A Monte-Carlo simulation and applications to published data. *British Journal of Mathematical and Statistical Psychology* 59:35–57. [155]

Horn, J. I. 1965. A rationale and test for the number of factors in factor analysis. *Psychometrika* 30:179–85. [156]

Horn, R. A. and C. R. Johnson. 1990. *Matrix analysis*. Cambridge: Cambridge University Press. [62]

Horst, P. 1935. Measuring complex attitudes. *Journal of Social Psychology* 6:369–74. [109, 129]

Hotelling, H. 1936. Relations between two sets of variables. *Biometrika* 28:321–77. [98]

Householder, A. S. 1958. Unitary triangularization of a nonsymmetric matrix. *Journal of the Association of Computing Machinery* 5:339–42. [35]

Hox, J. J. 1995. *Applied multilevel analysis*. Amsterdam: TT-Publikaties. [144]

Huang, J. Z., H. Shen, and A. Buja. 2009. The analysis of two-way functional data using two-way regularized singular value decompositions. *Journal of the American Statistical Association* 104:1609–20. [95]

Hunter, M. A. and Y. Takane. 1998. CPCA: A program for principal component analysis with external information on subjects and variables. *Behavior Research Methods, Instruments, and Computers* 30:506–16. [1, 203]

Hunter, M. A. and Y. Takane. 2002. Constrained principal component analysis: Various applications. *Journal of Educational and Behavioral Statistics* 27:41–81. [1, 203]

Hwang, H. and Y. Takane. 2002a. Generalized constrained multiple correspondence analysis, *Psychometrika* 67:215–28. [124]

Hwang, H. and Y. Takane, Y. 2002b. Structural equation modeling by extended redundancy analysis. In *Measurement and multivariate analysis*, ed. S. Nishisato, Y. Baba, H. Bozdogan, and K. Kanefuji, 115–24. Tokyo: Springer. [195]

Hwang, H. and Y. Takane. 2004. Generalized structured component analysis. *Psychometrika* 69:81–99. [198, 199]

Hwang, H. and Y. Takane. (2010). Nonlinear generalized structural component analysis. *Behaviormetrika* 37:1–14. [173, 199]

Hwang, H. and Y. Takane. (forthcoming). *Generalized structured component analysis*. Boca Raton, FL: Chapman and Hall/CRC Press. [199]

Hwang, H., W. S. DeSarbo, and Y. Takane. 2007a. Fuzzy clusterwise generalized structural component analysis. *Psychometrika*. 72:181-98. [199]

Hwang, H., M.-H. Ho, and J. Lee. 2010. Generalized structured component analysis with latent interaction. *Psychometrika* 75:228–42. [199]

Hwang, H., Y. Takane, and N. Malhotra. 2007b. Multilevel generalized structured component analysis. *Behaviormetrika* 34:95–109. [199]

Hwang, H., K. Jung, Y. Takane, and T. S. Woodward. 2012. Functional multiple-set canonical correlation analysis. *Psychometrika* 77:48–64. [123]

Hwang, H., K. Jung, Y. Takane, and T. S. Woodward. 2013. A unified approach to multiple-set canonical correlation analysis and principal components analysis. *British Journal of Mathematical and Statistical Psychology* 66:308–21. [12]

Hyvärinen, A. and E. Oja. 2000. Independent component analysis: Algorithms and applications. *Neural Networks* 13:411-30. [95]

Hyvärinen, A., J. Karhunen, and E. Oja. (2001). *Independent component analysis*. New York: Wiley. [95]

Ihara, M. and Y. Kano. 1986. A new estimator of the uniqueness in factor analysis. *Psychometrika* 51:563–6. [76]

Izenman, A. J. 1975. Reduced-rank regression for the multivariate linear model. *Journal of Multivariate Analysis* 5:248–64. [101]

Jennrich, R. I. and N. T. Trendafilov. 2005. Independent Component analysis as a rotation method: A very different solution to Thurstone's box problem. *British Journal of Mathematical and Statistical Psychology* 58:199–208. [96]

Johnston, J. 1984. *Econometric methods (third edition)*. New York: McGraw Hill. [50, 85]

Johnstone, I. M. and A. Y. Lu. 2009. On consistency and sparsity for principal components analysis in high dimensions. *Journal of the American Statistical Association* 104:682–93. [95]

Jolliffe, I. T. 2002. *Principal component analysis (second edition)*. New York: Springer. [93, 157]

Jolliffe, I. T. and M. Uddin. 2000. The simplified component technique: An alternative to rotated principal components. *Journal of Computational and Graphical Statistics* 9:689–710. [95]

Jolliffe, I. T., N. T. Trendafilov, and M. Uddin. 2003. A modified principal component technique based on lasso. *Journal of Computational and Graphical Statistics* 12:531–47. [95]

Jordan, C. (1874). Mémoire sur les formes bilineares. *Journal de Mathématiques Pures et Appliquées* 19:35–54. [62]

Jöreskog, K. G. 1967. Some contributions to maximum likelihood factor analysis. *Psychometrika* 32:443–82. [75]

Jöreskog, K. G. 1970. A general method for analysis of covariance structures. *Biometrika* 57:239–51. [201]

Jung, K., Y. Takane, H. Hwang, and T. S. Woodward. 2012. Dynamic GSCA (Generalized Structured Component Analysis) with applications to the analysis of effective connectivity in functional neuroimaging data. *Psychometrika* 77:827-48. [199]

Kaiser, H. F. 1976. Image and anti-image covariance matrices from a correlation matrix that may be singular. *Psychometrika* 41:295–300. [153]

Kalman, R. E. 1976. Algebraic aspects of the generalized inverse of a rectangular matrix. In *Generalized inverse and applications*, ed. M. Z. Nashed, 267–80. New York: Academic Press. [42]

Kempler, B. 1971. Stimulus correlates of area judgments: A psychological developmental study. *Developmental Psychology* 4:158–63. [189]

Khatri, C. G. 1966. A note on a MANOVA model applied to problems in growth curves. *Annals of the Institute of Statistical Mathematics* 18:75–86. [48, 83, 131, 142]

Khatri, C. G. 1990. Some properties of BLUE in a linear model and canonical correlations associated with linear transformations. *Journal of Multivariate Analysis* 34:211–26. [142]

Kiers, H. A. L. 1990. Majorization as a tool for optimizing a class of matrix functions. *Psychometrika* 55:417–28. [151, 191]

Kiers, H. A. L. 1991. Simple structure in component analysis techniques for mixtures of qualitative and quantitative variables. *Psychometrika* 56:197–212. [72]

Kiers, H. A. L. 1997. Weighted least squares fitting using ordinary least squares algorithm. *Psychometrika* 62:251–66. [148]

Kiers, H. A. L. and Y. Takane. 1993. Constrained DEDICOM. *Psychometrika.* 58:339–55. [181]

Kiers, H. A. L. and J. M. F. ten Berge. 1992. Minimization of a class of matrix trace functions by means of refined majorization. *Psychometrika* 57:371–82. [151, 191]

Kollo, T. and H. Neudecker. 1994. Asymptotics of eigenvalues and unit-length eigenvectors of sample variance and correlation matrices. *Journal of Multivariate Analysis* 47:283–300. [160]

Kollo, T. and H. Neudecker. 1997. Asymptotics of Pearson-Hotelling principal-component vectors of sample variance and correlation matrices. *Behaviormetrika* 24:51–69. [160]

Konishi, S. 1978. Asymptotic expansions for the distribution of statistics based on a correlation matrix. *Canadian Journal of Statistics* 6:49–56. [160]

Konishi, S. 1979. Asymptotic expansions for the distribution of statistics based on the sample correlation matrix in principal component analysis. *Hiroshima Mathematical Journal* 9:647–700. [160]

Koschat, M. A. and D. F. Swayne. 1991. A weighted procrustes criterion. *Psychometrika* 56:229–39. [151]

Kristof, W. 1970. A theorem on the trace of certain matrix products and some applications. *Journal of Mathematical Psychology* 7:515–30. [58]

Kruskal, J. B. 1964a. Multidimensional scaling by optimizing goodness of fit to a nonmetric hypothesis. *Psychometrika* 29:1–29. [104, 173]

Kruskal, J. B. 1964b. Nonmetric multidimensional scaling: A numerical method. *Psychometrika* 29:115–29. [104, 172, 173]

Kruskal, J. B. and R. N. Shepard. 1974. A nonmetric variety of linear factor analysis. *Psychometrika* 41:471–503. [173]

Lambert, Z. V., A. R. Wildt, and R. M. Durand. 1988. Redundancy analysis: An alternative to canonical correlation and multivariate multiple regression in exploring interset associations. *Psychological Bulletin* 104:282–9. [96]

Lauro, N. and L. D'Ambra. 1984. L'analyse non symétrique des correspondances. In *Data analysis and informatics III*, eds. E. Diday et al., 433–46, Amsterdam: Elsevier. [117]

Lawson, C. L. and R. J. Hanson. 1974. *Solving least squares problems*. Englewood Cliffs: Prentice Hall. [23, 36]

Lebart, L., A. Morineau, and K. M. Warwick. 1984. *Multivariate descriptive statistical analysis*. New York: Wiley. [87, 123]

Lee, M., H. Shen, J. Z. Huang, and J. S. Marron. 2010. Biclustering via sparse singular value decomposition. *Biometrics* 66:1087–95. [95]

Legendre, P. and L. Legendre 1998. *Numerical ecology*. Amsterdam: Elsevier. [156]

Light, R. J. and B. H. Margolin. 1971. An analysis of variance for categorical data. *Journal of the American Statistical Association* 66:534–44. [117, 118]

Linting, M., J. J. Meulman, P. J. F. Groenen, and J. J. Van der Kooij. 2007a. Nonlinear principal components analysis: Introduction and application. *Psychological Methods* 12:336–58. [171]

Linting, M., J. J. Meulman, P. J. F. Groenen, and J. J. Van der Kooij. 2007b. Stability of nonlinear principal components analysis: An empirical study using the balanced bootstrap. *Psychological Methods* 12: 359–79. [160]

Liu, L., D. M. Hawkins, S. Ghosh, and S. S. Young. 2003. Robust singular value decomposition analysis of microarray data. *National Academy of Science* 100:13167–72. [169]

Magnus, J. R. and H. Neudecker. 1988. *Matrix differential calculus*. Chichester: Wiley. [31]

Manton, J. H., R. Mahony, and Y. Hua. 2003. The geometry of weighted low-rank approximations. *IEEE Transactions on Signal Processing* 51:500–514. [148]

Markovsky, I., D. Sima, and S. van Huffel. 2010. Total least squares methods. *Advanced Review* 2:212–7. [148]

Marsaglia, G. and G. P. A. Styan. 1974. Equalities and inequalities for ranks of a matrices. *Linear and Multilinear algebra* 2:269–92. [24]

Maung, K. 1941a. Measurement of association in a contingency table with special reference to the pigmentation of hair and eye colors of Scottish school children. *Annals of Human Genetics* 11:189–223. [109]

Maung, K. 1941b. Discriminant analysis of Tocher's eye color data. *Annals of Human Genetics* 11:64–76. [109]

McDonald, R. P. 1978. A simple comprehensive model for the analysis of covariance structures. *British Journal of Mathematical and Statistical Psychology* 31:59–72. [87]

McKeen, J. W. 2004. Robust analysis of linear model. *Statistical Science* 19:562–70. [167]

Meredith, W. 1964. Rotation to achieve factorial invariance. *Psychometrika* 29:187–206. [121]

Meredith, W. and R. E. Millsap. 1985. On component analysis. *Psychometrika* 50:494–507. [75]

Merola, G. M. and B. Abraham. 2001. Dimensionality reduction approach to multivariate prediction. *Canadian Journal of Statistics* 29:191–200. [197]

Metzak, P., E. Feredoes, Y. Takane, L. Wang, S. Weinstein, T. Cairo, E. T. C. Ngan, and T. S. Woodward. 2011. Constrained principal component analysis reveals functionally connected load-dependent networks involved in multiple stages of working memory. *Human Brain Mapping* 32:856–71. [1]

Meulman, J. J. 1982. *Homogeneity analysis of incomplete data*. Leiden: DSWO Press. [165]

Mezzich, J. E. 1978. Evaluating clustering methods for psychiatric diagnosis. *Biological Psychiatry* 13:265–81. [2]

Mezzich, J. E. and H. Solomon, (1980). *Taxonomy and behavioral science: Comparative performance of grouping methods*. New York: Academic Press. [2]

Micceri, T. 1989. The unicorn, the normal curve, and other improbable creatures. *Psychological bulletin* 105:156–66. [201]

Mirsky, L. 1960. Symmetric gage functions and unitarily invariant norms. *Quarterly Journal of Mathematics* 11:50–9. [24, 62]

Mitra, S. K. 1968. A new class of g-inverse of square matrices. *Sankhyā, Series A* 30:323–330. [142]

Mitra, S. K. and C. R. Rao. 1974. Projections under seminorms and generalized inverse of matrices. *Linear Algebra and Its Applications* 9:155–67. [30]

Mooijaart, A. and J. J. F. Commandeur. 1990. A general solution of the weighted orthogonal procrustes problems. *Psychometrika* 55:657–63. [151]

Mulaik, S. A. 1972. *The foundation of factor analysis*. New York: McGraw-Hill. [95]

Nishisato, S. 1980. *Analysis of categorical data: Dual scaling and its applications*. Toronto: University of Toronto Press. [12, 83, 109, 123]

Nishisato, S. 1984. Dual scaling by reciprocal medians. *Proceedings of the 32nd Scientific Conference of the Italian Statistical Society*, 141–7. Rome, Italy: Societa Italiana di Statistica. [169]

Nishisato, S. 1987. Robust techniques for quantifying categorical data. In *Foundations of statistical inference*, ed. I. B. MacNeil and G. J. Umphrey, 209–17. Dordrecht, Holland: Reidel. [169]

Nishisato, S. 1988. Dual scaling: Its development and comparisons with other quantification methods. In *Deutsche Gesselleschaft für Operation Research Proceedings*, ed. H. D. Pressman et al., 376–89. Berlin: Springer. [83, 169]

Nishisato, S. and D. R. Lawrence 1989. Dual scaling of multiway data matrices: Several variants. In *Multiway data analysis*, ed. R. Coppi and S. Bolasco, 317–26. Amsterdam: North Holland. [83, 86]

Ogasawara, H. 2000. Standard errors of principal component loadings for unstandardized and standardized variables. *British Journal of Mathematical and Statistical Psychology* 53:155-74. [160]

Ogasawara, H. 2002. Concise formulas for the standard errors of component loading estimates. *Psychometrika* 67:289–97. [160]

Ogasawara, H. 2004. Asymptotic biases of the unrotated/rotated solutions in principal component analysis. *British Journal of Mathematical and Statistical Psychology* 57:353–76. [160]

Ogasawara, H. 2006. Higher-order asymptotic standard error and asymptotic expansion in principal component analysis. *Communications in Statistics - Simulation and Computation* 35:201–23. [160]

Okamoto, M. 1972. Four techniques of principal component analysis. *Journal of Japanese Statistical Society* 2:63–9. [83]

Okamoto, M. and M. Kanazawa. 1968. Minimization of eigenvalues of a matrix and optimality of principal components. *Annals of Mathematical Statistics* 39:859–63. [63]

Overall, J. E. and D. R. Gorham. 1962. The brief psychiatric rating scale. *Psychological Report* 10:799–812. [1]

Paige, C. C. 1985. The general linear model and the generalized singular value decomposition. *Linear Algebra and Its Applications* 70:269–84. [66]

Paige, C. C. 1986. Computing the generalized singular value decomposition. *SIAM Journal on Scientific Computing* 7:1126–46. [66]

Paige, C. C. and M. A. Saunders. 1981. Towards a generalized singular value decomposition. *SIAM Journal on Numerical Analysis* 18:398–405. [66]

Pearson, K. 1900. On the criterion that a given system of deviations from the probable in the case of a correlated system of variables is such that it can be reasonably supposed to have arisen from random sampling. *Philosophy Magazine* 50:157–72. [15, 93, 108]

Pearson, K. 1901. On lines and planes of closest fit to systems of points in space. *Philosophical Magazine, Series B* 2:559–72. [93]

Pearson, K. 1904. Mathematical contributions to the theory of evolution XIII: On the theory of contingency and its relation to association and normal correlation. *Drapers' Company Research Memoirs, Biometric Series 1*. [93, 109]

Pearson, K. 1906. On certain points connected with scale order in the case of a correlation of two characters which for some arrangement give a linear regression line. *Biometrika* 5:176–8. [93, 109]

Penrose, R. 1955. A generalized inverse of matrices. *Proceedings of Cambridge Philosophical Society* 51:406–13. [41]

Potthoff, R. F. and S. N. Roy. 1964. A generalized multivariate analysis of variance model useful especially for growth curve problems. *Biometrika* 51:313–26. [76, 83, 130, 132]

Puntanen, S. and G. P. H. Styan. 1989. The equality of the ordinary least squares estimator and the best linear unbiased estimator. *American Statistician* 43:153–64. [50]

Puntanen, S., G. P. H. Styan, and J. Isotalo. 2011. *Matrix tricks for linear statistical models: Our personal top twenty*. Berlin: Springer. [49]

Ramsay, J. O. 1982. When data are functions. *Psychometrika* 47:378–96. [74, 174]

Ramsay, J. O. 1988. Monotone regression splines in action. *Statistical Science* 3:425–61. [174]

Ramsay, J. O. and C. J. Dalzell. 1991. Some tools for functional data analysis. *Journal of the Royal Statistical Society, Series B* 53:539–72. [74]

Ramsay, J. O. and B. W. Silverman. 2005. *Functional data analysis (second edition)*. New York: Springer. [74, 123, 174]

Rao, C. R. 1962. A note on a generalized inverse of a matrix with applications to problems in mathematical statistics. *Journal of the Royal Statistical Society, Series B* 24:152–158. [49]

Rao, C. R. 1964. The use and interpretation of principal component analysis in applied research. *Sankhyā A* 26:329–58. [75, 83, 97, 202]

Rao, C. R. 1965. The theory of least squares when parameters are stochastic and its application to the analysis of growth curves. *Biometrika* 52:447–58. [83]

Rao, C. R. 1966. Covariance adjustment and related problems in multivariate analysis. In *Multivariate analysis, Vol. II*, ed. P. R. Krishnaiah, 97–103. New York: Academic Press. [83]

Rao, C. R. 1967. Least squares theory using an estimated dispersion matrix and its application to measurement of signals. In *Proceedings of the Fifth Berkeley Symposium on Mathematical Statistics and Probability*, Vol. 1, ed. L. M. Le Cam and J. Neyman, 335–72. Berkeley: University of California Press. [49, 83, 132]

Rao, C. R. 1973. *Linear statistical inference and its applications (second edition)*. New York: Wiley. [21]

Rao, C. R. 1979. Separation theorem for singular values of matrices and their applications in multivariate analysis. *Journal of Multivariate Analysis* 9:362–377. [63]

Rao, C. R. 1980. Matrix approximations and reduction of dimensionality in multivariate statistical analysis. In *Multivariate analysis V*, ed. P. R. Krishnaiah, 3–22. Amsterdam: North Holland. [23, 24, 29, 58, 63, 80, 83]

Rao, C. R. 1985. Tests for dimensionality and interactions of mean vectors under general and reducible covariance structures. *Journal of Multivariate Analysis* 16:173–84. [134]

Rao, C. R. and S. K. Mitra. 1971. *Generalized inverse of matrices and its applications*. New York: Wiley. [21,38, 39, 44, 66]

Rao, C. R. and H. Yanai. 1979. General definition and decomposition of projectors and some applications to statistical problems. *Journal of Statistical Planning and Inference* 3:1–17. [53, 84]

Reinsel, G. C. and R. P. Velu. 1998. *Multivariate reduced-rank regression: Theory and applications*. New York: Springer. [132, 139]

Reinsel, G. C. and R. P. Velu. 2003. Reduced-rank growth curve models. *Journal of Statistical Planning and Inference* 114:107–29. [84]

Revankar, N. S. 1974. Some finite sample results in the context of two seemingly unrelated regression equations. *Journal of the American Statistical Association* 69:187–90. [140]

Romano, J. P. 1990. On the behavior of randomization tests without a group invariance assumption. *Journal of the American Statistical Association* 85:686–92. [156]

Rousseeuw, P. J. and A. M. Leroy. 1987. *Robust regression and outlier detection*. New York: Wiley. [167]

Sachs, J. and H. Kong. 1994. Robust dual scaling with Tukey's biweight. *Applied Psychological Measurement* 18:301–9. [169]

Sands, R. and F. W. Young. 1980. Component models for three-way data: An alternating least squares algorithm with optimal scaling features. *Psychometrika* 45:39–67. [107]

Schmidt, E. (1907). Zur Theorie der linearen und nichtlinearen Integralgleichungen. I. Teil: Entwickelung willkürlichen Funktionen nach Vorgeschriebener. *Mathematische Annalen* 63:433–76. [62]

Schölkopf, B., C. J. C. Burges, and A. J. Smolla. 1997. *Advances in kernel method*. Cambridge, MA: MIT Press. [98]

Schönemann, P. H. 1966. A generalized solution of the orthogonal procrustes problems. *Psychometrika* 31:1–10. [150]

Schönemann, P. H. 1970. On metric multidimensional unfolding. *Psychometrika* 35:349–66. [105]

Schönemann, P. H. and J. H. Steiger. 1976. Regression component analysis. *British Journal of Mathematical and Statistical Psychology* 29:175–89. [202]

Schuermans, M., I. Markovsky, P. D. Wentzell, and S. van Huffel. 2005. On the equivalence between total least squares and maximum likelihood PCA. *Analytica Chimica Acta* 544:254–67. [148]

Searle, S. R. 1966. *Matrix algebra for the biological sciences*. New York: Wiley. [21]

Searle, S. R. 1971. *Linear models*. New York: Wiley. [49]

Seber, G. F. A. 1984. *Multivariate observations*. New York: Wiley. [53]

Segi, M. 1979. *Age-adjusted death rates for cancer for selected sites (A-classification) in 51 countries in 1974 (1979)*. Nagoya, Japan: Segi Institute of Cancer Epidemiology. [6, 7]

Shen, H. and J. Z. Huang. 2008. Sparse principal component analysis via regularized low rank matrix approximation. *Journal of Multivariate Analysis* 99:1015–34. [95]

Shepard, R. N. 1962. The analysis of proximities: Multidimensional scaling with an unknown distance function, I and II. *Psychometrika* 27:125–40, 219–46. [104, 173]

Shibayama, T. 1995. A linear composite method for test scores with missing values. *Memoirs of the Faculty of Education, Niigata University* 36:445–55. [163]

Shimizu, S., A. Hyvärinen, H. O. Hoyer, and Y. Kano. 2006. Finding a causal ordering via independent component analysis. *Computational Statistics and Data Analysis* 50: 3278–93. [95]

Siotani, M., T. Hayakawa, and Y. Fujikoshi. 1985. *Modern multivariate statistical analysis: A graduate course handbook*. Columbus, OH: American Sciences Press. [131]

Slater, P. 1960. The analysis of personal preferences. *British Journal of Mathematical and Statistical Psychology* 13:119–35. [125]

Stanek, E. J. III and G. G. Koch. 1985. The equivalence of parameter estimates from growth curve models and seemingly unrelated regression models. *American Statistician* 39:149–52. [139]

Stewart, D. and W. Love. 1968. A general canonical correlation index. *Psychological Bulletin* 70:160–3. [97]

Stewart, G. W. 1993. On the early history of the singular value decomposition. *SIAM Review* 35:551–66. [62]

Stewart, G. W. and J. Sun. 1990. *Matrix perturbation theory*. Boston: Academic Press. [62, 63, 160]

Stone, M. 1976. An asymptotic equivalence of model selection by cross validation and Akaike's criterion. *Journal of the Royal Statistical Society, Series B* 38:44–7. [157]

Takane, Y. 1980a. Analysis of categorizing behavior by a quantification method. *Behaviormetrika* 8:57–67. [124]

Takane, Y. 1980b. Maximum likelihood estimation in the generalized case of Thurstone's model of comparative judgment. *Japanese Psychological Research* 22:188–96. [126]

Takane, Y. 1984. The weighted additive model. In *Research methods for multimode data analysis in the behavioral sciences*, ed. H. G. Law et al., 470–516. New York: Praeger. [188, 190]

Takane, Y. 1987. Analysis of covariance structures and probabilistic binary choice data. *Communication and Cognition* 20:45–61. (A slightly expanded version is also in *New developments in psychological choice modeling*, (eds.) G. De Soete et al., 139–160, Amsterdam: North Holland, 1989.) [202]

Takane, Y. 1994. A review of applications of AIC in psychometrics. In *Proceedings of the first US/Japan conference on the frontiers of statistical modeling: An informational approach*, ed. H. Bozdogan, 379–403, Dortrecht: Kluver. [202]

Takane, Y. 2003. Relationships among various kinds of eigenvalue and singular value decompositions. In *New developments in psychometrics*, ed. H. Yanai et al., 45–56. Tokyo: Springer. [70]

Takane, Y. 2004. Matrices with special reference to applications in psychometrics. *Linear Algebra and Its Applications* 388C:341–61. [104]

Takane, Y. 2005. Optimal scaling. In *Encyclopedia of Statistics for Behavioral Sciences*, ed. B. S. Everitt and D. C. Howell, 1479-82. Chichester: Wiley. [170]

Takane, Y. 2007a. Applications of multidimensional scaling in psychometrics. In *Handbook of Statistics (Vol. 26): Psychometrics*, ed. C. R. Rao and S. Sinharay, 359-400. Amsterdam: Elsevier. [104]

Takane, Y. 2007b. More on regularization and (generalized) ridge operators. In *New Trends in Psychometrics*, ed. K. Shigemasu, et al., 443–452. Tokyo: University Academic Press. [57, 90]

Takane, Y. and M. A. Hunter. 2001. Constrained principal component analysis: A comprehensive theory. *Applicable Algebra in Engineering, Communication, and Computing* 12:391–419. [1, 71, 81, 82, 130]

Takane, Y. and M. A. Hunter. 2002. Dimension reduction in hierarchical linear models. In *Measurement and multivariate analysis*, ed. S. Nishisato et al., 145–154. Tokyo: Springer. [144]

Takane, Y. and M. A. Hunter. 2011. A new family of constrained principal component analysis (CPCA). *Linear Algebra and Its Applications* 434:2539–55. [1, 48, 71, 88]

Takane, Y. and H. Hwang. 2002. Generalized constrained canonical correlation analysis. *Multivariate Behavioral Research* 37:163–95. [100, 156]

Takane, Y. and H. Hwang. 2005. An extended redundancy analysis and its applications to two practical examples. *Computational Statistics and Data Analysis* 49:785–808. [9, 195]

Takane, Y. and H. Hwang. 2006. Regularized multiple correspondence analysis. In *Multiple correspondence analysis and related methods*, ed. J. Blasius and M. J. Greenacre, 259–79. London: Chapman and Hall. [124]

Takane, Y. and H. Hwang. 2007. Regularized linear and kernel redundancy analysis. *Computational Statistics and Data Analysis* 52:394–405. [97, 98]

Takane, Y. and S. Jung. 2008. Regularized partial and/or constrained redundancy analysis. *Psychometrika* 73:671–90. [97]

Takane, Y. and S. Jung. 2009a. Regularized nonsymmetric correspondence analysis. *Computational Statistics and Data Analysis* 53:3159–70. [119]

Takane, Y. and S. Jung. 2009b. Tests of ignoring and eliminating in nonsymmetric correspondence analysis. *Advances in Data Analysis and Classification* 3:315–40. [85, 116, 119, 120]

Takane, Y. and Y. Oshima-Takane. 2002. Nonlinear generalized canonical correlation analysis by neural network models. In *Measurement and multivariate analysis*, ed. S. Nishisato, Y. Baba, H. Bozdogan, and K. Kanefuji, 183–90. Tokyo: Springer. [120, 123]

Takane, Y. and Y. Oshima-Takane. 2003. Relationships between two methods of dealing with missing data in principal component analysis, *Behaviormetrika* 30:145–54. [163, 165, 174]

Takane, Y. and T. Shibayama. 1991. Principal component analysis with external information on both subjects and variables. *Psychometrika* 56:97–120. [1, 71, 128, 202]

Takane, Y. and H. Yanai. 1999. On oblique projectors. *Linear Algebra and Its Applications* 289:297–310. [39, 55, 56, 85]

Takane, Y. and H. Yanai. 2005. On the Wedderburn–Guttman theorem. *Linear Algebra and Its Applications* 410:267–78. [85, 141]

Takane, Y. and H. Yanai. 2007. Alternative characterizations of the extended Wedderburn–Guttman theorem. *Linear Algebra and Its Applications* 422:701–11. [57, 70, 85, 141]

Takane, Y. and H. Yanai. 2008. On ridge operators. *Linear Algebra and Its Applications* 428:1778–90. [90]

Takane, Y. and H. Yanai. 2009. On the necessary and sufficient condition for the extended Wedderburn–Guttman theorem. *Linear Algebra and Its Applications* 430:2890–95. [85, 141]

Takane, Y. and Z. Zhang. 2009. Algorithms for DEDICOM: Acceleration, deceleration, or neither? *Journal of Chemometrics* 23:364–70. [150]

Takane, Y. and L. Zhou. 2011. Exploratory multilevel redundancy analysis. IPS042.04 (http://www.isi2011.ie/content/access-congress-proceedings.html). [144]

Takane, Y. and L. Zhou. 2012. On two expressions of the MLE for a special case of the extended growth curve models. *Linear Algebra and Its Applications* 436:2567–77. [48, 137]

Takane, Y. and L. Zhou. in press. Anatomy of Pearson's chi-square statistic in three-way contingency tables. In R. Millsap (Ed.), Proceeding of the IMPS-2012. New York: Springer. [85]

Takane, Y., H. Hwang, and H. Abdi. 2008. Regularized multiple-set canonical correlation analysis. *Psychometrika* 73:753–75. [101, 120, 122, 124]

Takane, Y., K. Jung, and H. Hwang. 2010. An acceleration technique for ten Berge et al.'s algorithm for orthogonal INDSCAL. *Computational Statistics* 25:409–28. [107, 132]

Takane, Y., K. Jung, and H. Hwang. 2011. Regularized growth curve models. *Computational Statistics and Data Analysis* 55:1041–52. [84]

Takane, Y., S. Jung, and Y. Oshima-Takane. 2009. Multidimensional scaling. In *Handbook of quantitative methods in psychology*, ed. R. E. Millsap and A. Maydeu-Olivares, 219–42. London: Sage Publications. [104]

Takane, Y., H. A. L. Kiers, and J. de Leeuw. 1995. Component analysis with different constraints on different dimensions. *Psychometrika* 60:259–80. [72, 135, 177, 181, 186–189, 191]

Takane, Y., Y. Tian, and H. Yanai. 2007. On constrained *g*-inverses of matrices and their properties. *Annals of the Institute of Statistical Mathematics* 59:807–20. [44]

Takane, Y., H. Yanai, and H. Hwang. 2006. An improved method for generalized constrained canonical correlation analysis. *Computational Statistics and Data Analysis* 50:221–41. [100]

Takane, Y., H. Yanai, and S. Mayekawa. 1991. Relationships among several methods of linearly constrained correspondence analysis. *Psychometrika* 56:667–84. [85, 110, 112]

Takane, Y., F. W. Young, and J. de Leeuw. 1980. An individual differences additive model: An alternating least squares method with optimal scaling features. *Psychometrika* 45:183–209. [188, 190, 191]

Takeuchi, K., H. Yanai, and B. N. Mukherjee. 1982. *The foundations of multivariate analysis*. New Delhi: Wiley Eastern. [93]

Tanaka, Y. 1984. Sensitivity analysis in Hayashi's third method of quantification. *Behaviormetrika* 16:31–44. [160]

Tanaka, Y. 1988. Sensitivity analysis in principal component analysis: Influence on the subspace spanned by the principal components. *Communications in Statistics - Theory and Methods* 17:3157–75. [160]

Tanaka, Y. and T. Tarumi. 1986. Sensitivity analysis in Hayashi's second method of quantification. *Journal of Japan Statistical Society* 16:32–52. [160]

Tanaka, Y. and T. Tarumi. 1989. Sensitivity analysis in canonical factor analysis. *Journal of the Japanese Society of Computational Statistics* 2:9–20. [160]

Tarumi, T. 1986. Sensitivity analysis of descriptive multivariate methods formulated by the generalized singular value decomposition. *Mathematica Japonica* 31:957–77. [160]

Telser, L. G. 1964. Iterative estimation of a set of linear regression equation. *Journal of the American Statistical Association* 59:845–62. [140]

ten Berge, J. M. F. 1977. *Optimizing factorial invariance*. Unpublished Doctoral Dissertation, University of Groningen. [151]

ten Berge, J. M. F. 1979. On the equivalence of two oblique congruence rotation methods and orthogonal approximations. *Psychometrika* 44:359–64. [122]

ten Berge, J. M. F. 1983. A generalization of Kristof's theorem on the trace of certain matrix products. *Psychometrika* 48:519–23. [57, 58]

ten Berge, J. M. F. 1985. On the relationship between Fortier's simultaneous prediction and van den Wollenberg's redundancy analysis. *Psychometrika* 50:121–2. [97]

ten Berge, J. M. F. 1993. *Least squares optimization in multivariate analysis*. Leiden: DSWO Press. [57, 61, 151]

ten Berge, J. M. F. and D. L. Knol. 1984. Orthogonal rotations to maximal agreement for two or more matrices of different column orders. *Psychometrika* 49:49–55. [151]

ten Berge, J. M. F. and K. Nevels, K. 1977. A general solution to Mosier's oblique procrustes. *Psychometrika* 42:593–600. [151]

ter Braak, C. J. F. 1986. Canonical correspondence analysis : A new eigenvector technique for multivariate direct gradient analysis. *Ecology* 67:1167–79. [86, 109]

ter Braak, C. J. F. 1988. Partial canonical correspondence analysis. In *Classification and related methods of data analysis*, ed. H. H. Bock, 551–8. Amsterdam: North Holland. [115]

ter Braak, C. J. F. and S. de Jong. 1998. The objective function of partial least squares. *Journal of Chemometrics* 12:41–54. [47]

Tian, Y. and Y. Takane. 2009. On *V*-orthogonal projectors associated with a semi-norm. *Annals of the Institute of Statistical Mathematics* 61:517–30. [46]

Tibshirani, R. J. 1996. Regression shrinkage and selection via the lasso. *Journal of the Royal Statistical Society, Series B* 58:267–88. [95]

Timm, N. H. 1975. *Multivariate analysis with applications in education and psychology*. Belmont, CA: Wadsworth. [100]

Timm, N. H. and J. E. Carlson. 1976. Part and bipartial canonical correlation analysis. *Psychometrika* 41:159–176. [100]

Tipping, M. E. and C. M. Bishop. 1999. Probabilistic principal componentprobabilistic principal component analysis (PCA). *Journal of the Royal Statistical Society, Series B* 61:611–22. [175]

Torgerson, W. S. 1958. *Theory and methods of scaling*. New York: Wiley. [104]

Trendafilov, N. T. and I. T. Jolliffe. 2006. Projected gradient approach to the numerical solution of the SCoTLASS. *Computational Statistics and Data Analysis* 50:242–53. [95]

Tso, M. K. 1981. Reduced rank regression and canonical analysis. *Journal of the Royal Statistical Society, Series B* 43:183–9. [101]

Tucker, L. R. 1959. Intra-individual and inter-individual multidimensionality. In *Psychological Scaling*, ed. H. Gullicksen and S. Messick, 155–67. New York: Wiley. [125]

Tucker, L. R., L. G. Cooper, and W. Meredith. 1972. Obtaining squared multiple correlation from a correlation matrix which may be singular. *Psychometrika* 37:143–8. [153]

van de Velden, M. and Y. Takane. 2012. Generalized canonical correlation analysis with missing values. *Computational Statistics* 27:551–71. [163]

van den Wollenberg, A. L. 1977. Redundancy analysis: An alter native for canonical correlation analysis. *Psychometrika* 42:207–19. [83, 97]

van der Heijden, P. G. M. and J. de Leeuw. 1985. Correspondence analysis used complementary to loglinear analysis. *Psychometrika* 50:429–47. [116]

van der Heijden, P. G. M. and F. Meijerink. 1989. Generalized correspondence analysis of multi-way contingency tables and multi-way (super-) indicator matrices. In *Multiway data analysis*, ed. R. Coppi and S. Bolasco, 185–202. Amsterdam: Elsevier. [116]

van der Heijden, P. G. M., A. de Falguerolles, and J. de Leeuw. 1989. A combined approach to contingency table analysis with correspondence analysis and log-linear analysis. *Applied Statistics* 38:249–92. [116]

van der Leeden, R. 1990. *Reduced rank regression with structured residuals*. Leiden: DSWO Press. [97]

van Loan, C. F. 1976. Generalizing the singular value decomposition. *SIAM Journal on Numerical Analysis* 13:76–83. [66, 70]

Velicer, W. F. and D. N. Jackson. 1990. Component analysis versus common factor analysis: Some issues in selecting an appropriate procedure. *Multivariate Behavioral Research* 25:1–28. [201]

Velu, R. P. 1991. Reduced rank models with two sets of regressors. *Applied Statistics* 40:159–70. [178]

Verboon, P. 1994. *A robust approach to nonlinear multivariate analysis*. Leiden: DSWO Press. [167]

Verbyla, A. P. and W. N. Venables. 1988. An extension of the growth curve model. *Biometrika* 75:129–38. [135, 136]

Vines, S. K. 2000. Simple principal components. *Applied Statistics* 49:441–51. [95]

von Neumann, J. 1937. Some matrix inequalities and metrization of matrix space. *Tomsk University Review* 1:286–300. (Reprinted in A. H. Taub (Ed.), *John von Neumann collected works, Vol. IV*. New York: Pergamon Press. [58]

von Rosen, D. 1989. Maximum likelihood estimators in multivariate linear normal models. *Journal of Multivariate Analysis* 31:187–200. [135, 137]

von Rosen, D. 1995. Residuals in the growth curve model. *Annals of the Institute of Statistical Mathematics* 47:129–36. [134]

Wedderburn, J. H. M. 1934. *Lectures on matrices*. Colloquium Publication, Vol. 17, Providence: American Mathematical Society. [70]

Wentzell, P. D., D. T. Andrews, D. C. Hamilton, K. Faber, and B. R. Kowalski. 1997. Maximum likelihood principal component analysis. *Journal of Chemometrics* 11:339–66. [148]

Werner, H. J. 1992. G-inverses of matrix products. In *Data analysis and statistical inference*, ed. S. Schlach and G. Trenkler, 251–60. Berlin: Eul-Verlag. [39]

Weyl, H. 1912. Das asymptotische Verteilungsgesetz der Eigenwertlinearer partieller Differentialgleichungen (mit einer Anwendung auf der Theorie der Hohlrumstrahlung). *Mathmatische Annalen* 71:441–79. [62, 63]

Winsberg, S. 1988. Two techniques: Monotone spline transformations for dimension reduction in PCA and easy-to-generate metrics for PCA of sampled functions. In *Component and correspondence analysis*, ed. J. L. A. van Rijckevorsel and J. de Leeuw, 115–35. New York: Wiley. [75]

Winsberg, S. and J. O. Ramsay. 1983. Monotone spline transformations for dimension reduction. *Psychometrika* 48:575–95. [174]

Witten, D. M., R. J. Tibshirani, and T. Hastie. 2009. A penalized matrix decomposition, with applications to sparse principal components and canonical correlation analysis. *Biostatistics* 10:515–34. [95]

Woodward, T. S., E. Feredoes, P. D. Metzak, Y. Takane, and D. S. Manoach. 2013. Epoch-specific networks involved in working memory: A functional connectivity analysis. *NeuroImage* 65:529–39. [1]

Woodward, T. S., T. A. Cairo, C. C. Ruff, Y. Takane, M. A. Hunter, and E. T. C. Ngan. 2006. Load dependent neural systems underlying encoding and maintenance in verbal working memory. *Neuroscience* 139:317–25. [1]

Yanai, H. 1970. Factor analysis with external criteria. *Japanese Psychological Research* 12:143–53. [80, 83]

Yanai, H. 1980. A proposition of generalized method for forward selection of variables. *Behaviormetrika* 7:95–107. [157]

Yanai, H. 1988. Partial correspondence analysis and its properties. In *Recent developments in clustering and data analysis*, ed. C. Hayashi et al., 259–66. Boston: Academic Press. [116]

Yanai, H. 1990. Some generalized form of least squares g-inverse, minimum norm g-inverse and Moore-Penrose inverse matrices. *Computational Statistics and Data Analysis* 10:251–60. [46]

Yanai, H. and B. N. Mukherjee. 1987. A generalized method of image analysis from an intercorrelation matrix which may be singular. *Psychometrika* 52:555–64. [153]

Yanai, H. and Y. Takane. 1992. Canonical correlation analysis with linear constraints. *Linear Algebra and Its Applications* 176:75–89. [48, 85, 100]

Yanai, H., K. Takeuchi, and Y. Takane. 2011. *Projection matrices, generalized inverse matrices, and singular value decomposition*. New York: Springer. [viii, 21, 93]

Young, F. W. 1981. Quantitative analysis of qualitative data. *Psychometrika* 46:347–88. [170, 173]

Young, F. W., Y. Takane, and J. de Leeuw. 1978. The principal components of mixed measurement level multivariate data. *Psychometrika* 43:279–81. [171, 173]

Young, G. and A. S. Householder. 1938. Discussion of a set of points in terms of their mutual distances. *Psychometrika* 3:19–22. [104]

Zellner, A. 1962. An efficient method of estimating seemingly unrelated regressions and test for aggregation bias. *Journal of the American Statistical Association* 57:348–68. [138]

Zellner, A. 1963. Estimators for seemingly unrelated regression equations: Some exact finite sample results. *Journal of the American Statistical Association* 58:977–92. [138]

Zha, H. 1991. The restricted singular value decomposition of matrix triplets. *SIAM Journal on Matrix Analysis and Applications* 12:172–94. [70]

Zou, H., T. Hastie, and R. J. Tibshirani. 2006. Sparse principal component analysis. *Journal of Computational and Graphical Statistics* 15:265–86. [95]

Index

Printed and bound by CPI Group (UK) Ltd, Croydon, CR0 4YY
24/10/2024
01778277-0008